SYMPOSIA OF THE ZOOLOGICAL SOCIETY OF LONDON NUMBER 58

Mammal Population Studies

The Proceedings of a Symposium
held at the Zoological Society of London
on 28th and 29th November 1986

Edited by Stephen HARRIS
Department of Zoology, University of Bristol

Published for THE ZOOLOGICAL SOCIETY OF LONDON
by CLARENDON PRESS · OXFORD
1987

Oxford University Press, Walton Street, Oxford OX2 6DP

Oxford New York Toronto
Delhi Bombay Calcutta Madras Karachi
Petaling Jaya Singapore Hong Kong Tokyo
Nairobi Dar es Salaam Cape Town
Melbourne Auckland

and associated companies in
Beirut Berlin Ibadan Nicosia

Oxford is a trade mark of Oxford University Press

Published in the United States
by Oxford University Press, New York

British Library Cataloguing in Publication Data

Mammal population studies: the proceedings of a symposium held at the Zoological Society of London on 28th and 29th November 1986—(Symposia of the Zoological Society of London; No. 58).
1. Mammal populations—Research
I. Harris, S II. Zoological Society of London
III. Series
599.05'248'072 QL708.6
ISBN 1–19–854006–X

Library of Congress Cataloging in Publication Data

Mammal population studies.
(Symposia of the Zoological Society of London; no. 58)
Includes bibliographies and index.
1. Mammal populations—Congresses. 2. Mammals—Ecology—Congresses. I. Harris, Stephen, 1950– II. Zoological Society of London. III. Series.
QL1.Z733 no. 58 591 s 87–25709
[QL708.6] [599'.05248]
ISBN 0–19–854006–X

Set by Promenade Graphics Ltd., Cheltenham
Printed by St Edmundsbury Press,
Bury St Edmunds, Suffolk

Contents

Contributors xv

Organizer of symposium and Chairmen of sessions xvii

Population studies of mammals
R. J. BERRY

Text 1
References 4

Assessment of trends in size and structure of a colony of the greater horseshoe bat
R. E. STEBBINGS & H. R. ARNOLD

Synopsis 7
Introduction 7
Methods 8
History of the Dorset greater horseshoe bats 9
Distribution of roosts 10
Historical population changes 11
Results 12
Colony size 1970–1986 12
Sex ratio 14
Survival 15
Recruitment 18
Colony age structure 20
Discussion 21
Acknowledgements 23
References 23
Note added in proof 24

The application of Capture-Mark-Recapture methods to the enumeration of small mammal populations
W. I. MONTGOMERY

Synopsis 25
Introduction 26
The advantages of CMR 27
The use of CMR in small mammal ecology 28
The diversity of CMR models and their underlying assumptions 28
Two samples from a closed population: the Lincoln Index 29
Multiple samples from a closed population: a maximum likelihood approach 30

Increase in the fraction marked in a mark-and-release sequence 31
Frequency-of-capture models 33
Multiple samples from an open population: the Leslie-Chitty model 34
Multiple samples from an open population: the Jolly-Seber model 35
Statistical propriety and biological reality 35
Population closure 36
The loss of marks 37
The effect of marking: trap response 37
Heterogeneity in capture probability 38
Age-dependent survival 41
Paucity of data 42
The robustness of CMR models when assumptions are invalidated: fitting a model 43
CMR and MNA: some further problems 44
The design of CMR programmes: the light of experience 47
The limits of CMR in the revelation of the dynamics and regulation of small mammals 50
Conclusions 50
Acknowledgements 51
References 51

Patterns of mortality in a free-living rabbit (*Oryctolagus cuniculus*) population

D. P. COWAN

Synopsis 59
Introduction 59
Methods 60
Results 61
The breeding season and changes in population size 61
Variation in mortality with age and sex 64
Variation in juvenile mortality with time of birth 65
Variation in juvenile mortality between years 66
Causes of mortality 68
Discussion 70
Variation in mortality with age 70
Variation in mortality with sex 71
Variation in juvenile mortality with time of birth 71
Variation in juvenile mortality between years 72
Conclusions 74
Acknowledgements 74
References 75

Cycles in game-bag records of hares and rabbits in Britain

STEPHEN TAPPER

Synopsis 79
Historical background 79

Sources of data and methods 81
Results 84
Mountain hare 84
Brown hares 84
Rabbit 87
Discussion 87
Acknowledgements 96
References 96

Planning and monitoring an attempt to eradicate coypus from Britain

L. M. GOSLING & S. J. BAKER

Synopsis 99
Introduction 99
Materials and methods 100
The sample 100
Population reconstruction and simulation 101
Planning the eradication campaign 102
Trapper numbers 102
Equipment and strategic planning 103
The incentive bonus scheme 104
Mortality and recruitment during the eradication campaign 104
Trapping intensity and adult mortality rates 104
Recruitment rates 105
Population changes during the campaign and the effect of density-dependent change in reproductive success 106
Discussion 109
References 109
Appendix: Inputs to the simulation 111
Proportion of females littering 111
Litter size 111
Recruitment 111
Adult mortality 112
Population size 113
Structure of the simulation 113

Group stability and the structure of a capybara population

EMILIO A. HERRERA & DAVID W. MACDONALD

Synopsis 115
Introduction 115
Study area 117
Methods 117
Results 117
Group stability 118
Group size 118

Interannual group stability 120
Relations between groups 121
Group formation 125
Group size and the reproductive success of females 126
Discussion 126
Acknowledgements 129
References 129

Estimating the abundance of red deer (*Cervus elaphus* L.) and roe deer (*Capreolus capreolus* L.) and their current status in Great Britain

BRAIN W. STAINES & PHILIP R. RATCLIFFE

Synopsis 131
Introduction 132
Habitats 132
Methods 132
Direct methods 134
Mortality data 139
Indirect assessments 140
Comparison of techniques 143
Status and trends of deer populations 144
Red deer 144
Roe deer 147
Conclusions 148
Acknowledgements 149
References 149

Red deer population changes and the independent assessment of population size

PHILIP R. RATCLIFFE

Synopsis 153
Introduction 153
The Galloway red deer population 1965–1980 154
The Galloway red deer population 1981–1986 156
Independent assessment of deer density 157
Vantage-point counts 157
Cohort analysis 158
Management options in Galloway 160
The underestimation of deer numbers 160
Discussion 163
Acknowledgements 164
References 164

Long-term changes in New Forest deer populations aid correlated environmental change

R. J. PUTMAN & S. K. SHARMA

Synopsis 167
Introduction 167
  General introduction 167
  The New Forest, its vegetation and the larger herbivores 168
Methods 169
Results 172
  Deer numbers in relation to populations of other herbivores 172
  Deer numbers and vegetational change 173
Discussion 174
Acknowledgements 178
References 178

Census techniques for grey seal populations

A. J. WARD, D. THOMPSON & A. R. HIBY

Synopsis 181
Introduction 181
Census methodology 182
  Ground counting technique 182
  Aerial survey technique 183
  Subsampling versus complete enumeration 183
Estimating pup production from aerial survey data 183
  Empirical multiplication factors 185
  Maximum likelihood estimation 186
Conclusions 190
References 190

Census techniques for small populations, with special reference to the Mediterranean monk seal

A. R. HIBY & J. S. JEFFERY

Synopsis 193
Introduction 193
Objectives and choice of field techniques 194
Development of camera/detector equipment 197
  Detector 197
  Camera 198
  Housing and attachment systems 199
Results and qualitative inferences available 201
Methods of quantitative analysis 202
  Mark–recapture technique for small populations 202
  Problems unique to mark-recognition data 206
  Open population analysis for monk seals 208

Concluding remarks 208
Acknowledgements 209
References 209

Estimating pup production of elephant seals at South Georgia

P. ROTHERY & T. S. McCANN

Synopsis 211
Introduction 211
Methods 212
Mathematical model 212
Analysis of detailed counts 214
Analysis of the 1985 survey data 217
  Beach-to-beach variation 217
  Estimating pup production 219
Discussion 222
Acknowledgements 223
References 223

Techniques for estimating the size of whale populations

P. S. HAMMOND

Synopsis 225
Introduction 225
The Southern Hemisphere minke whale 227
  Method of estimation 227
  Application of method 229
The North Atlantic humpback whale 231
  Method of estimation 232
  Application of method 233
The California gray whale 235
  Method of estimation and its application 236
The Western Arctic bowhead whale 237
  Method of estimation and its application 238
Discussion 240
Acknowledgements 242
References 242

The use of spraints to monitor populations of otters (*Lutra lutra* L.)

J. W. H. CONROY & D. D. FRENCH

Synopsis 247
Introduction 247
The study area 249
Methods 249
  General 249

The monitoring programme 250
Results 252
Seasonal trends and preferences for sprainting in the detailed study area 252
The relationship between numbers of spraints and numbers of otters seen 253
The monitoring programme 255
Discussion 257
Conclusions 259
Acknowledgements 260
References 260

Seasonal reproduction, mortality and food of otters (*Lutra lutra* L.) in Shetland

H. KRUUK, J. W. H. CONROY & A. MOORHOUSE

Synopsis 263
Introduction 263
Study area and climate 264
Methods 265
Results 266
Birth seasons, in Shetland and elsewhere 266
Litter sizes 268
Condition and mortality 269
Diet and food supply 270
Feeding observations 271
Discussion 273
Acknowledgement 276
References 277

Badger population dynamics in a high-density area

C. L. CHEESEMAN, J. W. WILESMITH, J. RYAN & P. J. MALLINSON

Synopsis 279
Introduction 280
Methods 280
Social group territories 281
Capture of badgers 281
Population estimates 282
Age structure of population and estimation of mortality rates 282
Life tables and survivorship curves 283
Breeding potential and survival of adult females 283
Seasonal variation in body weight 283
Results 283
Social group territories 283
Population density 284
Age structure 285
Mortality rates 286
Breeding potential and survival of adult females 287

Seasonal variation in body weight 290
Discussion 291
Acknowledgements 293
References 294

Dynamics of a suburban badger (*Meles meles*) population

STEPHEN HARRIS & W. J. CRESSWELL

Synopsis 295
Introduction 295
Methods 296
The study area and field techniques 296
Population estimates 297
Estimation of group size by radio-isotope labelling of faeces 298
Age structure and mortality rates 298
Life tables and survivorship curves 299
Modelling the population 299
Results 300
Population estimates 300
Estimation of group size by radio-isotope labelling of faeces 301
Age structure and mortality rates 302
Life tables and survivorship curves 304
Proportion of females breeding 305
Modelling the population 305
Discussion 306
Acknowledgements 310
References 310

The use of sociological data to explain the distribution and numbers of urban foxes (*Vulpes vulpes*) in England and Wales

STEPHEN HARRIS & G. C. SMITH

Synopsis 313
Introduction 313
Materials and methods 315
The fox surveys 315
Statistical techniques 315
Results 318
Discussion 321
Acknowledgements 327
References 328

Population dynamics of mammalian diseases

DENIS MOLLISON

Synopsis 329
Introduction 330

Models 331
Epidemics 333
Endemics 334
Barriers and control 339
References 340

Summary 343
M. L. GORMAN

Index 347

Contributors

ARNOLD, H. R., Institute of Terrestrial Ecology, Monks Wood Experimental Station, Abbots Ripton, Huntingdon PE17 2LS, UK.

BAKER, S. J., Coypu Research Laboratory, Ministry of Agriculture, Fisheries and Food, Jupiter Road, Norwich NR6 6SP, UK.

BERRY, R. J., Department of Zoology, University College London, Gower Street, London WC1E 6BT, UK.

CHEESEMAN, C. L., Ministry of Agriculture, Fisheries and Food, Worplesdon Laboratory, Tangley Place, Worplesdon, Guildford, Surrey GU3 3LQ, UK.

CONROY, J. W. H., Institute of Terrestrial Ecology, Banchory, Kincardineshire, Scotland AB3 4BY, UK.

COWAN, D. P., Mammals and Birds Department, Ministry of Agriculture, Fisheries and Food, Worplesdon Laboratory, Worplesdon, Guildford GU3 3LQ, UK.

CRESSWELL, W. J., Department of Zoology, University of Bristol, Woodland Road, Bristol BS8 1UG, UK.

FRENCH, D. D., Institute of Terrestrial Ecology, Banchory, Kincardineshire, Scotland AB3 4BY, UK.

GORMAN, M. L., Culterty Field Station, University of Aberdeen, Newburgh, Ellon, Aberdeenshire AB4 0AA, UK.

GOSLING, L. M., Coypu Research Laboratory, Ministry of Agriculture, Fisheries and Food, Jupiter Road, Norwich NR6 6SP, UK.

HAMMOND, P. S., Sea Mammal Research Unit, c/o British Antarctic Survey, High Cross, Madingley Road, Cambridge CB3 0ET, UK.

HARRIS, S., Department of Zoology, University of Bristol, Woodland Road, Bristol BS8 1UG, UK.

HERRERA, E. A., Animal Behaviour Research Group, Department of Zoology, University of Oxford, South Parks Road, Oxford OX1 3PS. *Present address*: Smithsonian Tropical Research Institute, Apdo 2072, Balboa, Panama.

HIBY, A. R., Sea Mammal Research Unit, c/o British Antarctic Survey, Natural Environment Research Council, High Cross, Madingley Road, Cambridge CB3 0ET, UK.

JEFFERY, J. S., Sea Mammal Research Unit, c/o British Antarctic Survey, High Cross, Madingley Road, Cambridge CB3 0ET, UK.

KRUUK, H., Institute of Terrestrial Ecology, Banchory, Kincardineshire, Scotland AB3 4BY, UK.

MCCANN, T. S., British Antarctic Survey, Natural Environment Research Council, High Cross, Madingley Road, Cambridge CB3 0ET, UK.

MACDONALD, D. W., Animal Behaviour Research Group, Department of Zoology, University of Oxford, South Parks Road, Oxford OX1 3PS, UK.

MALLINSON, P. J., Ministry of Agriculture, Fisheries and Food, Worplesdon Laboratory, Tangley Place, Worplesdon, Guildford, Surrey GU3 3LQ, UK.

MOLLISON, D., Department of Actuarial Mathematics and Statistics, Heriot-Watt University, Riccarton, Edinburgh EH14 4AS, UK.

MONTGOMERY, W. I., Department of Zoology, Queen's University of Belfast, Belfast BT7 1NN, UK.

MOORHOUSE, A., Institute of Terrestrial Ecology, Banchory, Kincardineshire, Scotland AB3 4BY, UK.

PUTMAN, R. J., Department of Biology, The University, Southampton SO9 5NH, UK.

RATCLIFFE, P. R., Forestry Commission, Wildlife and Conservation Research Branch, Alice Holt Lodge, Wrecclesham, Farnham, Surrey GU10 4LH, UK.

ROTHERY, P., British Antarctic Survey, Natural Environment Research Council, High Cross, Madingley Road, Cambridge CB3 0ET, UK.

RYAN, J., Ministry of Agriculture, Fisheries and Food, Epidemiology Unit, Central Veterinary Laboratory, New Haw, Weybridge, Surrey KT15 3NB, UK.

SHARMA, S. K., Department of Biology, The University, Southampton SO9 5NH, UK.

SMITH, G. C., Department of Zoology, University of Bristol, Woodland Road, Bristol BS8 1UG, UK.

STAINES, B. W., Institute of Terrestrial Ecology, Hill of Brathens, Banchory, Scotland AB3 4BY, UK.

STEBBINGS, R. E., Institute of Terrestrial Ecology, Monks Wood Experimental Station, Abbots Ripton, Huntingdon PE17 2LS, UK.

TAPPER, S. C., The Game Conservancy, Fordingbridge, Hampshire, UK.

THOMPSON, D., Sea Mammal Research Unit, c/o British Antarctic Survey, Natural Environmental Research Council, High Cross, Madingley Road, Cambridge CB3 0ET, UK.

WARD, A. J., Sea Mammal Research Unit, c/o British Antarctic Survey, Natural Environment Research Council, High Cross, Madingley Road, Cambridge CB3 0ET, UK.

WILESMITH, J., Ministry of Agriculture, Fisheries and Food, Epidemiology Unit, Central Veterinary Laboratory, New Haw, Weybridge, Surrey KT15 3NB, UK.

Organizer of symposium

STEPHEN HARRIS, Department of Zoology, University of Bristol, Bristol BS8 1UG, UK.

Chairmen of sessions

R. J. BERRY, Department of Zoology, University College London, Gower Street, London WC1E 6BT, UK.

B. C. R. BERTRAM, Zoological Society of London, Regents Park, London NW1 4RY, UK.

J. P. CROXALL, British Antarctic Survey, Natural Environment Research Council, High Cross, Madingley Road, Cambridge CB3 0ET, UK.

M. L. GORMAN, Culterty Field Station, University of Aberdeen, Newburgh, Ellon, Aberdeenshire, Scotland AB4 0AA, UK.

Symp. zool. Soc. Lond. (1987) No. 58: 1–5

Population studies of mammals

R. J. BERRY

Department of Zoology
University College London
Gower Street, London WC1E 6BT

The most basic part of biology is taxonomy—being able to recognize and distinguish different sorts of organisms. The next phase is quantitative: describing and analysing variations in population size, numbers, reproductive success, mortality factors, habitat preferences, and so on. In other words, the proper study of biology inevitably involves an investigation of the processes which affect populations. Seventeen years ago, H. N. Southern and I organized a symposium for the Mammal and Zoological Societies on *Variation in mammalian populations* (Berry & Southern 1970). This symposium on *Mammal population studies* complements the earlier one, and presents a series of case histories illustrating the methodology and problems of population work in mammals.

If the first necessity in biological education is accepting the interdependence of investigations on individuals and on populations (a link which is not as natural as it should be, in large part owing to the persisting practice of teaching through a limited number of 'types', as perpetrated by Thomas Henry Huxley; he wrote in his *Autobiography* 'There is very little of the genuine naturalist in me. I never collected anything, and species work was a burden to me; what I cared for was the architectural and engineering end of the business' (Huxley 1983: 103)), the second need is acquiring a battery of techniques to analyse population processes. The conceptual bridge between these two stages seems to be difficult for many. Begon (1979) begins his monograph on *Investigating animal abundance* by pointing out that 'Biologists all too often act as if mathematical techniques possess magical powers, transforming incomprehensible raw data into clear and precise scientific conclusions. Nothing could be further from the truth. Far from being magical, such techniques are, in fact, essentially sterile. They become potent only in the light of the biological problem to which they are applied, and the biological context in which they are interpreted.' Elton (1930:35) made a related point: 'the ecological study of animal numbers is to many biologists a pretty dull subject, and not only dull but a trifle forbidding, since the thought of it makes the word "statistics" float ominously into their minds. It

ZOOLOGICAL SYMPOSIUM No. 58
ISBN 0-19-854006-X

is felt that the regulation of numbers is eminently a subject for specialists, and that the control of numbers, although it presents economic problems of great magnitude, can still be treated as a special and rather isolated branch of biology.' These words were written more than half a century ago, but they are still unfortunately true in part today.

Charles Elton can be regarded as the founder of mammal population biology, particularly from the data he assembled from his post as 'biological consultant' to the Hudson's Bay Company (1925–31). Elton entered this subject just as Raymond Pearl was resurrecting Verhulst's mathematical equations for human populations, and Lotka and Volterra were applying them to animal populations (McIntosh 1985: 90; Kingsland 1985). His interest in population numbers was stimulated by reading a Norwegian book on lemmings by Robert Collett, which Elton bought whilst returning from one of his Spitzbergen expeditions in the early 1920s. Collett was an Englishman, who worked in the University Museum in Trondheim. Elton's (1942) book on *Voles, mice, and lemmings* was subtitled 'Problems in population dynamics'.

Collett was clearly a 'scientific naturalist' in the same tradition as Elton himself. In a paper published in the *Proceedings of the Linnean Society* in 1877*, he discusses many of the topics that occupied Elton in his later work, and which repeatedly recur in this symposium: recruitment, dispersal, age-structure, density-dependence, proximate and ultimate mortality factors, models and predictability. Collett's conclusion about lemming plagues is still unproven, yet still a viable hypothesis: 'The immediate reason for the large increase of lemmings is doubtless owing to exceptional circumstances repeated for some years, and consequently favouring their pairing and bringing up of their young; the result of which is far greater numbers of families and larger numbers of young in each set than in ordinary years.'

Elton constantly linked population traits with their genetic basis, and it is important to emphasize this insistence. He wrote in his 1930 book: 'There is undoubtedly a strong tendency among many of the younger biologists to break away from the too rigid beliefs of the Darwinian school. They are dissatisfied, not so much because they believe that Darwin's theories of Natural Selection are wrong, but because these theories, which are obviously true as far as they go, do not assist in the least to explain a very large number of phenomena which are constantly encountered in the course of field work upon wild animals . . . . The disadvantage of this state of affairs is obvious,

* Collett's paper was written largely to correct information published the previous year by Crotch. For example, Crotch (1876) records that 'in November 1868 (quoted by Lilljeborg) a ship sailed for fifteen hours through a swarm of Lemmings which extended as far over the Trondhjems fjord as the eye could reach.' Collett comments, 'The notice which I gave Dr. Lilljeborg, and which is quite correctly quoted in his 'Fauna', does not state fifteen hours, but a quarter of an hour; and it was not a sailing ship, but a steamer.'

for a hearty belief in the theories of Darwin always acted as a very potent stimulus to naturalists, and supplied a co-ordinating principle for the ecological data that were then available. Without any such biological belief, the modern ecologist runs a risk of becoming a mere beaver in search of scientific lumber, or of falling back upon a severely mechanistic view of animals, based on the laws of physics and chemistry, solid in themselves, but unsatisfactory as a complete explanation of the life and mind of animals' (Elton 1930: 8, 9).

In the 1960s, there appeared to be a recognition that both population ecology and population genetics were becoming limited because their particular insights had not been integrated (Hedrick 1984). Birch (1960) wrote 'The ecological problems of populations have to do with the numbers of animals and what determines these numbers. The genetical problem of populations has to do with the kind of animals and what determines kind. These two disciplines meet when the questions are asked, how does the kind of animal (i.e. genotype) influence the numbers and how does the number of animals influence the kind, that is, the genetical composition of the population?' The previous mammalian population symposium in this series addressed these questions (Berry & Southern 1970). Unfortunately they have not been pursued as assiduously as they might; population genetics and population ecology are still separate disciplines, and Birch's questions are still not properly answered (Berry 1979).

In the context of mammalian population studies, it is pertinent (and sobering) to go back again to points raised by Elton in 1930. His words have an uncomfortably modern ring: 'It is suggested that if we knew enough about the ecological relations of the animals we could predict the effect of any interference, just as a clockmaker can work out the ultimate effect of the twirling of one wheel upon the rate of revolution of any of the others. At the same time it is assumed that an undisturbed natural animal community lives in a certain harmony, referred to as "the balance of nature", and that although rhythmical changes may take place in this balance, yet that these are regular and essentially predictable, and, above all, nicely fitted into the environmental stresses without . . . . This picture has the advantage of being an intelligible and apparently logical result of natural selection in producing the best possible world for each species. It has the disadvantage of being untrue. "The balance of nature" does not exist, and perhaps never has existed. The numbers of wild animals are constantly varying to a greater or less extent, and the variations are usually irregular in period and always irregular in amplitude. Each variation in the numbers of one species causes direct and indirect repercussions on the numbers of the others, and since many of the latter are themselves independently varying in numbers the resultant confusion is very remarkable. The simile of the clockwork mechanism is only true if we imagine that a large proportion of the cog-wheels have their own mainsprings, which do not unwind at a constant speed. There is also the difficulty that each wheel retains the right to arise and migrate and settle down in another clock . . . .'

'The frequency of apparently spontaneous acts on the parts of animals is not in harmony with prevailing ideas about their behaviour, for according to these ideas they are expected to manifest only the characteristics of a very complicated motor car, which sometimes goes wrong. The Darwinian theory, which I take to be mainly true as far as it goes, regards animals as stationary units, or at any rate units with regular ecological habits, and acted upon by a number of selective eliminating agents. Darwinism, in fact, implies the selection of helpless bundles of flesh by an environment which lacks any traces of biological purposiveness. Now what we know of the real, as opposed to the textbook, behaviour of wild animals establishes the fact that they do not remain still to be selected. In periods of stress it is a common thing for the animals to change their habits, and usually this involves migration . . . . In fact we are face to face with a process which may be called the SELECTION OF THE ENVIRONMENT BY THE ANIMAL, as opposed to the NATURAL SELECTION OF THE ANIMAL BY THE ENVIRONMENT' (Elton 1930: 16–17, 50–51).

It would be wrong to suggest that we are no further forward than when Elton originally wrote those ideas. Southwood (1977) has traced the interactions of the factors defining the relationships between animal and habitat in both time and space, and derived a 'reproductive success matrix'; and I have sought to show the capriciousness and variability of selection in moulding genomes (Berry 1985, 1986). We have advanced a long way since Collett wrote about lemmings or Elton first formulated his ideas on ecology and evolution, but there is still much work to be done.

References

Begon, M. (1979). *Investigating animal abundance*. Edward Arnold, London.

Berry, R.J. (1979). Genetical factors in animal population dynamics. In *Population dynamics*: 53–80. (Eds Anderson, R.M., Turner, B.D. & Taylor, L.R.). Blackwell, Oxford.

Berry, R.J. (1985). The processes of pattern: genetical possibilities and constraints in coevolution. *Oikos* **44**: 222–8.

Berry, R.J. (1986). Genetics of insular populations of mammals, with particular reference to differentiation and founder effects in British small mammals. *Biol. J. Linn. Soc.* **28**: 205–30.

Berry, R.J. & Southern, H.N. (Eds). (1970). *Variation in mammalian populations*. Academic Press, London & New York. (*Symp. zool. Soc. Lond.* No. 26).

Birch, L.C. (1960). The genetic factor in population ecology. *Am. Nat.* **94**:5–24.

Collett, R. (1877). On *Myodes lemmus* in Norway. *Proc. Linn. Soc. Lond.* **13**: 327–34.

Crotch, D. (1876). On the migration and habits of the Norwegian lemming. *Proc. Linn. Soc. Lond.* **13**: 27, 83, 157.

Elton, C. (1930). *Animal ecology and evolution*. Clarendon Press, Oxford.

Elton, C. (1942). *Voles, mice and lemmings. Problems in population dynamics*. Clarendon Press, Oxford.

Hedrick, P.W. (1984). *Population biology*. Jones and Barlett, Boston.

Huxley, T.H. (1983). *Autobiography*. (Oxford Paperback Edition). (Ed. De Beer, G.R.). University Press, Oxford.

Kingsland, S.E. (1985). *Modelling nature*. University Press, Chicago.

McIntosh, R.P. (1985). *The background of ecology*. University Press, Cambridge.

Southwood, T.R.E. (1977). Habitat, the templet for ecological strategies? *J. Anim. Ecol.* **46**: 337–65.

Symp. zool. Soc. Lond. (1987) No. 58: 7–24

Assessment of trends in size and structure of a colony of the greater horseshoe bat

R. E. STEBBINGS and
H. R. ARNOLD

Institute of Terrestrial Ecology
Monks Wood Experimental Station
Abbots Ripton
Huntingdon
England
PE17 2LS

Synopsis

In the last 100 years the greater horseshoe bat has become extinct over half its previous range in Britain. The causes are probably many, including the killing of colonies or their exclusion from roosts as well as changes in habitat, particularly those which cause reduction in their food.

There is no accurate documentation of the past history of colony distribution or size of populations in Britain. Case history results are presented here for one colony which formerly occupied about 3,000 km^2 of Dorset and parts of Wiltshire and Hampshire. The colony was almost eliminated by a remedial timber treatment in 1953 involving the pesticide Lindane. The size of the colony before the timber was treated is estimated to have been about 15,000 bats (0.05/ha) and there may have been 1,500 surviving in 1957 when a previous mark–recapture study began. Between 1972 and 1986 the colony has declined from an estimated 140 to 90 bats living in an area of 1,500 km^2 (6.0×10^{-4} bats/ha). The sex ratio at birth is unity but the ratio in the colony has remained constant at 38.5% males to 61.5% females. Study of cohorts of known age shows that juveniles in their first year have a survival rate of 0.65 and survival rises steadily to 0.85 at nine years of age. Of the total annual mortality, most (50.7%) occurs between January and March, with 26.7% from April to June, 18.6% in July and 4.0% from August to December.

Introduction

Once large declines in numbers of animals in a population have been recorded, periodic estimation of the subsequent changes is a necessity in the development and monitoring of a conservation strategy for an endangered species. The very fact that a species is threatened or endangered often severely limits the way in which information can be gathered and hence may

ZOOLOGICAL SYMPOSIUM No. 58
ISBN 0-19-854006-X

greatly affect the accuracy of results. An ironic situation develops in which the procedures necessary to gather good results may cause animals additional stress, thereby increasing mortality—but without those results the case for conservation may be weak and we may not know whether the conservation measures are successful. The greater horseshoe bat, *Rhinolophus ferrumequinum*, in Britain appears to be a good example of this problem. It appears, mostly from anecdotal information, that this species has declined substantially in the past century and observations suggest these animals are particularly sensitive to disturbance by observers. The aims of this work were, first, to find methods of estimating the size of a greater horseshoe bat population, and secondly to examine annual and seasonal changes in population size. Clearly, the major constraint was how to find out how many bats there were without greatly influencing survival and hence numbers. Excessive disturbance in hibernation could lead to deaths. However, greater horseshoe bats can live up to 30 years (Caubère, Gaucher & Julien 1984), are colonial animals and spend varying time in succession in traditional roosts. Because of this behaviour there may be many temporally and geographically widely spaced opportunities to catch them.

Methods

A detailed population study began in Dorset in December 1970 and involved marking all greater horseshoe bats with magnesium alloy C-shaped forearm rings. Observations by one of us (RES) had begun on this population ten years earlier in November 1960, having developed from work by several others which started in the early 1950s. A large ringing study had started in 1957 and continued until 1963. Several naturalists were involved at various times and, through their groundwork, knowledge was built up of the many different roosts used by the bats.

The population study reported here depended upon the annual core of consistent search of the major accessible roosts, and a small number of opportunistic visits, both to already known and also to new sites. Core visits were made in February–March, May, June, July, September–October and December and amounted to about 18 days per year. Other visits averaged six days annually. Generally bats were caught in roosts no more than twice per year; some sites were visited once and others two or three times each year. It was considered important to leave some known sites relatively undisturbed as sanctuaries for the bats. However, there are likely to be many other roosts which remain unknown to us, as often only half the known population could be found at one time.

Roost sites were found in a variety of ways. Most were notified to us by people who saw bats, some were located by following bats marked with radio-transmitters and some were found by searching. By mark–recapture a picture emerged of the kinds of movements these animals made seasonally, but, of course, at least some movements may have been induced by our disturbance.

All bats were given a temporary mark at birth by clipping toe claws. From 1978 each bat was given a distinctive combination so that growth pattern and survival could be related to date of birth. Ringing was possible for bats aged 60 days or more and efforts were made to mark the juveniles before they dispersed from their nursery. Subsequently bats of the year usually could be recognized by their toe claw clips or grey pelage. Females were identified as having successfully given birth if the pubic false teats were 1 mm or more in length in August or September. Females were categorized as immature if their pubic teats were less than 0.2 mm long. In fact, our category of immatures probably included animals which were sexually mature but which had not given birth and weaned a baby. The pubic teats are used by the baby as anchor points, with the baby holding one of them in its mouth for hours daily.

Having learned from other peoples' mistakes, we began ring-marking of the colony in 1970 cautiously, using newly developed rings and not disturbing any individual in hibernation more than twice per winter. This level of disturbance, which was considered unlikely to affect the bats significantly, meant there was a high chance of bats not being caught each year. Attempts were made to estimate the colony size using various standard mark–recapture statistics but none appeared to give reasonable results. This was to be expected because of the sampling technique used in this study and because marked bats do not mix randomly with the whole population. A better approach was likely to be retrospective census by examination of the case histories of each bat.

History of the Dorset greater horseshoe bats

No significant documentary or observational evidence has been found which gives any indication of the size or distribution of the greater horseshoe bat population in Dorset before about 1950. After that time several naturalists became aware of some sites frequented by the bats. An increasing number of visits were made annually and, in 1957, a substantial study began which involved people searching for roosts and ring-marking the bats. Over 37 visits were made to some sites in the first year, and bats were caught on each occasion. Some bats were taken for photography and others for homing experiments and released elsewhere. Bats were greatly disturbed in both hibernation and summer roosts. By 1960 a picture had developed of the distribution of greater horseshoe bats in Dorset, west Hampshire, south Wiltshire and south-east Somerset (Fig. 1). The bats had apparently been born in and centred on the two nursery sites known at that time, 25 km

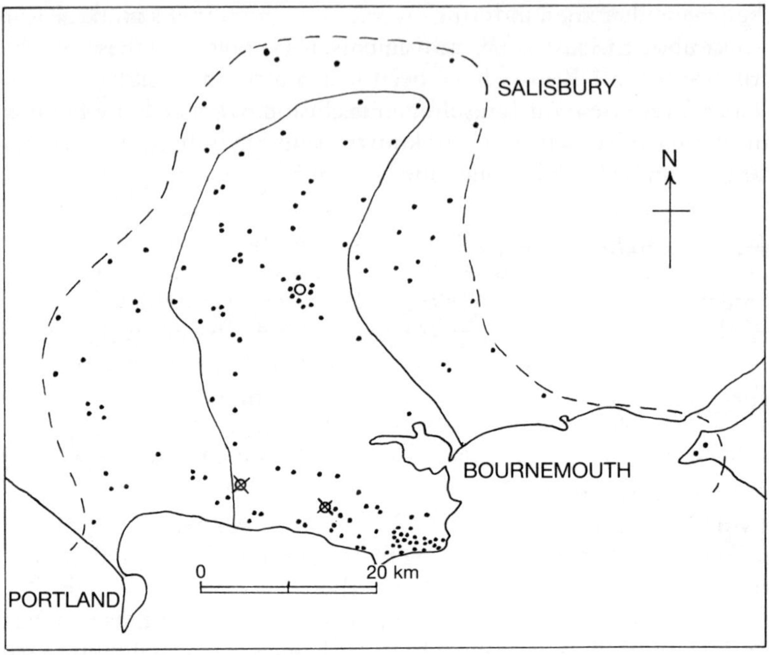

Fig. 1. Distribution of roost sites for the greater horseshoe bat colony (some have been omitted in the south). The circles with crosses indicate former nursery roosts and the open circle is the remaining nursery. The broken line encompasses the roosts used regularly until 1970 and the continuous line shows the area covered by roosts occupied since 1970.

apart in central and southern Dorset. This initial study showed there to be a discrete population in Dorset, members of which gathered in differing places at various times. From this our concept of a colony was born.

Colony : A definition

A colony of bats is a group of animals of one species in an area, which normally associate and breed exclusively with each other. At various times of the year, some members of the colony will join other members, and sometimes a substantial number (perhaps over half the members of the colony) may occupy one roost simultaneously. The rest of this paper will use the word 'colony' to imply interrelated animals born in Dorset.

Distribution of roosts

Greater horseshoe bats in the Dorset colony were found in many different kinds of roosts, from caves, mines, tunnels, cellars and castle walls, to barns and houses.

In 1963, when the early ringing study stopped, over 170 different roost sites had been identified, covering an area of about 3,000 km^2 (Fig. 1). Since 1970 only about 35 (20%) of the original sites survive and these occur in an area of 1,500 km^2. Roosts have been lost mostly by infilling of mines or renovation of houses which resulted in exclusion of bats. Undoubtedly there remain many roosts which are unknown to us, as only part of the known population can be found at one time.

Historical population changes

The principal aims of the naturalists in the 1950s were to record the movements and longevity of the bats. Attention was not given to estimating population size, but from their records an indication of bat numbers may be gleaned.

On 12 July 1961, one of us (RES) interviewed a gardener of a country house lying west of Corfe Castle in Dorset, and was told of a large colony of bats which used to live there. The bats seasonally occupied a range of 18th century outbuildings and annually bat guano was dug out as fertilizer for the gardens. In late May 1953 these buildings were sprayed with a 1% solution of 'Gammexane' or gamma HCH (Lindane). This was at a time of year when the females were gathering at the site which they used as a nursery.

It is difficult to be sure of the number of bats killed: a minimum figure of many thousand is indicated by the gardener who, eight years later, reported removing seven large barrow loads of dead bats. From the volume, we have estimated a minimum of 15,000 were involved.

For several years subsequently, large numbers of bats died after they entered the building. For instance, in 1961, 109 dead bats were removed, but over 80 had been dead one or more years. Bats continue to visit the building to the present but, despite attempts, no successful breeding is known to have occurred in the last 26 years.

Concurrent with this major breeding site were two subsidiary nurseries, a small one about 10 km to the west, and another larger one 25 km to the north. The former no longer exists because the building was converted to a house; the latter is the only remaining breeding site.

The fact that a substantial number of greater horseshoe bats were still alive in Dorset between 1955 and 1958 was revealed by two kinds of observations. First, naturalists made estimates of the numbers of bats in three large clusters which simultaneously occupied cellars and mines, and estimated that each cluster contained between 200 and 350 bats. Because we found these large clusters to consist mostly of young animals it can be assumed that several hundred other bats—mostly the older breeding animals—would have been elsewhere.

Secondly, 401 bats were ringed in the first year from 1957, this being at a time when many roosts were unknown. Our current knowledge of roost

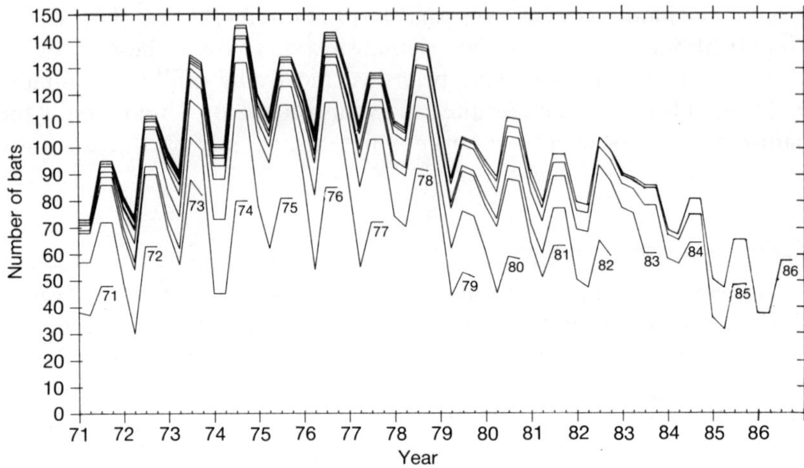

Fig. 2. Annual progressive improvement in knowledge of the colony size (sexes combined) derived by mark–recapture and from case histories.

sites allows us to capture in the first year an average of 0.523 (n=14, $s.d.$=0.068) of the total population subsequently shown to be alive at that time (see below for a full explanation). If the initial capture rate in 1957 had been similar, it would have implied a total population of about 767 ± 84 bats. However, since in 1957 approximately that number of bats were estimated in three known hibernating clusters and since the adults tend to be solitary or roost in small groups elsewhere, it is more likely that the colony totalled at least 1,500 bats in 1957.

Results

Colony size 1970–1986

Observed

Figure 2 shows the number of bats of both sexes known to be alive (from case histories) each year as the study progressed. The reason for the retrospective improvement in number of bats known to be alive is that not all bats are caught every year and some bats remain uncaught for several years. Females have had maximum intervals between captures of 14 years and males seven years.

The pattern of retrospective improvement in knowledge of the colony size as bats are recaptured is used to estimate current population size. Although a similar annual search effort is made, the proportion of the population caught each year varies. As a result, estimates for the most recent years will be the least reliable.

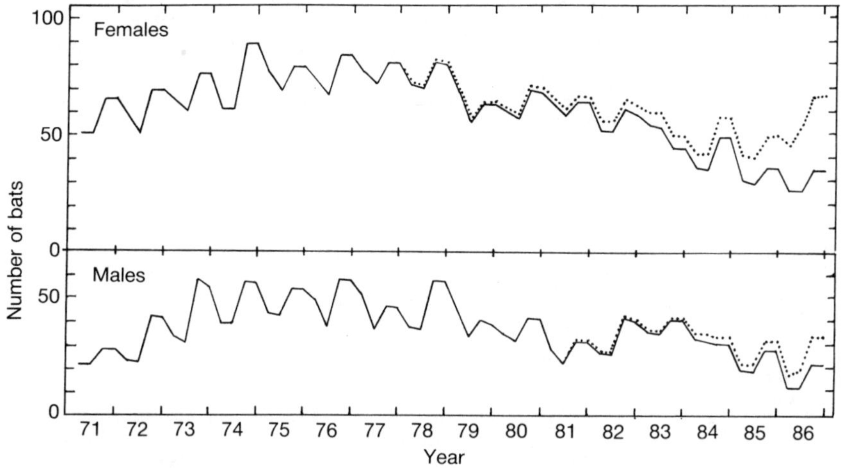

Fig. 3. Changing number of males and females in the colony from 1971 to 1986. Continuous line shows population size to December 1986 known from mark–recapture and retrospective census. Dotted line represents population estimates.

Estimate

The case histories were examined, and the total captures of individuals were counted for each year as if it was the first of the study. Then, for each subsequent year, bats which were caught and were known to be old enough to have been alive the previous year were added to that year's total. In this way, the progressive discovery of each year's total was tabulated. Then, the number of bats known in each second year was divided by the first, and similarly the third year by the second, *et seq.*, to give a value of the change every year. The values for each year category were averaged so we had an average value for the annual proportional change in number from the first year of study to the second, second to the third, and so on to the fifteenth to sixteenth year. These average figures were then applied in the following way.

The total number of bats caught in 1986 was multiplied by each of the 15 values. The figures for 1985 was multiplied by the last 14 values, omitting the figure for the first to second year. The value for 1984 was multiplied by the last 13 values and so on until the 1972 total was corrected using the last figure only. These estimates (Fig. 3) are produced four times each year: in March (at the end of hibernation), in June and again in July (before and after parturition) and in December (early hibernation).

The major fault in this estimate is that it will still tend to underestimate the true number of bats alive at any particular time. Inevitably there will be bats which are not recaptured perhaps for several years, and which die

Table 1. Differences between the estimates made on 31 December 1979, 1980 and 1981, and (a) an estimate made on 31st December 1986; (b) numbers, derived from case histories, of bats actually known to be alive in those years.

(a)	Date of estimate					
	1979		1980		1981	
	Diff.	%	Diff.	%	Diff.	%
1973	−2	1.5	−4	3.0	−2	1.5
1974	−1	0.7	−4	2.7	−4	2.7
1975	+1	0.7	−2	1.5	−1	0.7
1976	−5	3.5	−10	7.7	−5	3.5
1977	+1	0.8	−8	6.2	−1	0.8
1978	−9	6.5	−25	17.8	−14	10.1
1979	−7	6.8	−13	12.6	0	0
1980			−6	5.3	−7	6.2
1981					−3	3.0

(b)	Date of estimate					
	1979		1980		1981	
	Diff.	%	Diff.	%	Diff.	%
1973	−2	1.5	−4	3.0	−2	1.5
1974	−1	0.7	−4	2.7	−4	2.7
1975	+1	0.7	−2	1.5	−1	0.7
1976	−5	3.5	−10	7.7	−5	3.5
1977	+1	0.8	−8	6.2	−1	0.8
1978	−8	5.8	−24	17.4	−13	9.4
1979	−6	5.9	−12	11.8	−1	1.0
1980			−3	2.7	−4	3.6
1981					0	0

before recapture. It is believed these bats tend to be older animals which generally roost singly in widely scattered sites.

We compared the estimates made on 31st December in 1979, 1980 and 1981 with a similar estimate made in 1986 to see how accurate the predictions were. We also compared those former estimates with the number of animals subsequently known to have been alive in those years (Table 1). In 1979 and 1981 the estimates differed by averages of 2.9 and 3.1% from the 1986 estimate and by 2.7 and 2.6% from the number of animals known to have been alive. The differences for 1980 were greater at 7.1 and 6.6%. Almost all figures underestimated subsequent discoveries.

Sex ratio

Sex ratio at birth is not significantly different from unity (282 births 1978–1986; 149 males, 133 females; χ^2 0.91) but the slight preponderance of males (52.8%) is in contrast to the sex ratio of the entire colony, which is significantly biased and averages 38.5% males (χ^2 5.76, $p>0.02$, 1 *d.f.*; n = 60, range 29–48%). Figure 4 shows the changing ratio seasonally; throughout the study the proportion of males has remained more or less constant.

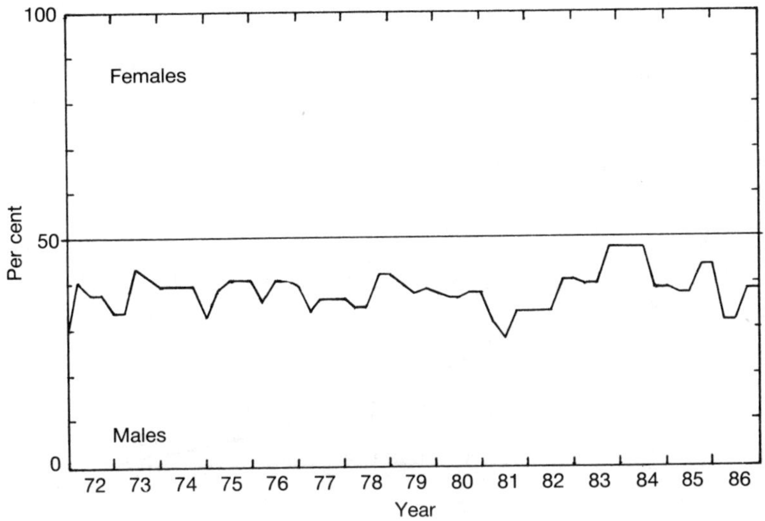

Fig. 4. Seasonal changes in sex ratio of the greater horseshoe bat colony.

These observations suggest that any changes there may have been to colony size have affected both sexes equally. Hence results for colony size presented below are for combined sexes.

Survival

Time-specific survival curves were constructed by plotting the sum of the proportion of marked bats surviving after one, two, three years and so on (Fig. 5). Log curves plotted separately for either sex were similar: males, $y = 0.58 - 0.25 \ln x$ ($r^2 = 0.92$) and females, $y = 0.54 - 0.21 \ln x$ ($r^2 = 0.93$).

Since the survival rate of older animals was likely to be higher than that of juveniles, we plotted the survival of cohorts of known age from year to year for the first nine years (Fig. 6). After that the number of surviving animals (22) in the sample becomes too small to produce reliable results. Surprisingly, the survival rate of animals in their second year (0.62) appears slightly lower than in their first (0.65), but thereafter there is a steady improvement until the ninth year (0.86). From the second to the ninth year the survival rate increases by 0.031 per year ($r^2 = 0.92$). Survival for the separate sexes is similar except in the first four years, when the rates for males were 0.66, 0.66, 0.58, 0.70 and for females 0.63 (−0.03), 0.60 (−0.06), 0.77 (+0.19) and 0.82(+0.12). Average survival for the first six years is 67.2% for males and 74.6% for females. This accounts for the consistent female bias in the sex ratio of the whole colony.

There is an obvious annual cycle in the changing colony size due to recruitment (births) in July and deaths occurring at other times (Fig. 2). Only

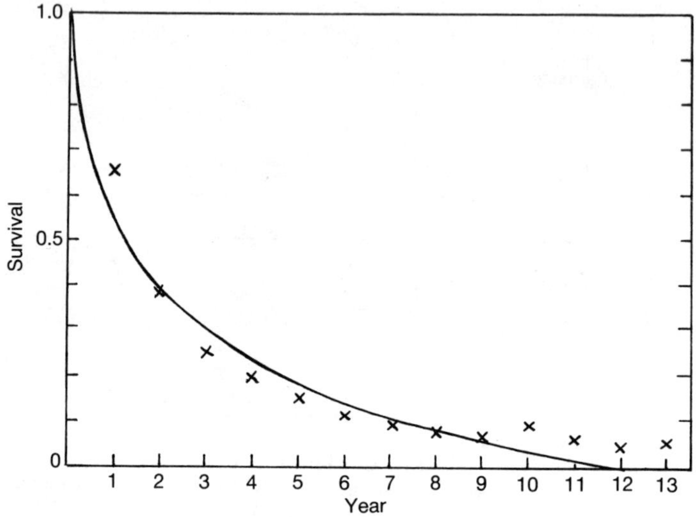

Fig. 5. Survival of Dorset greater horseshoe bats. Sexes and all years combined. Fitted log curve. $y = 0.54 - 0.22\ ln\ x, r^2 = 0.91$.

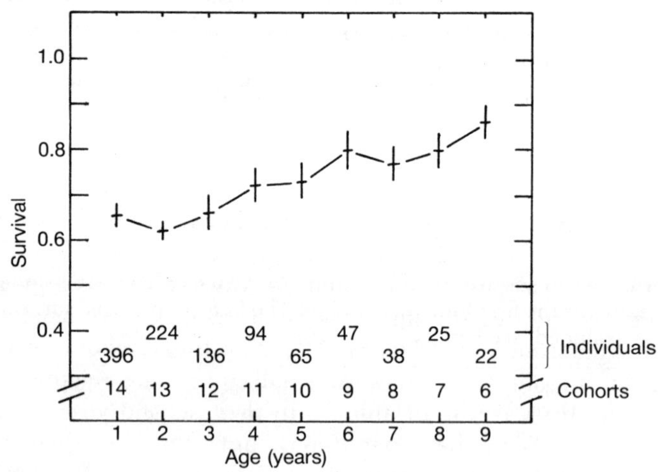

Fig. 6. Survival of cohorts of known age for bats aged from 1–9 years (sexes combined).

16 dead marked bats have been found and in almost all cases the date of death could not be determined.

The mean seasonal survival for the first four years of life of 12 annual cohorts born in the years 1972–1982 is shown in Fig. 7 and Table 2. The lowest survival rate (0.73) occurred within the first year of life, from birth (July) to December, whereas in the three succeeding years the highest survi-

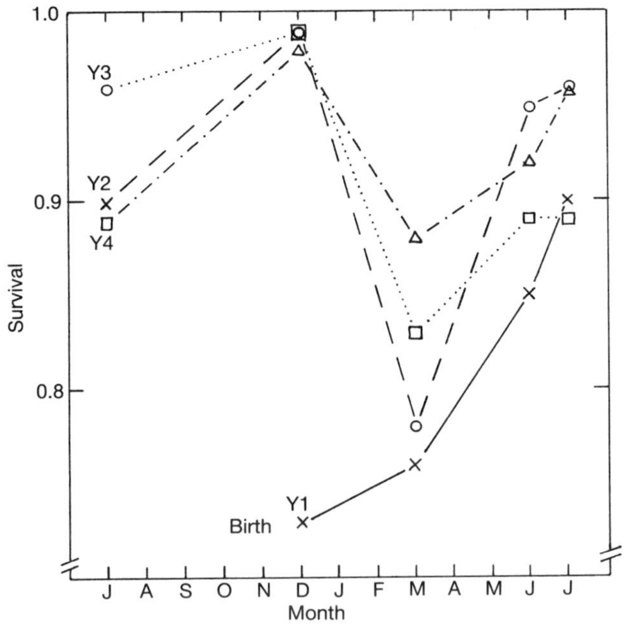

Fig. 7. Mean seasonal survival of 12 annual cohorts born in years 1972–1983, for the first four years from birth (sexes combined). The figure for December of the birth year is calculated from the known number of bats weaned. This figure is omitted from the averages shown in Table 2.

× = bats from 0–12 months
○ = bats from 13–24 months
□ = bats from 25–36 months
△ = bats from 37–48 months

Table 2. Mean survival of year cohorts for years 1972–1986 summed seasonally for the first four years of life. The first six months' survival after birth is excluded (see Fig. 8).

	August–December	January–March	April–June	July
n	35	46	46	46
$\bar{x}$	0.985	0.810	0.898	0.927
s.e. 95%	0.004	0.034	0.030	0.052

val rates, 0.99, 0.99 and 0.98 respectively, were recorded for that season. Most deaths occurred in the period from January to March with survivorship of 0.76, 0.78, 0.83 and 0.88 for the first four years respectively, showing a steady improvement with increasing age. Survival at the end of hibernation from April to June was higher than expected, as bats were considered vulnerable to starvation at this time.

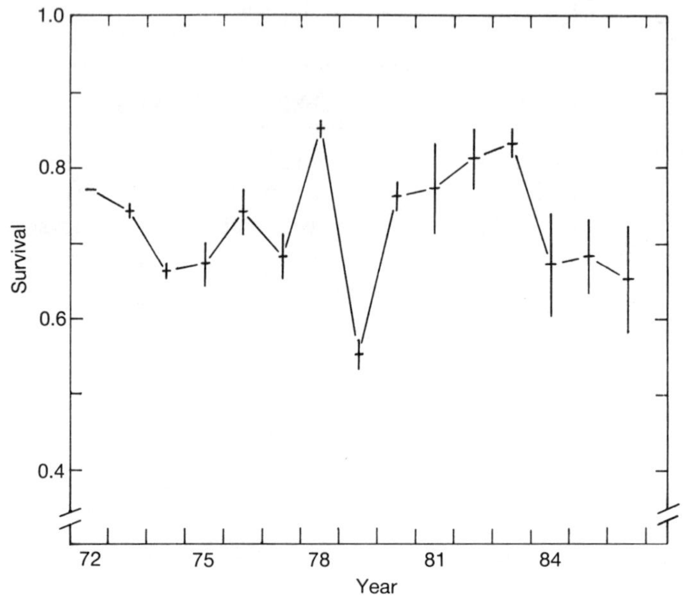

Fig. 8. Survival, year effects. Survival of all marked bats of all age groups from year to year. Sexes combined. Weanlings excluded unless subsequently ringed.

Survival of all marked bats of all age classes was examined from year to year. It was found to vary considerably, from 0.85 (1978) to only 0.55 (1979) (Fig. 8). The high mortality in 1979 had already been suspected (Fig. 2). Possible causes of this will be discussed below.

Recruitment

There are roost sites on the periphery of the Dorset colony's range which two or more colonies share for hibernation. But there is little evidence to suggest that bats from these peripheral colonies or from colonies elsewhere in Britain are entering the Dorset colony. More than 1,200 bats have been ringed in Dorset and several thousand in surrounding areas over 40 years, yet in 30 years only two bats have been found within the Dorset colony that came from the Somerset colonies 60 km to the north, and neither stayed. Only one bat is known to have emigrated westward to the area of south Devon (120 km), but that died in its first winter. Effectively, therefore, the only recruitment to the Dorset colony is by birth.

Age of females at first breeding

In Dorset, greater horseshoe bats usually give birth in July to a single young, with births spread over a three-week period which sometimes begins in late

June or extends into the first half of August. But how old are the females before they first reproduce? In practice, it is usually necessary to recapture individual bats between June and September each year following marking, if the age at which they first breed is to be recorded with precision. It is, however, also possible to determine age at first breeding for bats recaptured after an interval of a year or more that are found still to be immature (not having bred) and the following year are seen to have bred.

Animals of known birth date that were followed through until they first gave birth (Fig. 9a) first bred at a mean age of 3.96 years (n = 26).

However, other bats, also of known birth date (Fig. 9c), still had not given birth after three to seven years, with one bat having reached 11 years of age without having bred. The average age of these bats, which could be considered old enough to breed, was 4.53 years (n = 28). Many would probably first breed in their next year.

The last category (Fig. 9b) is bats which were immature females of unknown age upon initial capture, but which bred after a recorded time span. The minimum average breeding age of these was 4.17 years (n = 23), and one did not breed until it was at least nine years old.

It appears from these results that some females first breed at three years old but many are five to seven years old and occasionally some may be older. It is also known that some bats, especially young animals, did not breed every year, so recruitment will be slow.

Births

Recording the number of bats born to the colony annually may give a measure of the colony's success. Early in the study it was uncertain whether all births occurred in one place or if other sites might be used, as was known to be the case in former times. We now know that only one site is used because we mark each baby at birth—when the mothers have flown off to feed after dusk leaving the babies hanging in a crèche. Other unmarked juveniles are not found within the area occupied by the colony.

The precise number of bats born annually from 1977 to 1986 is shown in Fig. 10. Regressions were plotted for the years 1972–1976 and 1977–1986. In the former period the number of births were increasing at 1.4 per year (r^2 = 0.98) but in the latter there has been a decline of −2.0 per year (r^2 = 0.71).

The breeding site is a derelict building, the roof of which was partially vandalized in 1969. Water ingress caused decay and by 1977 holes had appeared in the roof covering. In 1978 a 5 × 3 m section collapsed, letting in wind and rain. Ambient temperatures in the roof were lower and this may have contributed to the decline in the number of births. The building continued to deteriorate and in the middle of the 1980–1981 winter, when the bats were away in hibernation, the rest of the roof was removed and

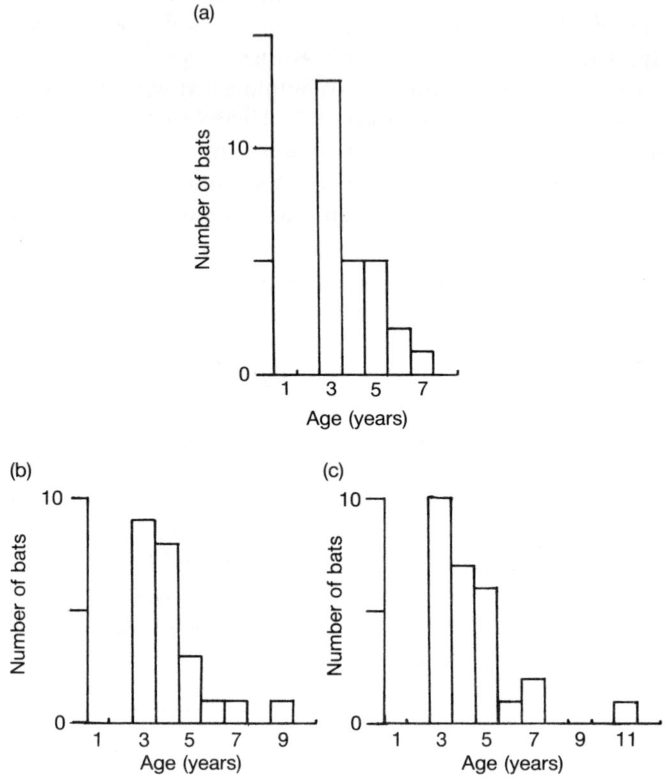

Fig. 9. Females. Age at first breeding. (a) Bats of known birth date; (b) Bats of unknown age upon initial capture: minimum age at first breeding shown; (c) Bats of known birth date which have not bred. (See text for explanation.)

replaced with a purpose-built structure. The bats did not return in the summer of 1981 until towards the end of weaning, and we did not discover how many bats were born that year. Electric heating has been provided in succeeding years and the bats have roosted on the heaters and given birth successfully each year. Details will be published elsewhere.

Colony age structure

As the study progressed, the proportion of animals of known age increased (Table 3), and with these true figures we can begin to observe further discrepancies between known and estimated colony size. Table 4 shows the number of births (one per female) with the known number of females alive in July each year when the births occur. In the first few years, 1972–1976, there were clearly more adult females alive than were caught, showing that those early years were underestimating actual colony size by at least six to seven animals. In the years 1977–1980 our knowledge of the colony size is probably at its best, and we know our more recent knowledge is still relatively poor as a larger proportion of animals have yet to be caught (Fig. 3). The 1977–1980 period had an average of 18% surplus females of potential breeding age over the number of bats born. Even this may be an underestimate, because in June 1978 (Table 3) there were 54 bats that were three years old (i.e., they would have been four years old in July) or older.

Discussion

Greater horseshoe bats in hibernation are often conspicuous, roosting in large clusters where they may be easily counted. In many studies bats have been counted in hibernacula and population statistics derived from the results (e.g. Bezem, Sluiter & van Heerdt 1960; Bárta *et al.* 1981). This type of data refers to populations which may include bats from many colonies. Bats in hibernacula seem particularly sensitive to disturbance by observers and frequent observations will not only cause bats to abandon the site but probably contribute significantly to mortality (Stebbings 1969).

Also, different sex and age groups of one species show differential preferences for hibernation sites, based primarily on temperature (Harmata 1969; Ransome 1968; R. E. Stebbings unpublished), and hence hibernation sites

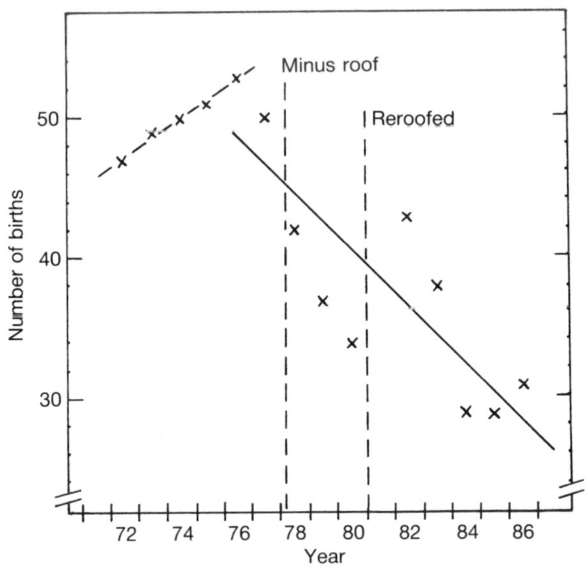

Fig. 10. Number of births per year (see text for explanation). Mean annual increase 1972–1976, +1.4 ($r^2 = 0.98$) and mean annual decrease 1977–1986, −2.0 ($r^2 = 0.71$).

Table 3. Age composition of the colony in June for two selected years—1978 and 1983.

1978	Age (years)	1	2	3	4	5	6	7	8	9	10	11	12	13	Total
Males	Known	7	5	5	3	3	3	1							27
	Estimated		1	1			7					1			10
Females	Known	7	9	3	12	4	7	1	2						45
	Estimated				2	0	10					13			25
															107
1983	**Age (years)**	**1**	**2**	**3**	**4**	**5**	**6**	**7**	**8**	**9**	**10**	**11**	**12**	**13**	**Total**
Males	Known	15	7	5	1	1	2		1	1	1				34
	Estimated														0
Females	Known	7	4	3	2	4	1	2	1	5	2	3		2	36
	Estimated			1	1	1	5					8			16
															86

Known = bats of known age
Estimated = minimum age—bats which were of unknown age when first caught

Table 4. Precise number of births related to known number of females of breeding age.

	Number of births	Number of females 4+ years old in July	Discrepancy
1972	47	41	−6
1973	49	43	−6
1974	50	43	−7
1975	51	44	−7
1976	53	49	−4
1977	50	55	+5
1978	42	51	+9
1979	37	46	+9
1980	34	47	+13
1981	no record	48	—
1982	43	43	0
1983	38	31	−7
1984	29	27	−2
1985	29	19	−10
1986	31	19	−12

do not usually contain an unbiased sample. Often males predominate, e.g. 60.7% of *Myotis daubentonii* in Denmark (Egsbaek & Jensen 1963) and 56.3% in England (Stebbings 1965). Therefore it is important to obtain information on a colony as a whole.

Although we will never be able to establish exactly how many greater horseshoe bats there were in Dorset before the timber treatment in 1953, our guess is that there must have been at least 15,000, which would give a density of 0.05 per ha over the then known range. This is a lower density than for eight European vespertilionid species (0.1 to 1.1 per ha) reported by Gaisler (1979). The 1986 density in Dorset is 6.0×10^{-4} per ha or one bat in 167 km^2, assuming a colony size of about 90 bats. However, the bats are not evenly spread throughout the range, but instead are clumped in favoured areas. Retrospective census from case histories has the major fault of underestimating true colony size because bats die at various time spans after the last recapture. With a constant annual search effort trends do eventually become apparent, but the most accurate information is for some years before the present.

The same argument is true for survival statistics, which depend on the accumulation of several years' results before conclusions can be reached. Gaisler (1979) suggests that, for hibernating monoestrous bats, around 30% of juveniles and 60% of adults survive annually. However, we have shown that juvenile and second-year greater horseshoe bats average 64% survival, immatures about 70% and adults about 80%. Even these figures we consider to be underestimates because of the presumed existence of the uncaught bats. Gaisler's (1979) figures will certainly be too low for female greater horseshoe bats which on average are over four years old before they first breed. Most other European bats first breed in their first or second years, and even these have been shown to have high survival: for example, female *Plecotus auritus* had a 76% survival (Stebbings 1970).

We feel that the question has not yet been answered of whether, by counting the number of births, an adequate indication of the colony size is given, mainly because there are few years for which we are reasonably confident of the quality of results. We do not yet know whether there is a constant proportion of non-breeding adults, nor indeed do we know the factors that determine the age at which they first reproduce. The problem remains of a colony apparently close to extinction requiring research to aid judgements as to conservation priorities.

Acknowledgements

This work was funded in part by the Nature Conservancy Council. We are most grateful to Andrew Watson, Michael Blackmore and Sdeuard Bisserôt for their records and observations; to Ken Lakhani for statistical advice; to the many householders and landowners for access to their properties and especially to the National Trust and Ministry of Defence.

References

Bárta, Z., Červený, J., Gaisler, J., Hanák, P., Hanák, V., Horáček, I., Hůrka, L., Miles, P., Nevrlý, M., Rumler, Z., Sklenár, J. & Žalman, J. (1981). Výsledky zimniho sčítáni netopýrů v Československu : 1969–1979. *Sb. okresn. Muz. Mostě* (*Přirod.*) **3**: 71–116.

Bezem, J.J., Sluiter, J.W. & Heerdt, P.F. van (1960). Population statistics of five species of the bat genus *Myotis* and one of the genus *Rhinolophus*, hibernating in the caves of S. Limburg. *Archs néerl. Zool.* **13**: 511–39.

Caubère, B., Gaucher, P. & Julien, J.F. (1984). Un record mondial de longevité in natura pour un chiroptère insectivore? *Terre Vie* **39**: 351–3.

Egsbaek, W. & Jensen, B. (1963). Results of bat banding in Denmark. *Vidensk. Meddr. dansk naturh. Foren.* **125**: 269–96.

Gaisler, J. (1979). Ecology of bats. In *Ecology of small mammals*: 282–342. (Ed. Stoddart, D.M.). Chapman & Hall, London.

Harmata, W. (1969). The thermopreferendum of some species of bats (Chiroptera). *Acta theriol.* **14**: 49–62.

Ransome, R.D. (1968). The distribution of the Greater horse-shoe bat, *Rhinolophus ferrum-equinum*, during hibernation, in relation to environmental factors. *J. Zool., Lond.* **154**: 77–112.

Stebbings, R.E. (1965). Observations during sixteen years on winter roosts of bats in west Suffolk. *Proc. zool. Soc. Lond.* **144**: 137–43.

Stebbings, R.E. (1969). Observer influence on bat behaviour. *Lynx* (N.S.) No. 10: 93–100.

Stebbings, R.E. (1970). A comparative study of *Plecotus auritus* and *P. austriacus* (Chiroptera, Vespertilionidae) inhabiting one roost. *Bijdr. Dierk.* **40**: 91–4.

Note added in proof

In June 1987 an unexpected discovery supported our estimate that the Dorset greater horseshoe colony numbered at least 15,000 bats before 1953. A previously unknown former nursery was discovered only 2.7 km south-east of the present nursery. An estate worker remembers seeing clusters of bats, including babies, covering several square metres in cellars of a large house throughout the 1940s. The house and cellars were demolished in 1949. He further stated that bats lined the roof of the adjacent stables, measuring *c.* 50 × 8 m, which was cleared out in June 1945. In the winter of 1951–1952 a new owner removed a 'lorry load' of droppings, but does not remember any further roof cleaning.

An inspection on 16 June 1987 revealed only small quantities of droppings through most of the roof, but, in a small area which is difficult to enter, there were 1.9 m^3 of old disintegrated droppings. On top of these was a paper dated November 1962 which appeared to be contemporary. Lack of droppings on top suggested no significant number of bats had roosted there for over 20 years.

Interpretation: the estate worker's description suggests several thousand greater horseshoe bats roosted at the site in the 1930s and 1940s.

The fact that a large number of bats apparently disappeared from a traditional site in the early 1950s supports the hypothesis that most of the colony was killed by the Lindane treatment to another nursery 24 km to the south. It is known that greater horseshoe bats regularly visit other roosts and it seems likely that they died when they visited the treated roof.

Symp. zool. Soc. Lond. (1987) No. 58: 25–57

The application of Capture–Mark–Recapture methods to the enumeration of small mammal populations

W. I. MONTGOMERY

Department of Zoology
Queen's University of Belfast
Belfast BT7 1NN
Northern Ireland

Synopsis

A review of literature, 1970–1986, revealed that Capture–Mark–Recapture (CMR) methods of estimating population size were used in approximately 10% of studies of small mammals in which population size was assessed. Even when applied, the full potential of these techniques was seldom realized. The underutilization of CMR is explained by fears of violating the assumptions underlying many CMR methods, the failure to recognize that certain CMR models are sufficiently robust to withstand such deviations and the misapprehension that other methods suffer less from differential trappability and survival. The probability of initial capture may differ from that of subsequent capture. Probability of capture may also vary with species, sex and age, and from one individual to another. Age-dependent survival rates may confound the Jolly–Seber method but, again, this method can withstand minor departures from its statistical basis. Trap response and heterogeneity in capture probability may be accommodated by developments in the estimation of closed populations (Otis, Burnham, White & Anderson 1978). Age-dependent survival has been assumed in recent modifications of the Jolly–Seber method. Comparison of the Jolly–Seber model and enumeration methods suggests that Minimum Number Alive (MNA) suffers even more than CMR from heterogeneity in trappability and should not be used. MNA involves underestimation of population size to a varying and unknown extent. Following Pollock (1982), it is suggested that population size should be estimated over a sequence of samples from a single sampling period using an appropriate CMR model for a closed population, while survival and birth rates should be estimated from data pooled for each of a number of such sampling periods using the Jolly–Seber method. Where the assumption of population closure within sampling periods is invalid, Jolly–Seber estimates of population size based on data lumped for sampling periods may be used. The limitations of CMR methods in

ZOOLOGICAL SYMPOSIUM No. 58
ISBN 0-19-854006-X

elucidating the processes that determine the distribution and abundance of small mammals are discussed.

Introduction

The idea that information about an animal population may be obtained through marking some individuals may be traced back over 300 years, although Peterson did not describe the fundamental principle until the latter part of the 19th century (Cormack 1968; Otis *et al.* 1978; Blower, Cook & Bishop 1981). Lincoln's (1930) estimate of waterfowl population size using ringing returns represents the beginning of the intense, almost universal, application of Capture–Mark–Recapture (CMR) techniques in field ecology. Early developments, mainly in the context of entomology or fisheries, included the analyses of sequential CMR data by Jackson (1933), Fisher & Ford (1947) and Schnabel (1938). Deterministic estimates, where random fluctuations in population parameters such as survival are precluded, were later joined by the more rigorous mathematical treatments of the stochastic CMR models of population estimation employed in the Jolly–Seber (Jolly 1965; Seber 1965) and Manly & Parr (1968) methods, in which the probabilities of population events are defined in the initial assumptions. CMR techniques may also be broadly divided between those appropriate to 'closed' populations, where the animals have a survival rate of one and recruitment into the population is zero, and those pertaining to populations in which natality, mortality and migration operate. CMR models for 'open' populations are exemplified by the Jolly–Seber method in which the parameters of population size, survival rate, recruitment and capture probability are estimated. The advantage of closed models of population estimation is that they involve the estimation of fewer parameters. The advantage of open models of population estimation lies in greater reality and broader applicability.

During the 1940s various workers in North America and Australia applied CMR methods to the estimation of mammal populations, but not without misgivings (Otis *et al.* 1978). Dice (1941), for example, stated that 'the application of the proportional method of calculating mammalian populations may often require as much effort as the complete trapping or counting of the whole sample population'. Hayne (1949), following Schumacher & Eschmeyer (1943), suggested estimating the size of a closed population of small mammals using the linear relationship of the fraction of individuals marked in a sequence of samples and the cumulative number of marks. The most sophisticated suite of models for estimating the size of closed populations of small mammals (Otis *et al.* 1978; White, Anderson, Burnham & Otis 1982) incorporates heterogeneity in the probability of capture with time, with capture history and between individuals. A maximum likelihood (ML) method for the estimation of the death rate and size of open populations of small mammals was presented by Leslie & Chitty (1951) and Leslie (1952). Arnason & Baniuk (1980) described a computer system for the CMR analysis of open populations using the Jolly–Seber model. Le Boulengé (1984) presents a comprehensive package for the analysis of CMR data on small mammals. Since the mid-1960s, the Jolly–Seber method has been applied sporadically to rodent populations. In Begon (1983), for example, 25% of 53 published uses of the Jolly–Seber method involved rodents. Begon noted the general abuse of the Jolly–Seber method, with estimates lacking any indication of error, being based on insufficient data and showing a cavalier disregard of the assumptions the method makes. Elsewhere, Nichols & Pollock (1983) lamented the predominance of simple enumeration of populations using 'minimum number known alive' (MNA) and 'calendar-of-catches' (CC) in the literature on small mammal populations. These authors proposed that closed population models might be appropriate to estimate population size within a sampling period and that survival rates could be estimated from recapture rates between sampling periods using the Jolly–Seber model.

The objectives of this review are to report the current use of CMR in studies of small mammals, examine the validity of assumptions underlying CMR models, describe how the application of such methods might be enhanced to good effect and indicate their limitations in revealing fundamental processes in the temporal and spatial dynamics of small mammals. It is not intended to present detailed accounts of CMR analyses. For detailed, technical accounts of the most important methods the reader is referred to Cormack (1968), Otis *et al.* (1978) and Seber (1982). Less mathematically-involved accounts are presented by Parr, Gaskell & George (1968), Cormack (1973), Flowerdew (1976), Caughley (1977), Southwood (1978), Begon (1979), Davis & Winstead (1980), Blower *et al.* (1981) and White *et al.* (1982). Recently, Seber (1986) reviewed advances in estimating animal abundance during the first half of the 1980s.

The advantages of CMR

The argument for the application of CMR techniques in the study of field populations of small mammals is impressive. CMR methods permit the estimation of population size of elusive, mobile organisms that are seldom if ever amenable to direct counts. CMR methods have a clear advantage over removal methods of estimating small mammal population parameters, such as the Standard Minimum method (Grodziński, Pucek & Ryszkowski 1966), because the population remains intact and *in situ*. CMR methods do not suffer from either of the major drawbacks of CC or MNA methods of population estimation where estimates tail off towards the end of a series of samples and lack measurement of error. CC and MNA also require frequent, regular periods of intensive trapping effort (Flowerdew 1976). These advantages apply to even the most simple of CMR models. Recent CMR models for the estimation of closed populations (Otis *et al.* 1978) have the advantage of incorporating differential trappability and response to traps. The Jolly–Seber model for open populations allows variable levels of ingress into and egress from the population, permits estimation of population size and survival rates and their standard errors for the majority of occasions on which the population is sampled, and is fully stochastic (Begon 1983).

The use of CMR in small mammal ecology

Current methodology for the estimation of the abundance of five genera of small rodents is summarized in Table 1. The literature review spanned the period 1970 to 1986 and is representative rather than comprehensive. Studies in the survey included enclosed and unenclosed, field and laboratory, populations. Each study was counted only once even where several species were involved. Not all the papers involved estimation of population size. The proportion of studies on *Mus* in which population size featured was smaller than in the other genera. Methodology varied to some extent between the genera. The MNA method, for example, predominated among studies of *Microtus* and *Peromyscus* to a much greater extent than among studies of *Apodemus* and *Clethrionomys*. This may reflect a methodological difference between studies in North America and Europe. It is evident, however, that estimation of population size using CMR is infrequent. Approximately 10% (20) of the 204 studies in which population size was estimated involved CMR. Of these 20 studies, eight involved use of the Lincoln Index or Hayne's modified Lincoln Index method (Hayne 1949), seven employed the Jolly–Seber method and five the Leslie & Chitty ML method. Only four studies presented standard errors for estimations based on CMR models. These data suggest that a potentially powerful technique for the estimation of population size and other population parameters, CMR, is underutilized in the study of small mammals. Further, when CMR methods are applied, they are seldom fully exploited.

The diversity of CMR models and their underlying assumptions

This brief account of the bases and assumptions of CMR methods is restricted to those techniques that have been or might be applied more widely in the study of the ecology of small mammals. Basic models are described without reference to how they have been elaborated to include, for example, events such as loss of marked animals dying in traps. The methods

Table 1. Summary of literature survey of methods of estimating population size in the study of small rodents of specified genera between 1970 and 1986. For details see text.

	Microtus	*Clethrionomys*	*Peromyscus*	*Mus*	*Apodemus*	Total
Papers examined	128	53	126	29	55	391
Population estimated	82	26	50	8	28	204
Method						
Trap index	9	6	5	2	5	27
Removal method	3	3	1	0	4	11
Calendar of captures	1	9	0	0	4	14
Minimum number alive	55	5	31	1	7	99
Number caught	7	2	9	4	11	33
Capture-mark-recapture	7	1	4	1	7	20

are separated into those dealing with closed and those with open populations. This classification has a statistical rather than a biological basis (Pollock 1982).

Two samples from a closed population: the Lincoln Index

The basis of the Lincoln Index is that, in a static population, the ratio of the number of marked individuals released after one sample to total population size equals the ratio of marked to unmarked individuals in a subsequent sample. Caughley (1977) and Seber (1982) indicate the number of recaptures required to provide a standard error of a given magnitude for an estimate of a population of a given size. Blower *et al.* (1981) derive the Lincoln Index by three methods and show that population size and sampling effort may be estimated for the time of the first or the second sample. They demonstrate that the index gives an estimate of numbers for the first and second occasion only when there are neither losses nor gains, for the first occasion only when there is population loss but no gain (provided marked and unmarked animals are lost in the same proportion) and the second occasion only when there is gain but no loss. The problem, however, is to decide whether the population is subject to losses, gains or gains and losses. The form of the Lincoln Index most commonly used is

$$\hat{N}_{i+1} = n_i \, n_{i+1} / m_{i+1} \tag{1}$$

where n_i and n_{i+1} are the number of individuals captured in the first and second samples, m_i is the number of marked animals in the second sample and $\hat{N}_{i+1}$ is the population estimate. (See Blower *et al.* 1981 for a general discussion of the Lincoln Index). The Lincoln Index is generally regarded as dependent on the following assumptions (Seber 1982):

1. The population is closed.
2. All individuals have the same chance of being caught in the initial sample.
3. Capture in the first sample does not affect the probability of an individual being caught again.
4. Marked animals must intermingle evenly with unmarked ones.
5. The marks applied in the first sample must last until the second.
6. There must be no ambiguity in identifying marks in the second sample.

The Lincoln Index is biased upward by births and immigration and is best restricted to populations sampled between periods of recruitment (Caughley 1977). The effects of immigration may be kept to a minimum by minimizing the period between the first and second samples.

Multiple samples from a closed population; a maximum likelihood approach

The principal method in statistical estimation of population size over the last 50 years has been the maximum likelihood (ML) method (White *et al.* 1982). ML estimators are found using calculus techniques on the 'likelihood function'. In the present context, the likelihood function is a function of the parameters N, population size, and p, capture probability (Fig. 1), given a data set where capture of an individual, i, on a sampling occasion, j, is denoted by 1 and its non-capture by 0. Data so arranged are referred to as the X matrix (Otis *et al.* 1978; White *et al.* 1982). In CMR models, ML estimators rarely exist as simple formulae. Darroch (1958) developed a ML estimator for N based on four sampling occasions with variable capture probabilities over time. Here the approximate ML estimator is the unique value of N that satisfies

$$(1 - M_5/N) = (1 - n_1/N)(1 - n_2/N)(1 - n_3/N)(1 - n_4/N) \tag{2}$$

where n_1 to $_4$ is the number caught on occasions 1 to 4 and M_5 is the number of individuals caught during the trapping programme. The general form of Darroch's ML estimator is

$$(1 - M_{t+1}/N) = \Pi\ (1 - n_j\,/N) \tag{3}$$

where n_j is the number caught on occasion j, and M_{t+1} is the total number of individuals caught on all occasions. For values of t in excess of 2, Equation 3 cannot be solved algebraically and is solved iteratively.

Otis *et al.* (1978) demonstrated that estimators based on Equation 3 are only approximate and derive eight ML models. Three sources of variation in

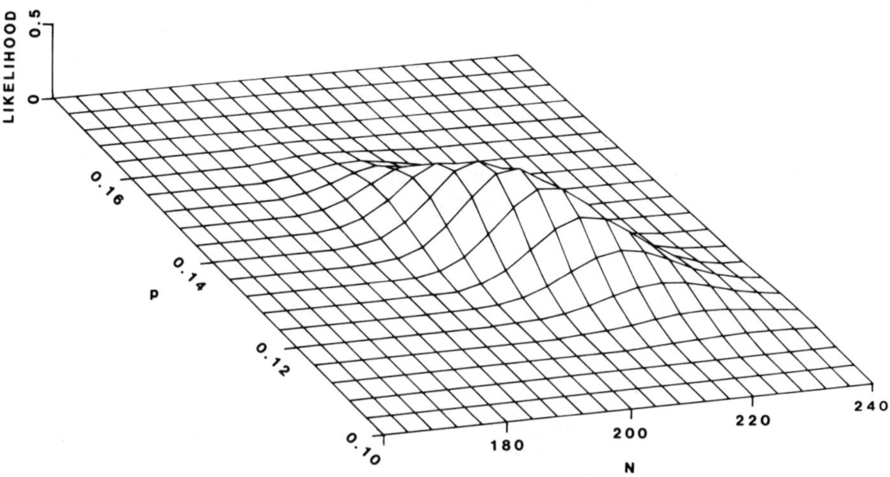

Fig. 1. A three-dimensional representation of population estimation for a given data set where there are two unknown variables; N = the population size and p = constant capture probability. The maximum likelihood estimates N and p are those which maximize the likelihood function ($N = 208, p = 0.136$) (After White *et al.* 1982).

capture probabilities are incorporated into seven of these models, the first being based on the assumption that capture probabilities are constant. Capture probabilities are allowed to vary with time (see above), response to capture and heterogeneity due to species, sex, age or individual differences. These factors may operate singly, in any combination by pairs or all together (Table 2). Relationships between the eight models are illustrated in Fig. 2. Otis *et al.* (1978) described estimation methods for only five of these models, the remainder being difficult to discriminate from their special cases (Fig. 2). Otis *et al.* (1978) emphasize a unified approach in choosing which model to employ in the estimation of population parameters according to the assumptions violated or upheld in the trap data. They suggest that the most appropriate trappability model should be selected through a rigorous examination of the data and that model should then be used to estimate further parameters.

Increase in the fraction marked in a mark-and-release sequence

The rationale of this method, suggested first by Schumacher & Eschmeyer (1943) and developed by Hayne (1949) for application to CMR data on small mammals, is that there is an increase in the fraction marked in a sequence of samples. The population is estimated from the slope of the line describing the relationship between recaptures as the proportion of successive samples, and the cumulative number of marks in the population (Fig. 3). The method assumes that the latter relationship is linear and passes through

Table 2. The eight models forming the basis for the estimation of closed populations proposed by Otis *et al.* (1978). Variation in capture probability with time indicates that trappability is equal for all individuals but varies between samples. Variation due to behaviour indicates a trap response such that the probability of catching an unmarked animal differs from that of catching a marked one. Heterogeneity in trappability indicates variation in capture probability among individuals which may or may not be of different sub-groups in the population.

Model	Sources of variation in capture probability	Population estimator
M_0	None	Null
M_t	Time	Darroch (1958)
M_b	Behaviour	Zippin (1956)
M_h	Heterogeneity	Jackknife (Otis *et al.* 1978)
M_{tb}	Time, behaviour	None
M_{th}	Time, heterogeneity	None
M_{bh}	Behaviour, heterogeneity	Generalized removal (Otis *et al.* 1978)
M_{tbh}	Time, behaviour, heterogeneity	None

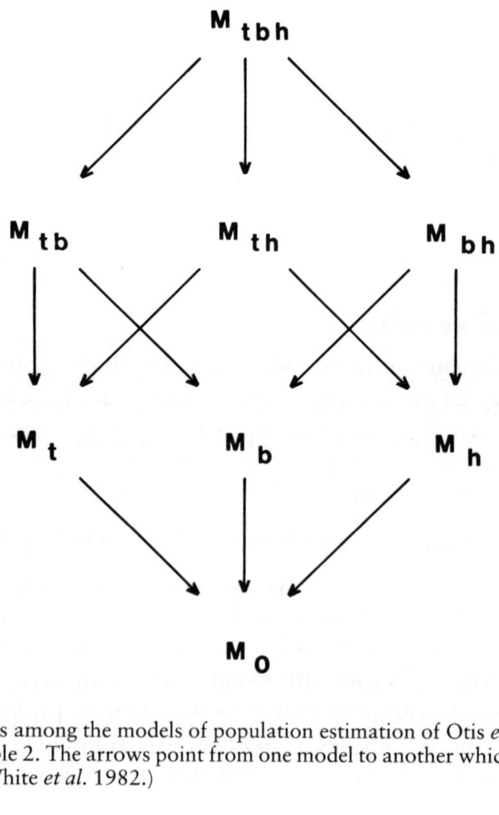

Fig. 2. Relationships among the models of population estimation of Otis *et al.* (1978). Models are defined as in Table 2. The arrows point from one model to another which is a special case of the former. (After White *et al.* 1982.)

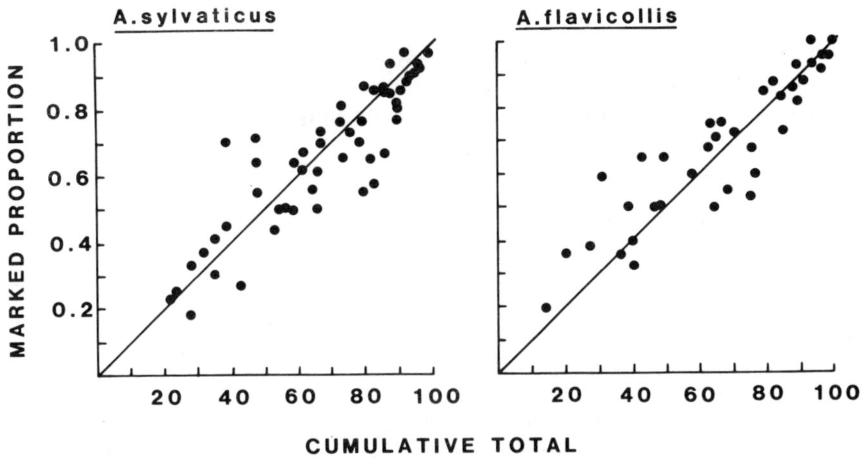

Fig. 3. The relationship between the proportion of each sample that is marked and the cumulative number of individuals marked previously in the population in *A. sylvaticus* and *A. flavicollis*. Lines have a slope of 1.0. (After Montgomery 1985.)

the origin, the probability of capture is constant and capture does not affect the chance of recapture. There must also be no ingress into or egress from the population. Seber (1982) discusses a number of CMR regression methods for the estimation of population size, including several where non-linearity in the relationship of the proportion of each sample that is marked and the cumulative number of marks in the population is assumed. Du Feu, Hounsome & Spence (1983) introduced a ML estimator of population size based on the number of individuals captured and the number of recaptures.

Frequency-of-capture models

These methods depend on the number of individuals captured once, twice, three times and so on during the course of trapping. These form a zero-truncated frequency distribution of captures where the missing zero class represents the unknown number of individuals never caught. Methods of population estimation attempt to fit the data to a theoretical distribution, such as the Binomial, Poisson, Negative Binomial or Geometric distributions, and hence evaluate the number of individuals not captured. Goodness-of-fit tests may be used to determine which of a number of distributions best describes the frequency-of-capture data. The Binomial and Poisson methods are dependent on the assumption that probability of capture is constant while the Negative Binomial and Geometric models permit unequal catchability (Caughley 1977; Seber 1982). Frequency-of-capture methods assume that the distribution of capture frequency can be described by one of a number of mathematical distributions. Seber (1986) suggests that these methods of estimation should be abandoned, since it is frequently difficult to distinguish between the models, because CMR data may fit more than one theoretical frequency distribution, and all models are particularly sensitive to heterogeneity in capture probability, even those designed to accommodate such variation.

Multiple samples from an open population: the Leslie–Chitty model

CMR methods based on Method B of organizating data (Leslie & Chitty 1951; Leslie 1952), in which animals are classified solely in accord with the date on which they were last captured, remain the only ones specifically designed for applications in the estimation of field populations of small mammals. A number of estimates of population size were described by Leslie (1952). The general model, where egress per unit time, $Q = 1 - P$, is assumed to be constant throughout the period of sampling but ingress, B_t, into the population is occurring, requires the population to be sampled at equidistant intervals. Losses are assumed to fall evenly on all classes in the population. The number of individuals entering the population during the interval t to $t + 1$, which are alive at $t + 1$, is B_tN_t. The change in population size over the interval t to $t + 1$ is

$$N_{t+1} = (P + B_t)\, N_t \tag{4}$$

This is evaluated by estimating N_t, N_{t+1} and P. The estimate of population size, $\hat{N}_t$, is based on the equation

$$\hat{N}_t = \O_t\, R_t/s_t \tag{5}$$

where R_t is the number of individuals captured, marked and returned to the population, s_t is the sum of the number of individuals last captured at time x where $x = 0,1,2 \ldots, (t - 1)$, and $\O_t$ is derived from the solution of a polynomial function of P for the expected number of recaptures of animals last released at time x. In practice, $\O_t$ is calculated as the total number of animals captured once or more, which are alive in the population as a whole at t (ψ_t). Leslie (1952) found that Equation 5 leads to overestimation of population size and suggested modifying the equation in a manner similar to Bailey's (1951) modification of the Lincoln Index so that

$$\hat{N}_t = \psi_t(R_t + 1)/(s_t + 1) \tag{6}$$

The assumptions underlying estimation by this method are that sampling the population is random and that marked and unmarked animals are captured with equal facility (Leslie 1952).

Multiple samples from an open population: the Jolly–Seber model

The Jolly–Seber method uses data in statistically the most efficient manner (Blower *et al.* 1981). It employs the form of the Lincoln Index that is unaffected by gain but must be corrected for loss (Equation 1). This correction is made by calculating the number of marked animals from all previous releases, M_i, on a given date from

$$R_i/n_i = z_i/(\hat{M}_i - m_i) \qquad (7)$$

where n_i is the total number of captures on date i, m_i is the number of recaptures on date i, R_i is the number of animals marked on i and recaptured subsequently and z_i is the number of animals marked previous to i, not caught on i but recaptured subsequently. Equation 7 can be solved for $\hat{M}_i$ and this substituted into the basic equation

$$\hat{N}_i = n_i\hat{M}_i/m_i \qquad (8)$$

to find the estimate of population size, $\hat{N}_i$. Equation 7 depends on the assertion that any class of marked animals should behave as any other class. Those caught on i, (n_i), and those marked previously but not recaptured on i, ($\hat{M}_i - m_i$), should be equally liable to capture after i. This relationship should persist even if sampling intensity varies after (i + 1) (Blower *et al.* 1981). The basic Jolly–Seber method is based on the following assumptions (Seber 1982):

1. Each animal in the population, marked and unmarked, has the same probability of being captured in the ith sample provided it survives and is in the population when this sample is taken.
2. After release, every marked animal in the population has the same probability of surviving to the (i + 1)th sample and remaining in the population. In practice, this requires that survival is independent of age.
3. All captured animals are equally likely to be returned to the population.
4. Marks are not lost.
5. The time required for sampling the population is short and animals are released immediately after the sample is taken.

Statistical propriety and biological reality

The task of the statistician is made easier by predetermination of the manner in which 'animals' behave. In the statistical population all individuals may be equally catchable, be unresponsive to confinement in traps and handling, and mingle thoroughly with unmarked animals before being liable to recapture. The application of CMR techniques to the estimation of biological populations, however, depends largely on characteristics of the population and the behaviour of individuals, over which the biologist has little control. Biological reality should determine which CMR models are applied (Otis *et al.* 1978). Most investigators, however, are unaware of the relevant biology of the organisms under study and the inadequacy of their models of population estimation, until data are investigated. Biological populations may violate the assumptions of statistical CMR models for the estimation of population parameters in a number of ways. These statistical anomalies and the manner in which they may be assessed are reviewed below. Biologists can rarely comment on animals they fail to catch at all and most of the discussion concerns individuals that constitute the marked portion of the population. It is uncertain whether the marked proportion of a population is representative of the population as a whole.

Population closure

The distinction between open and closed populations is fundamental in the construction of statistical models of CMR. Biological populations, however, are rarely completely closed and the biologist requires some method of determining whether CMR methods for the estimation of closed populations might be applied to his data. Robson & Flick (1965), Robson & Regier (1968), Pollock, Solomon & Robson (1974) and Otis *et al.* (1978) have proposed tests for ingress and/or egress. None is wholly satisfactory because they have little chance of rejecting closure unless samples are large and the departure from closure is substantial (Seber 1982). Also, certain tests are invalid when closure is true but the assumptions inherent in a particular CMR model are false (Otis *et al.* 1978; Seber 1982). These authors conclude that the best evidence for population closure is biological rather than statistical. White *et al.* (1982), for example, present data for an eight-day CMR census of cotton rats, *Sigmodon hispidus*. On Day 6, the number of new individuals was noticeably greater than on each of the previous five days, so that the authors were led to conclude that there had been a sudden influx of animals into the study area. Hence the assumption of closure was invalid. Montgomery (1985) conversely concluded that populations of *Apodemus sylvaticus* and *A. flavicollis* trapped over five days could be regarded as effectively closed given the observed patterns of capture (see below) and rates of predation and recruitment. Certain CMR methods permit rapid evaluation of population closure. For example, in the method of du Feu *et al.* (1983), population closure is indicated by the levelling out of the relationship between population size and number of recaptures (D. W. Yalden pers. comm.). Trapping programmes may also be designed in such a manner that the assumption of population closure is fulfilled as nearly as possible, by minimizing the sampling period and avoiding known times of migration and recruitment (Seber 1982).

The loss of marks

Loss of marks will lead to overestimation of population size and underestimation of survival rates (Begon 1979). This may be a problem in studies of small mammals employing ear-tags or leg-rings. Seber (1982) reviews several methods for estimating loss of marks using double-tagging. For example, estimates of population size and its variance based on the Lincoln Index may be made in double-tagging experiments provided the tagging methods are independent of each other (Seber & Felton 1981). Elsewhere, Arnason & Mills (1981) have shown that the population estimate in the full Jolly–Seber model (allowing both ingress and egress) is unbiased by tag loss but there is loss of precision and estimates of survival and births are biased. Correction is possible only with an estimate of the rate of losing marks. Where the Jolly–Seber model is amended to assume no ingress, tag loss may be accommodated without a measurement of tag loss rate (Arnason & Mills 1981).

The effect of marking: trap response

Marking may affect an animal's survival or probability of recapture. This is termed 'trap response'. It is possible to test for the effects of marking even though their causation may remain uncertain. Such tests commonly involve the comparison of the recapture rate of individuals marked in sample i with that of individuals marked before sample i, using contingency analysis (Robson 1969; Begon 1979; Blower *et al.* 1981; Seber 1982). They have only rarely been applied to CMR data for small mammals despite the frequent use of leg-ringing, toe-clipping and ear-punching (Twigg 1976). Bishop & Hartley (1976) found that marking did not affect the number of recaptures of *Rattus norvegicus*. Smal & Fairley (1982) and Fairley (1982) reported that newly marked *C. glareolus* and *A. sylvaticus* were less likely to be recaptured on the following day than individuals marked in an earlier trapping session. The effect of marking, however, is apparent only in certain species and may be short-lived (Table 3). Another approach to determining the effect of marking is to compare the frequencies of captures or latency to first capture between previously marked and freshly marked individuals (see Gurnell 1982). These tests usually reveal that trap-experienced individuals are more liable to capture in future trapping sessions than trap-naive animals (Fig. 4). In these analyses, variation in trappability due to sources other than initial capture is evened out by lumping data over many trapping sessions. This may sometimes be inadequate and may be overcome by the ML test of the equality of the probability of first capture and the probability of recapture presented by Otis *et al.*(1978). Phillips & Campbell (1970) suggested that a linear relationship between the proportion of marked individuals in a sample and the number of marked animals in the population

Table 3. Recapture data for *A. sylvaticus, A. flavicollis* and *C. glareolus* released on Day 1 of a five-day trapping period. Numbers of recaptured and non-recaptured individuals were determined after one day, four days and at any time. Data lumped for 17 separate trapping periods over 31 months.

	One day			Four days			Any time		
	Rec.	Non.	% rec.	Rec.	Non.	% rec.	Rec.	Non.	% rec.
A. sylvaticus									
New	49	47	51	80	16	83	89	7	93
Old	88	42	68	120	10	92	123	7	95
χ^2	5.735 $p < 0.05$			3.531 n.s.			0.095 n.s.		
A. flavicollis									
New	24	26	48	40	10	80	43	7	86
Old	40	29	59	60	9	87	63	6	91
χ^2	0.332 n.s.			0.591 n.s.			0.382 n.s.		
C. glareolus									
New	8	13	38	16	5	76	18	3	86
Old	12	17	41	23	6	79	27	2	93
χ^2	0.003 n.s.			0.007 n.s.			0.146 n.s.		

(Fig. 3) indicates that marked and unmarked animals are equally liable to capture. Roff (1973), however, stated that this test should only be used as a rough indication of a marking effect. Finally, comparison of toe-clipping and ear-tags in a study of *Microtus pennsylvanicus* suggests that the former technique may slightly shorten the duration of residency of males but not females (Pavone & Boonstra 1985). The available evidence, therefore, indicates that capture and survival probabilities may be affected by marking and that this in turn may affect estimates of population parameters. The nature and direction of bias in the estimates of population parameters depends on the CMR method employed and is discussed by Begon (1979) and Nichols, Hines & Pollock (1984).

Heterogeneity in capture probability

Heterogeneity in capture probability may occur over a sampling period, within a sampling area, among population sub-groups or among individuals. There are three main methods by which the assumption of equal catchability may be investigated. First, contingency analysis for the recapture and non-recapture of population sub-groups may reveal heterogeneity. This is similar to the method of testing for the effects of marking. E. G. White (1975), for example, conducted 2,337 contingency table analyses to identify homogeneous groupings in grasshopper data classified according to site, species, growth stage and sex. Skalski, Robson & Simmons (1983) present an exten-

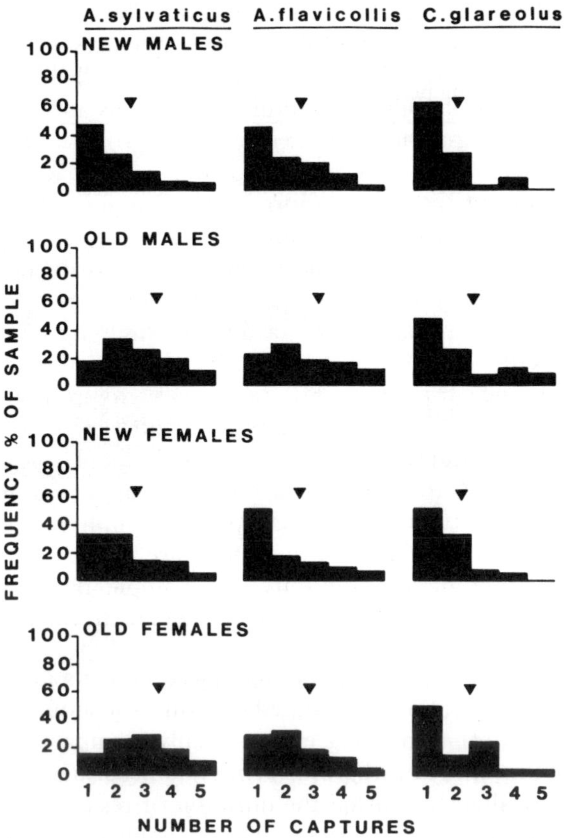

Fig. 4. Frequency of captures in five-day capture periods of freshly marked (new), previously marked (old), male and female *A. sylvaticus*, *A. flavicollis* and *C. glareolus*. Triangles indicate mean number of captures per sample. See text for statistical details. Data from Montgomery (1977).

sion of this technique illustrated by data for Nuttall's cottontail, *Sylvilagus nuttallii*. Secondly, the frequency distribution of captures may be tested against a random (Poisson) or non-random (Geometric or Negative Binomial) truncated null distribution. Young, Neess & Emlen (1952), for example, found captures of *M. musculus* exhibited significant deviations from expected Poisson distributions. David & Johnson (1952) present an approximate test for goodness-of-fit to a zero-truncated Poisson distribution. This has been used to test for heterogeneity in the frequency distribution of captures of freshly and previously marked, male and female *A. sylvaticus*, *A. flavicollis* and *C. glareolus* (Fig. 4). There was good agreement with the assumption of random distribution of capture frequency in six out of 12 data sets. Here, heterogeneity in capture frequency might be expected because of the collation of data over a large number of trapping periods. Tanton (1965) found a good fit of capture frequencies of male and female *A. sylvaticus* to a zero-truncated Negative Binomial distribution, suggesting heterogeneity in probability of capture. Again, this author lumped data for the whole of a long-term study. Leslie (1958) proposed a test for randomness in the frequencies of recaptures of individuals. Where probability of capture is constant these will conform to a Binomial distribution. Leslie's test has been applied widely to both closed and open populations (Seber 1982). Carothers (1971) proposed a modification of Leslie's test using a greater proportion of small data sets. This, however, has not been introduced widely as it requires some arbitrary trimming of data (Otis *et al.* 1978). The latter authors present an example of capture heterogeneity detected by Leslie's test using data for *Peromyscus* sp. Further examples of the use of Leslie's test are presented below. Roff (1973) concluded that frequency of capture and recapture methods, including Leslie's test, are unreliable because CMR data generally conform to one or more theoretical distributions with little chance of rejecting the null hypothesis or little indication of the better fit. Further, these tests are incapable of discriminating simple from compound distributions. For example, failure to detect heterogeneity in the probability of capture could be due to the congruence of two or more Poisson distributions.

The third method for detecting heterogeneity in CMR data is the direct comparison of observed probability of capture frequency or latency to first capture or recapture, among population sub-groups. This method is favoured by mammalogists although there are frequent statistical shortcomings in the literature. Variation in capture probability of small mammals has been associated with many variables including weather conditions (Sidorowicz 1960; Getz 1961; Gentry, Golley & McGinnis 1966), habitat (Perry, Pardue, Barkalow & Monroe 1977; Smal & Fairley 1982), the design of trapping programme (Andrzejewski, Bujalska, Ryszkowski & Ustyniuk 1966; Renzulli, Flowers & Tamarin 1980), species (Gurnell 1982), sex (Gentry, Golley & Smith 1968; Gliwicz 1970; Montgomery 1979), breeding condition (Jensen 1975; Reich & Tamarin 1984), age (Gliwicz 1970; Jensen 1975; Smal & Fairley 1982), social status (Andrzejewski, Petrusewicz & Walkova 1963), season (Tanton 1965; Crawley 1970; Green 1979), population density (Montgomery 1977) and presence of other species (Kalinowska 1971; Montgomery 1979; Boonstra, Rodd & Carleton 1982). Gurnell (1982) points out that trappability is a complex phenomenon and it is likely that variation in capture probability is determined by the interaction of several variables. Smith, Gardner, Gentry, Kaufman & O'Farrell (1975) present the most comprehensive review of this topic. Variation in capture rate over five-day trapping periods, comparing all individuals with first-time captives, for *Apodemus* spp. and *C. glareolus* is illustrated in Fig. 5. Mean

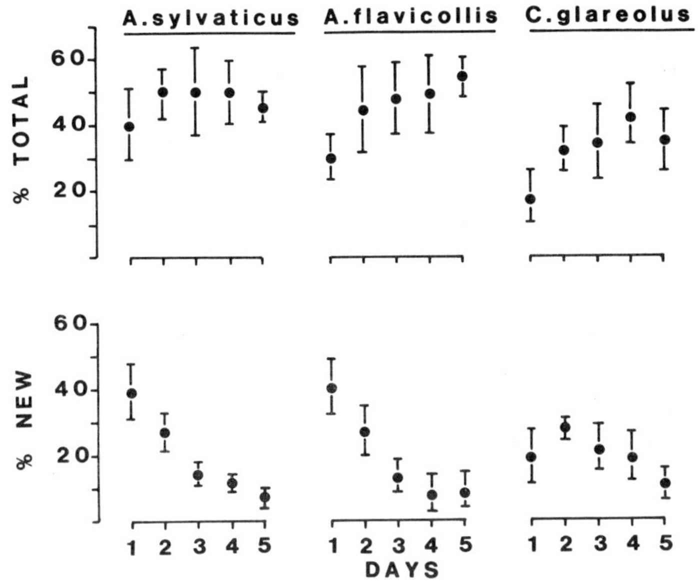

Fig. 5. Proportion of total number of individuals recaptured or captured for the first time on each day of a five-day trapping period. Vertical bars indicate 95 per cent confidence limits of the means. See text for statistical details. Data from Montgomery (1977).

rates of capture of *A. sylvaticus* and *A. flavicollis* differed significantly from that of *C. glareolus* ($F = 19.465$; *d.f.* 2,15; $p < 0.001$ and $F = 15.295$; *d.f.* 2,15; $p < 0.001$ respectively). Variation in capture frequency was significant over time ($F = 15.225$; *d.f.* 8,60; $p < 0.001$ for time over all species). It is also clear that new *C. glareolus* tend to appear later in the sampling period than new *Apodemus* (Fig. 5). Species differences and the difference in capture rate between newly captured individuals and those with an established trap history are likely to be the major sources of variation in CMR data (Gurnell 1982; Montgomery 1985).

Age-dependent survival

The Jolly–Seber method requires that all individuals, regardless of their age, have an equal chance of surviving from one sample period to the next. This can be tested using contingency analysis of the homogeneity of the number of individuals recaptured or not recaptured among two or more age classes. Bishop & Hartley (1976), for example, using this method, found no indication of age-dependent mortality among *R. norvegicus*. Alternatively, the relationship between age and survival may be determined from life tables. If survival is age-independent, loss should decline exponentially so that the relationship of the logarithm of frequency remaining in a class and time

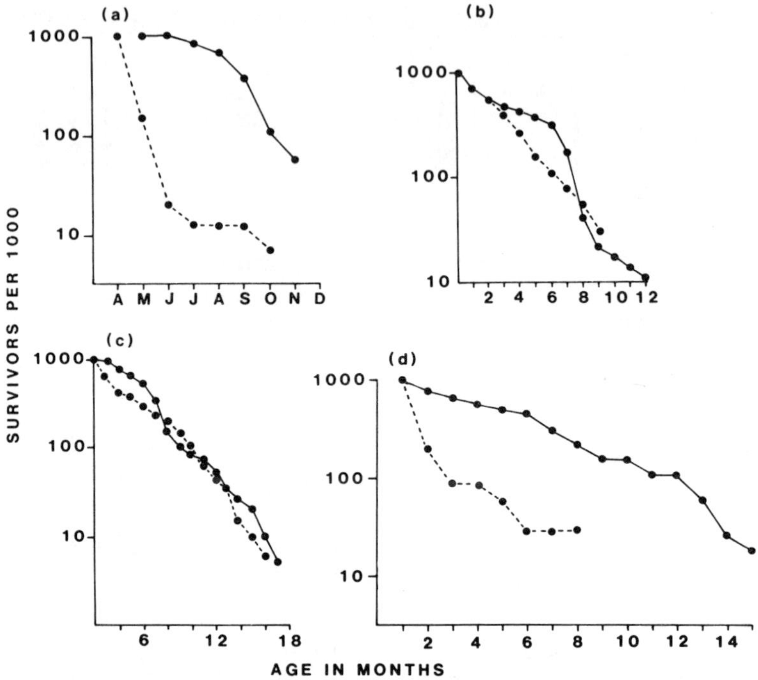

Fig. 6. Age-dependent survival rates for different cohorts of (a) *M. agrestis* (Myllymäki 1977), (b) *A. flavicollis* (Bobek 1969), (c) *C. glareolus* (Pucek, Ryszkowski & Zejda 1969) and (d) *Mus musculus* (Stickel 1979).

should be linear (Blower *et al.* 1981). This is clearly not the case among the examples illustrated in Fig. 6. Age-dependence of survival rates may be one of the most common violations of CMR assumptions in studies of small mammals.

Paucity of data

A major problem with complex statistical models is that a large amount of data is required for estimating a number of unknown parameters. In particular, where probabilities of capture are low, the variances of the estimates of population parameters are large (Seber 1986). The effects of Bailey's correction (Bailey 1951; see Equation 6) for scarcity of recaptures on CMR estimation by the Lincoln Index, Fisher–Ford and Jolly–Seber methods are discussed by Blower *et al.* (1981). If the number of recaptures in any sample falls below 20, then Bailey's correction should be included in the Jolly–Seber method (M. V. Hounsome pers. comm.). Where few animals are recaptured, perhaps fewer than 10 per sample, as a result of low survival and sampling intensity, Fisher–Ford estimates are more precise and accurate than those of Jolly–Seber.

The robustness of CMR models when assumptions are invalidated: fitting a model

The choice of CMR technique depends, to a large extent, on the degree to which the data conform to the various assumptions of CMR models. Even where assumptions are not substantiated, certain CMR methods may give reasonable results because of their inherent robustness. 'Statistical models make idealized assumptions, and are likely to be of limited use unless they are fairly robust under failure of their assumptions' (Buckland 1982). Following Carothers (1979), Buckland asserts that 'useful and reliable estimation can be possible even when a goodness-of-fit test indicates that the model employed is wrong.' The consequences of invalidation of the assumptions of CMR are best understood for the Jolly–Seber model. Here, discussion is restricted to this method, the most generally applicable. Gilbert (1973) and Carothers (1973) contended that unequal catchability is not a problem if all animals have a probability of capture above 0.5 and the distribution of capture probability is reasonably symmetrical. Only extreme trap-shyness renders the estimates inappropriate. Such problems may be alleviated by additional sampling days (Southwood 1978). Manly (1970) concluded that Jolly–Seber estimates can tolerate a modest degree of age-dependence in the rate of survival, although in cases of high infant mortality, population size and survival were overestimated. Bishop & Sheppard (1973) found that the Jolly–Seber method gave poorer results than the Fisher–Ford method, which assumes constant survival rate, when survival was low (0.5 or less) and probability of capture was low (0.05 or less). In studies of small mammals, such low probability of capture is unlikely. Later, Carothers (1979) found that the negative bias in survival rates due to heterogeneity in capture probability as estimated from the Jolly–Seber method was remarkably small, less than 1% of the actual value, and Buckland (1982) described how heterogeneity in trappability may be accommodated through an *ad hoc* procedure in the estimation of survival by his modified Jolly–Seber method. The robustness of the Jolly–Seber method is discussed more fully by Seber (1982, 1986).

Although heterogeneity in capture probability among sub-groups within the population may indicate that these should be analysed separately, this may not be possible because of shortage of data. This is a problem that affects all CMR methods. The effects of paucity of data are to produce widely fluctuating estimates with a lack of precision. In the Jolly–Seber method, problems in the estimation of population parameters can arise when the number of marked animals in a sample falls below 10 (Blower *et al.* 1981). These difficulties may be avoided where probability of capture is high (Seber 1986). Alternatively, the restricted Jolly–Seber models, which require the estimation of fewer parameters, may be satisfied by otherwise inadequate data sets. Jolly (1982) gave four models based on the Poisson formulation (Jolly 1979); A, the full Jolly–Seber method; B, constant survival rate; C, constant probability of capture, and D, constant survival and probability of capture. In the absence of methods that are robust to small sample sizes and the invalidation of underlying assumptions, the biologist is forced to compromise statistical propriety with biological reality.

CMR and MNA: some further problems

The use of live traps in conjunction with CMR to estimate population size of small mammals has long provoked scepticism. Different trap designs, for example, have been shown to sample different portions of the subject population unevenly (Chelkowska 1967; Hansson & Hoffmeyer 1973; K. D. Taylor, Hammond & Quy 1974; Pelikán, Zejda & Holišova 1977; Boonstra & Krebs 1978; Beacham & Krebs 1980; Boonstra & Rodd 1984). Trap odour may also affect the probability of catching a particular species or population sub-group (Mazdzer, Capone & Drickamer 1976; Boonstra & Krebs 1976; Daly, Wilson & Faux 1978; Montgomery 1979; Daly, Wilson & Behrends 1980; Stoddart & Smith 1986). Stoddart (1982), for example, found that as trapping continued, more unmarked animals were caught if the traps had been cleaned, and therefore suggested that cleaning traps increased population estimates. Some individuals may fail to encounter traps or investigate but do not enter them, while others may set off traps before entering or escape after capture (Gurnell 1978; Hurst & Berreen 1985). These phenomena are seldom taken into account in estimation of population parameters in studies of small mammals. Factors such as escape and premature operation of traps are probably rare, while differences in the catchability of individuals in certain traps may be offset by trap density or sampling intensity (Southern 1973; Gurnell 1976; Montgomery 1985). Further, large amounts of data are generally required to demonstrate differential trappability, suggesting that differences are not substantial (see, for example, Montgomery 1979, and Gurnell 1982). The consistency of population density, age structure, survival and so on among studies with different field methods and data treatment also suggests that CMR is a robust technique. For example, CMR studies of *A. sylvaticus* produce much the same picture of annual fluctuations in abundance as those employing kill traps or removal where many of the above problems would be irrelevant (Fig. 7).

The widespread application of MNA in the estimation of population size of small mammals (Table 1) is frequently justified by the assertion that both the Jolly–Seber method and MNA provide a 'good measure of numerical

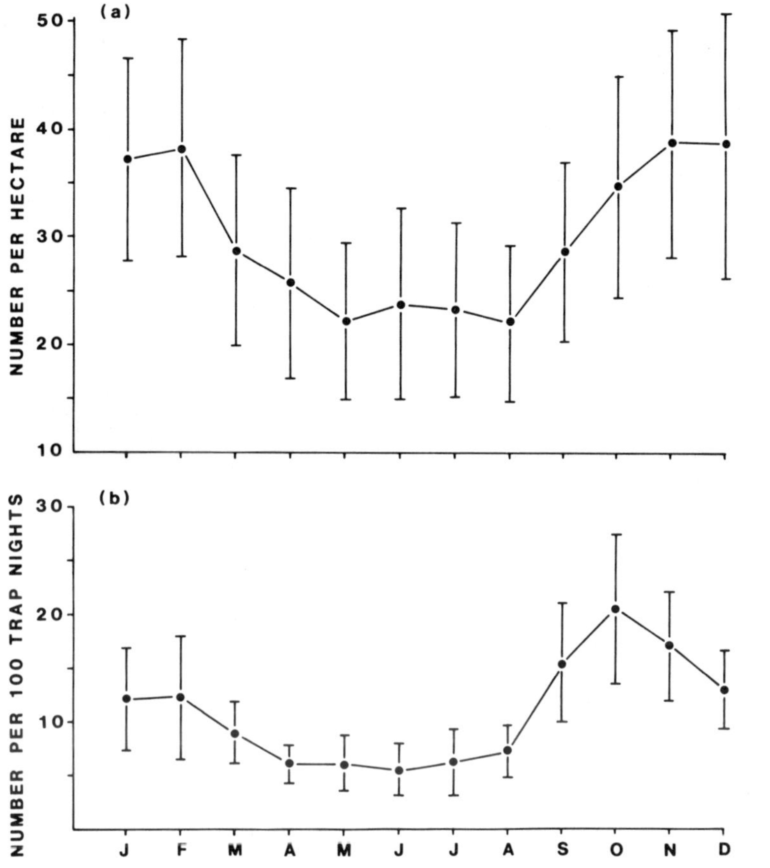

Fig. 7. Population dynamics of *A. sylvaticus* based on (a) CMR (number of years, 64) and (b) numbers per 100 trap nights where mice were killed or removed (number of years, 42). Vertical bars indicate 95% confidence limits of the means. Data after various published and unpublished sources.

changes in field populations' (Hilborn, Redfield & Krebs 1976). These authors found that MNA is insensitive to heterogeneity in trappability (cf. Jolly & Dickson 1983) but gives a poor estimate when average trappability falls below 0.5. Recently, the use of MNA estimators has been challenged. Jolly & Dickson (1983) discuss the relationship between MNA and the Jolly–Seber estimates of population size and survival, showing that population size calculated by the latter technique must always be greater than MNA (see, for example, Sadleir 1970; Lefebvre, Otis & Holler 1982; Sauer & Slade 1986; Fig. 8). Jolly & Dickson conclude that MNA is only useful as a lower limit to the Jolly–Seber estimate of population size and has a serious

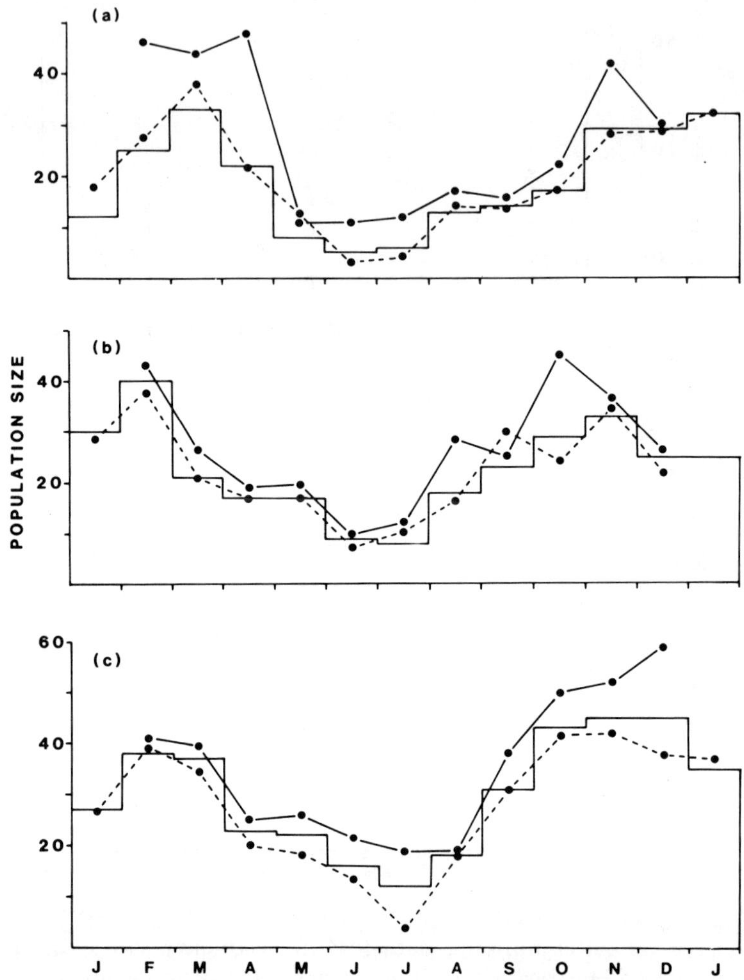

Fig. 8. Minimum Number Alive (histograms) and population size estimated after Hayne (1949) (broken line) and Jolly–Seber (solid lines) on (a) grid A, (b) grid B and (c) grid C in matched areas of woodland at Woodchester Park, Gloucestershire (data from Montgomery 1981 and Table 4).

negative bias under conditions of equal catchability. Further, under conditions of heterogeneity in capture probability, MNA has an even greater negative bias than the Jolly–Seber method. Jolly & Dickson take the extreme view that there is no conceivable situation where 'MNA could be preferable to the J–S estimate of population size, whether in an absolute sense or as an indicator of population trend'. Similarly, Nichols & Pollock (1983) reject the widespread use of enumeration estimators in the study of small mammals. However, where numbers are particularly low and recaptures infrequent, MNA may remain a useful estimate of population size (Krebs, Gilbert, Boutin, Sinclair & Smith 1986). Boonstra (1985) found that Jolly–Seber estimates of the population size of *M. pennsylvanicus* were 0–30% higher (average 10.6%) than the MNA. The population estimates differed most during and immediately after the breeding season. Estimates of survival by the Jolly–Seber and MNA methods were similar despite a number of occasions where survival rates calculated by the former method were more than 1.0. At these times, calculations based on MNA gave biologically more reasonable estimates of survival. Boonstra (1985) concluded that MNA must still be considered a robust method of population estimation but recommended the Jolly–Seber method on the grounds that 'it is usually less biased'. The fundamental failure of MNA is the underestimation of population size to an unknown and varying degree (Blower *et al.* 1981). These points are supported by the analysis of data for *A. sylvaticus* presented in Fig. 8 and Table 4.

The design of CMR programmes: the light of experience

There is a lamentable lack of practical guidance from statisticians for the estimation of population parameters by CMR methods (Seber 1986). The following guidelines for the estimation of population parameters of field populations of small mammals are based on the approximate method for the reduction of bias due to heterogeneity in trappability suggested by Pollock (1982). He defined primary sampling periods each of which is divided into a number of secondary sampling periods. Here, the former are referred to as sampling periods and the latter as days within a sampling period. Pollock (1982) suggested estimating population size using data from secondary samples (days) and assuming a closed population. Survival and other population parameters could then be estimated between primary sampling periods (months) after lumping data for the secondary sampling periods (days). Bishop & Hartley (1976) lumped data in just this manner to avoid bias introduced by heterogeneity in trappability of individual *R. norvegicus*.

1. Sampling period should be short in relation to the interval between samples (25% or less). The sampling period should consist of a minimum of five days. Where 'trap deaths' are a problem the sampling period may have to be shortened.

2. Density of traps should be as great as possible or commensurate with past experience of the density and mobility of the subject population. There should be at least two single-capture live traps for each animal known or expected to live in the study area. Multiple-capture traps should be avoided unless appropriate statistical techniques are employed.

3. Non-random placement of traps, prebaiting, two or more traps per point, adding traps to successful trap points and moving traps between days

Table 4. Estimates of population size ($\hat{N}$) calculated after Hayne (1949) and estimates of population size ($\hat{N}$), survival ($\hat{\varnothing}$) and ingress ($\hat{B}$) based on the Jolly-Seber model for three sets of CMR data concerning *A. sylvaticus* on three grids in matched areas of woodland at Woodchester Park, Gloucestershire, running concurrently over 13 months. For details see Montgomery (1981).

	Hayne		Jolly-Seber					
Sample	$\hat{N}_i$	Upper 95% c.l.	$\hat{N}_i$	s.e.	$\hat{\varnothing}_{i-1}$ to i	s.e.	$\hat{B}_{i-1}$ to i	s.e.
Grid A								
1	18.0	34.4	—	—	—	—	—	—
2	27.4	31.6	46.0	18.1	0.94	0.25	—	—
3	38.1	42.2	43.8	8.7	0.49	0.12	21.1	8.3
4	21.5	25.5	48.0	23.1	1.00	0.45	10.1	8.4
5	12.5	20.1	11.3	5.7	0.17	0.11	3.2	2.6
6	2.8	20.5	11.0	9.0	0.52	0.38	5.2	6.5
7	3.7	5.2	12.3	8.9	0.62	0.44	5.5	7.8
8	13.7	39.1	17.3	7.0	0.63	0.29	9.5	6.6
9	13.4	14.6	15.5	4.4	0.62	0.15	4.8	3.6
10	17.0	31.4	22.3	6.5	0.74	0.14	10.9	4.3
11	28.0	33.3	41.8	10.6	1.01	0.15	19.2	8.1
12	28.5	29.5	29.9	5.6	0.64	0.10	3.2	4.4
13	32.4	39.6	—	—	—	—	—	—
Grid B								
1	28.6	36.1	—	—	—	—	—	—
2	37.8	45.1	42.9	7.0	0.93	0.06	—	—
3	21.2	21.6	26.5	5.7	0.56	0.09	3.2	1.4
4	17.0	17.2	18.8	4.5	0.62	0.12	4.2	1.3
5	16.8	17.7	19.4	4.8	1.00	0.10	1.6	1.3
6	7.5	12.2	10.0	3.6	0.49	0.15	1.1	0.7
7	10.5	19.0	12.1	5.9	0.66	0.27	5.5	3.5
8	16.4	23.0	28.0	11.2	0.61	0.24	20.6	9.5
9	29.9	31.3	24.8	5.5	0.66	0.12	6.2	6.7
10	24.2	26.0	45.2	12.1	0.94	0.17	21.8	7.8
11	34.4	40.9	36.5	6.9	0.55	0.11	11.4	4.7
12	21.8	26.2	25.8	5.2	0.66	0.09	1.8	2.1
13	—	—	—	—	—	—	—	—
Grid C								
1	26.7	33.9	—	—	—	—	—	—
2	39.1	48.9	41.1	7.4	0.62	0.10	—	—
3	34.3	42.2	39.5	7.0	0.60	0.09	15.0	2.7
4	20.1	22.4	24.9	5.3	0.49	0.09	5.5	1.8
5	18.4	26.0	26.1	6.2	0.93	0.14	3.0	2.0
6	13.4	37.4	21.4	6.5	0.58	0.15	6.3	3.0
7	4.1	4.4	19.0	9.4	0.76	0.28	2.8	6.4
8	18.5	21.4	19.1	4.7	0.75	0.25	4.8	4.7
9	31.0	38.2	38.2	8.9	0.88	0.14	21.5	5.1
10	41.6	47.5	49.8	8.7	0.47	0.10	31.8	5.3
11	42.2	51.4	52.0	8.2	0.80	0.07	13.0	5.1
12	37.7	39.3	58.9	9.9	0.91	0.09	12.3	4.8
13	37.1	39.6	—	—	—	—	—	—

may increase the proportion of the population caught within a single sampling period. This will promote precision of estimates.

4. Estimation of population size should be based on data for a single sampling period. This should be conducted on the premise that the population is closed. If this is biologically unacceptable, estimates should be based on the Jolly–Seber model used in conjunction with MNA after lumping data for each of the sampling periods. The assumptions of the CMR model employed should be validated, or the nature of their violation determined, to elucidate the direction and magnitude of bias in the estimates of population parameters.

5. Data should be lumped within each sampling period for estimation of survival rates during the intervals between samples. The Jolly–Seber method remains statistically and biologically the most elegant and efficient use of CMR data.

6. Mark loss should be estimated in a double-tagging experiment.

7. Where there are sufficient data, analyses should be conducted separately for subgroups within the population.

8. Replication of the estimation of population parameters using data from alternate rows of a grid or duplicate grids in the same study area is desirable to evaluate the precision of field and statistical methodology.

Lefebvre *et al.* (1982) present an analysis of population dynamics of cotton rats, *Sigmodon hispidus*, where numbers were estimated by closed (Otis *et al.* 1978) and open (Jolly–Seber after Arnason & Baniuk 1980) population techniques. Despite heterogeneity in capture probabilities over eight-day capture periods, open and closed population estimates were similar. This approach has also been followed in Fig. 8 where population estimates were based on the assumption that samples were drawn from a closed population of *A. sylvaticus* over four days during each of 13 months. This census was replicated on three grids of identical dimensions in woodland of similar character (Montgomery 1981). As in many studies of non-microtine rodents, there were insufficient data to allow estimation of males and females or any age class separately. Leslie's test for random sampling (see above) was non-significant for males ($\chi^2 = 32.52$; *d.f.* 23; n.s.) and data lumped for males and females ($\chi^2 = 51.68$; *d.f.* 43; n.s.) but rejected the null hypothesis for females alone ($\chi^2 = 36.2$; *d.f.* 19; $p < 0.025$); females were recaptured more frequently than expected. This seems to have been offset by the converse tendency in males. It implies that any negative bias arising in population estimates through heterogeneity in capture probability will be slight when using male and female data together. Estimates were made for the ML method of Zippin (1956) and the modified Lincoln Index method of Hayne (1949). The former, recommended by Otis *et al.* (1978) for estimation where a trap response is present (see above), produced estimates with very erratic standard errors and results are not presented. The latter method produced estimates which in some cases were below the MNA value, presumably because the Hayne's estimate used only data for mice actually captured in a particular month. The Jolly–Seber method produced estimates of survival and ingress (Table 4) as well as population size. These were based on data lumped for each sampling period (month). The Jolly–Seber population estimates, although greater for most sampling periods, produced similar trends to MNA and the Hayne's estimate. The reliability of population parameters estimated using the Jolly–Seber method, judged by the relative magnitude of their standard errors, seemed considerable. The reliability of the CMR census techniques in general is supported by the uniformity of the three replicates.

The limits of CMR in the revelation of the dynamics and regulation of small mammals

All ecological techniques have limitations. Here, it is argued that MNA is more limited in elucidating the population biology of small mammals than CMR estimation. The limitations of CMR procedures because of non-fulfilment of underlying assumptions and the inadequacy of statistical models have already been discussed (see above). CMR suffers from two further drawbacks. First, it is inadequate for determining the interaction of spatial and temporal changes in population size. CMR studies open a window in time on populations of mobile animals. This window is usually small in relation to the overall population and the assumption of homogeneity in population dispersion is certainly unrealistic (L. R. Taylor, Woiwod & Perry 1978). For example, studies of the spatial and temporal changes in abundance of *A. sylvaticus* at Tollymore Forest Park, an area of some 700 ha of mixed woodland, over three years, revealed changes in space just as great as those over time (W. I. Montgomery unpublished). The second limitation to the study of population processes that is inherent in CMR estimations, especially those where ingress and egress are assumed, is the interdependence of estimates of survival, 'births' and population size. This prohibits direct evaluation of density-dependence using, for example, population size and survival estimates. Alternative techniques must be used for the comparison of population change and population size.

Conclusions

This review of the application and potential value of CMR methods of population estimation in the study of small mammals has four main findings:

1. The current underutilization of CMR, potentially a most useful

technique, results from failure to appreciate the robust nature of some CMR estimators, and the misapprehension that other techniques, chiefly MNA, are inherently less affected by factors such as heterogeneity in catchability.

2. The field biologist must evaluate trappability, age structure and survival, ingress and egress of a subject population in order to select an appropriate CMR model or amend methodology to approach more nearly a statistical ideal. Biological knowledge is as powerful a tool as the available statistical tests for validating the assumptions of CMR models. There is a need for efficient, easy to use, goodness-of-fit tests for CMR models.

3. The development of more realistic CMR models, incorporating, for example, heterogeneity in catchability and age structure, has continued rapidly throughout the 1970s and early 1980s. Methods for estimating closed populations (Otis *et al.* 1978; Pollock & Otto 1983) are more advanced than those for open populations in this respect. Among the latter, Pollock & Mann (1983) and Stokes (1984) have proposed modifications of the Jolly–Seber model allowing age-dependent survival. These recent models should be tested on field populations of small mammals.

4. In the absence of thoroughly robust and flexible models for the estimation of open populations, certain changes in current field practices permit wider application of the Jolly–Seber model to long-term studies of populations of small mammals.

There is a strong case for the re-evaluation of modern statistical models for the estimation of populations of small mammals using CMR. Experience has demonstrated that there are few general rules for successful application of CMR. The biologist must justify the use of particular CMR methods. Ultimately choice of method will depend on the aim of the study, the biology of the subject population and the behaviour of the statistical model.

Acknowledgements

I am grateful to Drs D. W. Yalden, J. Gurnell and E. Le Boulengé for their illuminating criticisms of an earlier draft of the MS, Dr M. V. Hounsome for his assistance in the Jolly–Seber calculations, Greg Forsythe for compilation of the wood mouse data for Fig. 7, and Miss Gemma Branagh for typing the manuscript.

References

Andrzejewski, R., Bujalska, G., Ryszkowski, L. & Ustyniuk, J. (1966). On a relation between the number of traps in a point of catch and trappability of small rodents. *Acta theriol.* **11**: 343–9.

Andrzejewski, R., Petrusewicz, K. & Walkowa, W. (1963). Absorption of newcomers by a population of white mice. *Ekol. pol.* (*A*) **11**: 223–40.

Arnason, A.N. & Baniuk, L. (1980). A computer system for mark–recapture analysis of open populations. *J. Wildl. Mgmt* **44**: 325–32.

Arnason, A.N. & Mills, K.H. (1981). Bias and loss of precision due to tag loss in Jolly–Seber estimates for mark–recapture experiments. *Can. J. Fish. aquat. Sci.* **38**: 1077–95.

Bailey, N.T.J. (1951). On estimating the size of mobile populations from recapture data. *Biometrika* **38**: 293–306.

Beacham, T.D. & Krebs, C.J. (1980). Pitfall versus live-trap enumeration of fluctuating population of *Microtus townsendii*. *J. Mammal.* **61**: 486–99.

Begon, M. (1979). *Investigating animal abundance: capture–recapture for biologists.* Edward Arnold, London.

Begon, M. (1983). Abuses of mathematical methods in ecology: applications of Jolly's capture–recapture method. *Oikos* **40**: 155–8.

Bishop, J.A. & Hartley, D.J. (1976). The size and age structure of rural populations of *Rattus norvegicus* containing individuals resistant to the anticoagulant Warfarin. *J. Anim. Ecol.* **45**: 623–46.

Bishop, J.A. & Sheppard, P.M. (1973). An evaluation of two capture–recapture models using the technique of computer simulation. In *The mathematical theory of the dynamics of biological populations*: 235–52. (Eds Bartlett, M.S. & Hiorns, R.W.). Academic Press, London.

Blower, J.G., Cook, L.M. & Bishop, J.A. (1981). *Estimating the size of animal populations.* George Allen & Unwin, London.

Bobek, B. (1969). Survival, turnover and production of small rodents in a beech forest. *Acta theriol.* **14**: 191–210.

Boonstra, R. (1985). Demography of *Microtus pennsylvanicus* in Southern Ontario: enumeration versus Jolly–Seber estimation compared. *Can. J. Zool.* **63**: 1174–80.

Boonstra, R. & Krebs, C.J. (1976). The effect of odour on trap response in *Microtus townsendii*. *J. Zool., Lond.* **180**: 467–76.

Boonstra, R. & Krebs, C.J. (1978). Pitfall trapping of *Microtus townsendii* *J. Mammal.* **59**: 136–48.

Boonstra, R. & Rodd, F.H. (1984). Efficiency of pitfalls versus live traps in enumeration of populations of *Microtus pennsylvanicus*. *Can. J. Zool.* **62**: 758–65.

Boonstra, R., Rodd, F.H. & Carleton, D.J. (1982). Effect of *Blarina brevicauda* on trap response of *Microtus pennsylvanicus*. *Can. J. Zool.* **60**: 438–42.

Buckland, S.T. (1982). A mark–recapture survival analysis. *J. Anim. Ecol.* **51**: 833–47.

Carothers, A.D. (1971). An examination and extension of Leslie's test of equal catchability. *Biometrics* **27**: 615–30.

Carothers, A.D. (1973). The effects of unequal catchability on Jolly–Seber estimates. *Biometrics* **29**: 79–100.

Carothers, A.D. (1979). Quantifying unequal catchability and its effect on survival estimates in an actual population. *J. Anim. Ecol.* **48**: 863–9.

Caughley, G. (1977). *Analysis of vertebrate populations.* Wiley, Chichester.

Chelkowska, H. (1967). An attempt at comparing two methods of trapping small rodents (in pitfalls and live traps). *Ekol. pol. (A)* **15**: 779–85.

Cormack, R.M. (1968). The statistics of capture–recapture methods. *Oceanogr. mar. Biol. Ann. Rev.* **6**: 455–506.

Cormack, R.M. (1973). Commonsense estimates from capture–recapture studies. In *The mathematical theory of the dynamics of biological populations*: 225–34. (Eds Bartlett, M.S. & Hiorns, R.W.). Academic Press, New York.

Crawley, M.C. (1970). Some population dynamics of the Bank vole, *Clethrionomys glareolus* and the Wood mouse, *Apodemus sylvaticus* in mixed woodland. *J. Zool., Lond.* **160**: 71–89.

Daly, M., Wilson, M.I. & Behrends, P. (1980). Factors affecting rodents' responses to odours of strangers encountered in the field: experiments with odour-baited traps. *Behav. Ecol. Sociobiol.* **6**: 323–9.

Daly, M., Wilson, M.I. & Faux, S.F. (1978). Seasonally variable effects of conspecific odors upon capture of deer mice (*Peromyscus maniculatus gambelii*). *Behav. Biol.* **23**: 254–9.

Darroch, J.N. (1958). The multiple recapture census: I. Estimation of a closed population. *Biometrika* **45**: 343–59.

David, F.N. & Johnson, N.L. (1952). The truncated Poisson. *Biometrics* **8**: 275–85.

Davis, D.E. & Winstead, R.L. (1980). Estimating the numbers of wildlife populations. In *Wildlife management techniques manual* (4th edn): 221–45. (Ed. Schemnitz, S.D.). The Wildlife Society, Washington, D.C.

Dice, L.R. (1941). Methods for estimating populations of mammals. *J. Wildl. Mgmt* **5**: 398–407.

du Feu, C., Hounsome, M. & Spence, I. (1983). A single-session mark-recapture method of population estimation. *Ringing Migr.* **4**: 211–26.

Fairley, J.S. (1982). Short-term effects of ringing and toe-clipping on the recapture of Wood mice (*Apodemus sylvaticus*). *J. Zool., Lond.* **197**: 295–7.

Fisher, R.A. & Ford, E.B. (1947). The spread of a gene in natural conditions in a colony of the moth *Panaxia dominula* (L.). *Heredity, Lond.* **1**: 143–74.

Flowerdew, J.R. (1976). Techniques in mammalogy. Chapter 4. Ecological methods. *Mammal Rev.* **6**: 123–59.

Gentry, J.B., Golley, F.B. & McGinnis, J.T. (1966). Effect of weather on captures of small mammals. *Am. Midl. Nat.* **75**: 526–30.

Gentry, J.B., Golley, F.B. & Smith, M.H. (1968). An evaluation of the proposed International Biological Program census method for estimating small mammal populations. *Acta theriol.* **13**: 313–27.

Getz, L.L. (1961). Responses of small mammals to live traps and weather conditions. *Am. Midl. Nat.* **66**: 160–9.

Gilbert, R.O. (1973). Approximations of the bias in the Jolly–Seber capture-recapture model. *Biometrics* **29**: 501–26.

Gliwicz, J. (1970). Relation between trappability and age of individuals in a population of the bank vole. *Acta theriol.* **15**: 15–23.

Green, R (1979). The ecology of Wood mice (*Apodemus sylvaticus*) on arable farmland. *J. Zool., Lond.* **188**: 357–77.

Grodziński, W., Pucek, Z. & Ryszkowski, L. (1966). Estimation of rodent numbers by means of prebaiting and intensive removal. *Acta theriol.* **11**: 297–314.

Gurnell, J. (1976). Studies on the effect of bait and sampling intensity on trapping and estimating Wood mice, *Apodemus sylvaticus*. *J. Zool., Lond.* **178**: 91–105.

Gurnell, J. (1978). Observations on trap response in confined populations of Wood mice, *Apodemus sylvaticus*. *J. Zool., Lond.* **185**: 279–87.

Gurnell, J. (1982). Trap response in woodland rodents. *Acta theriol.* **27**: 123–37.

Hansson, L. & Hoffmeyer, I. (1973). Snap and live trap efficiency for small mammals. *Oikos* **24**: 477–9.

Hayne, D.W. (1949). Two methods for estimating population from trapping records. *J. Mammal.* **30**: 399–411.

Hilborn, R., Redfield, J.A. & Krebs, C.J. (1976). On the reliability of enumeration for mark and recapture census of voles. *Can. J. Zool.* **54**: 1019–24.

Hurst, J.L. & Berreen, J. (1985). Observations on the trap-response of wild House mice, *Mus domesticus* Rutty in poultry houses. *J. Zool., Lond. (A)* **207**: 619–22.

Jackson, C.H.N. (1933). On the true density of tsetse flies. *J. Anim. Ecol.* **2**: 204–9.

Jensen, T.S. (1975). Trappability of various functional groups of the forest rodents *Clethrionomys glareolus* and *Apodemus flavicollis*, and its application in density estimation. *Oikos* **26**: 196–204.

Jolly, G.M. (1965). Explicit estimates from capture–recapture data with both death and immigration – stochastic model. *Biometrika* **52**: 225–47.

Jolly, G.M. (1979). A unified approach to mark–recapture stochastic models, exemplified by a constant survival rate model. In *Sampling biological populations*: 277-82. (Eds Cormack, R.M., Patil, G.P. & Robson, D.S.). International Co-operative Publishing House, Fairland, Maryland. (*Statist. Ecol. Ser.* **5**).

Jolly, G.M. (1982). Mark–recapture models with parameters constant in time. *Biometrics* **38**: 301–21.

Jolly, G.M. & Dickson, J.M. (1983). The problem of unequal catchability in mark–recapture estimation of small mammal populations. *Can. J. Zool.* **61**: 922–7.

Kalinowska, A. (1971). Trapping of *Apodemus flavicollis* and *Clethrionomys glareolus* into a double trap. *Acta theriol.* **16**: 73–8.

Krebs, C.J., Gilbert, B.S., Boutin, S., Sinclair, A.R.E. & Smith, J.N.M. (1986). Population biology of snowshoe hares I. Demography of food-supplemented populations in the southern Yukon, 1976-84. *J. Anim. Ecol.* **55**: 963–82.

Le Boulengé, E. (1984). Computer package for the analysis of capture–recapture data. *Acta zool. fenn.* **173**: 69–72.

Lefebvre, L.W., Otis, D.L. & Holler, N.R. (1982). Comparison of open and closed models for cotton rat population estimates. *J. Wildl. Mgmt* **46**: 156–63.

Leslie, P.H. (1952). The estimation of population parameters from data obtained by means of the capture–recapture method: I. The maximum likelihood equations for estimating the death rate. *Biometrika* **38**: 368–88.

Leslie, P.H. (1958). Statistical appendix. *J. Anim. Ecol.* **27**: 84–6.

Leslie, P.H. & Chitty, D. (1951). The estimation of population parameters from data obtained by means of the capture–recapture method: II. An example of the practical application of the method. *Biometrika* **40**: 137–69.

Lincoln, F.C. (1930). Calculating waterfowl abundance on the basis of banding returns. *Circ. U.S. Dep. Agric.* No. 118: 1–4.

Manly, B.F.J. (1970). A simulation study of animal population estimation using the capture–recapture method. *J. appl. Ecol.* **7**: 13–39.

Manly, B.F.J. & Parr, M.J. (1968). A new method of estimating population size, survivorship, and birth rate from capture–recapture data. *Trans. Soc. Br. Ent.* **18**: 81–9.

Mazdzer, E., Capone, M.R. & Drickamer, L.C. (1976). Conspecific odors and trappability of deer mice (*Peromyscus leucopus noveboracensis*). *J. Mammal.* **57**: 607–9.

Montgomery, W.I. (1977). *Studies on the ecology of two sympatric species of* Apodemus (*Rodentia: Muridae*). Ph.D. thesis: University of Manchester.

Montgomery, W.I. (1979). An examination of interspecific, sexual and individual biases affecting rodent captures in Longworth traps. *Acta theriol.* **24**: 35–45.

Montgomery, W.I. (1981). A removal experiment with sympatric populations of *Apodemus sylvaticus* (L.) and *A. flavicollis* (Melchior) (Rodentia: Muridae). *Oecologia* **51**: 123–32.

Montgomery, W.I. (1985). Interspecific competition and the comparative ecology of two congeneric species of mice. In *Case studies in population biology*: 126–87. (Ed. Cook, L.M.). Manchester University Press, Manchester.

Myllymäki, A. (1977). Demographic mechanisms in the fluctuating populations of the field vole *Microtus agrestis*. *Oikos* **29**: 468–93.

Nichols, J.D., Hines, J.E. & Pollock, K.H. (1984). Effects of permanent trap response in capture probability on Jolly–Seber capture–recapture model estimates. *J Wildl. Mgmt* **48**: 289–94.

Nichols, J.D. & Pollock, K.H. (1983). Estimation methodology in contemporary small mammal capture–recapture studies. *J. Mammal* **64**: 253–60.

Otis, D.L., Burnham, K.P., White, G.C. & Anderson, D.R. (1978). Statistical interference from capture data on closed animal populations. *Wildl. Monogr.* **62**: 1–135.

Parr, M.J., Gaskell, T.J. & George, B.J. (1968). Capture–recapture methods of estimating animal numbers. *J. biol. Educ.* **2**: 95–117.

Pavone, L.V. & Boonstra, R. (1985). The effects of toe clipping on the survival of the meadow vole (*Microtus pennsylvanicus*). *Can. J. Zool.* **63**: 499–501.

Pelikán, J., Zejda, J. & Holišová, V. (1977). Efficiency of different traps in catching small mammals. *Fol. zool.* **26**: 1–13.

Perry, H.R., Jr, Pardue, G.B., Barkalow, F.S., Jr & Monroe, R.J. (1977). Factors affecting trap responses of the gray squirrel. *J. Wildl. Mgmt* **41**: 135–43.

Phillips, B.F. & Campbell, N.A. (1970). Comparison of methods of estimating population size using data on the whelk *Dicathais aegrota* (Reeve). *J. Anim. Ecol.* **39**: 753–9.

Pollock, K.H. (1982). A capture–recapture design robust to unequal probability of capture. *J. Wildl. Mgmt* **46**: 752–7.

Pollock, K.H. & Mann, R.H.K. (1983). Use of an age-dependent mark–recapture model in fisheries research. *Can. J. Fish. aquat. Sci.* **40**: 1449–55.

Pollock, K.H. & Otto, M.C. (1983). Robust esimation of population size in closed animal populations from capture–recapture experiments. *Biometrics* **39**: 1035–49.

Pollock, K.H., Solomon, D.L. & Robson, D.S. (1974). Tests for mortality and recruitment in a K-sample tag–recapture experiment. *Biometrics* **30**: 77–87.

Pucek, Z., Ryszkowski, L. & Zejda, J. (1969). Estimation of average length of life in bank vole, *Clethrionomys glareolus* (Schreber, 1780). In *Energy flow through small mammal populations*: 137–201. (Eds Petrusewicz, K. & Ryszkowski, L.). Polish Scientific Publications, Warsaw.

Reich, L.M. & Tamarin, R.H. (1984). Multiple capture trap association of meadow voles (*Microtus pennsylvanicus*). *J. Mammal.* **65**: 85–90.

Renzulli, G.B., Flowers, J.F. & Tamarin, R.H. (1980). The effects of trapping design on demographic estimates in the meadow vole, *Microtus pennsylvanicus*. *Am. Midl. Nat.* **104**: 397–401.

Robson, D.S. (1969). Mark–recapture methods of population estimation. In *New developments in survey sampling*: 120–40. (Eds Johnson, M.L. & Smith, H., Jr). Wiley Interscience, New York.

Robson, D.S. & Flick, W.A. (1965). A non-parametric statistical method for culling recruits from a mark–recapture experiment. *Biometrics* **21**: 936–47.

Robson, D.S. & Regier, H.A. (1968). Estimation of population number and mortality rates. In *Methods for assessment of fish production in fresh waters*: 124–58. (Ed. Ricker, W.E.). Blackwell Sci. Publ., Oxford. (*Int. biol. Progm Handb*. No. 3).

Roff, D.A. (1973). An examination of some statistical tests used in the analysis of mark–recapture data. *Oecologia* **12**: 35–54.

Sadleir, R.M.F.S. (1970). Population dynamics and breeding of the deermouse (*Peromyscus maniculatus*) on Burnaby Mountain, British Columbia. *Syesis* **3**: 67–74.

Sauer, J.R. & Slade, N.A. (1986). Size-dependent population dynamics of *Microtus ochrogaster*. *Am. Nat.* **127**: 902–8.

Schnabel, Z.E. (1938). The estimation of the total fish population of a lake. *Am. Math. Mon.* **45**: 348–52.

Schumacher, F.X. & Eschmeyer, R.W. (1943). The estimation of fish population in lakes or ponds. *J. Tenn. Acad. Sci.* **18**: 228–49.

Seber, G.A.F. (1965). A note on the multiple-recapture census. *Biometrika* **52**: 249–59.

Seber, G.A.F. (1982). *The estimation of animal abundance and related parameters*. (2nd edn). Charles Griffin & Company Ltd., London.

Seber, G.A.F. (1986). A review of estimating animal abundance. *Biometrics* **42**: 267–92.

Seber, G.A.F. & Felton, R. (1981). Tag loss and the Peterson mark–recapture experiment. *Biometrika* **68**: 211–9.

Sidorowicz, J. (1960). Influence of the weather on capture of micromammalia. I. Rodents (Rodentia). *Acta theriol.* **4**: 139–58.

Skalski, J.R., Robson, D.S. & Simmons, M.A. (1983). Comparative census procedures using single mark–recapture methods. *Ecology* **64**: 752–60.

Smal, C.M. & Fairley, J.S. (1982). The dynamics and regulation of small rodent populations in the woodland ecosystems of Killarney, Ireland. *J. Zool., Lond.* **196**: 1–30.

Smith, M.H., Gardner, R.H., Gentry, J.B., Kaufman, D.W. & O'Farrell, M.H. (1975). Density estimation of small mammal populations. In *Small mammals:*

their productivity and population dynamics: 25–53. (Eds Golley, F.B., Petrusewicz, K. & Ryszkowski, L.). Cambridge University Press, Cambridge. (*Int biol. Progm* 5).

Southern, H.N. (1973). A yardstick for measuring populations of small rodents. *Mammal Rev.* **3**: 1–10.

Southwood, T.R.E. (1978). *Ecological methods with particular reference to insect populations.* (2nd edn). Chapman & Hall, London.

Stickel, L.F. (1979). Population ecology of house mice in unstable habitats. *J. Anim. Ecol.* **48**: 871–87.

Stoddart, D.M. (1982). Does trap odour influence estimation of population size of the short-tailed vole, *Microtus agrestis*? *J. Anim. Ecol.* **51**: 375–86.

Stoddart, D.M. & Smith, P.A. (1986). Recognition of odour-induced bias in the live-trapping of *Apodemus sylvaticus*. *Oikos* **46**: 194–9.

Stokes, S.L. (1984). The Jolly–Seber method applied to age-stratified populations. *J. Wildl. Mgmt* **48**: 1053–9.

Tanton, M.T. (1965). Problems of live trapping and population estimation for the Woodmouse, *Apodemus sylvaticus* (L.). *J. Anim. Ecol.* **34**: 1–22.

Taylor, K.D., Hammond, L.E. & Quy, R.J. (1974). The reactions of common rats to four types of live-capture trap. *J. appl. Ecol.* **11**: 453–60.

Taylor, L.R., Woiwod, I.P. & Perry, J.N. (1978). The density-dependence of spatial behaviour and the rarity of randomness. *J. Anim. Ecol.* **47**: 383–406.

Twigg, G.I. (1976). Techniques in mammalogy. Chapter 3. Marking mammals. *Mammal Rev.* **6**: 101–15.

White, E.G. (1975). Identifying population units that comply with capture–recapture assumptions in an open community of alpine grasshoppers. *Res. Popul. Ecol.* **16**: 153–87.

White, G.C., Anderson, D.R., Burnham, K.P. & Otis, D.L. (1982). *Capture–recapture and removal methods for sampling closed populations.* Los Alamos National Laboratory, Los Alamos.

Young, H., Neess, J. & Emlen, J.T., Jr (1952). Heterogeneity of trap response in a population of house mice. *J. Wildl. Mgmt* **16**: 169–80.

Zippin, C. (1956). An evaluation of the removal method of estimating animal populations. *Biometrics* **12**: 163–89.

Symp. zool. Soc. Lond. (1987) No. 58: 59–77

Patterns of mortality in a free-living rabbit (*Oryctolagus cuniculus*) population

D. P. COWAN

Mammals and Birds Department, Ministry of Agriculture, Fisheries and Food, Worplesdon Laboratory, Worplesdon, Nr Guildford, Surrey GU3 3LQ

Synopsis

Patterns of mortality within a free-living rabbit population in southern England were investigated over six consecutive years. Juvenile mortality was high (only 6.5% survived from 4–8 to more than 52 weeks of age). Adult mortality increased amongst adults more than three years old. Mortality was higher for males than females amongst juveniles up to 16 weeks of age. Annual mortality of adult males (64%) was higher than that of females (53%) as a result of lower survival outside the breeding season. Mortality increased amongst successive cohorts of juveniles born during the course of the breeding season. Variation in juvenile mortality between years was partly explicable in terms of differences in population density. Hence, numbers of emergent young were not directly related to the numbers recruited into the subsequent breeding populations, and the sizes of breeding populations showed less variation than the peak summer populations. Myxomatosis and coccidiosis were strongly implicated as agents of mortality influencing these trends.

Introduction

The rabbit is once again a serious pest of agriculture in Britain (Rees, Ross, Cowan, Tittensor & Trout 1985), having increased in both distribution and abundance in recent years (Trout, Tapper & Harradine 1986). Changes in patterns of mortality are likely to underlie changes in the sizes of rabbit populations, both in the past and in the future, and will influence the effectiveness of management strategies. However, no detailed estimates of mortality for any free-living populations in Britain have become available since the pioneering work of Southern (1940). This paper examines variations in mortality with respect to age, sex and time of birth both within and between

ZOOLOGICAL SYMPOSIUM No. 58
ISBN 0-19-854006-X

years in a free-living population and attempts to identify some of the causes of these differences.

Methods

The rabbit population living in a 10 ha bowl of chalk downland, which forms part of Aston Rowant National Nature Reserve, Oxfordshire, was studied during the period October 1976 to March 1983. Cage traps (Henderson 1979), baited with sliced carrots and set outside burrow entrances, were used to catch rabbits during the period March to October each year, except in 1977 when trapping did not commence until June. Seventy to 100 traps were set for at least four consecutive days during each 28-day period between the above dates. In addition, ferrets were used to bolt rabbits from their burrows into nets during December and January each winter, when all warrens (multiple-entrance burrow systems) were ferreted once, except for 1977/1978 when they were ferreted twice. Details of the method are given in Cowan (1984).

Captured rabbits were weighed, sexed and, on the basis of the extent of apophyseal fusion, divided into two age categories, above and below nine months old (Watson & Tyndale-Biscoe 1953). Individuals weighing more than 650 g when caught (approximately ten weeks of age) were given individually recognizable, coloured eartags; those below this weight when caught were given numbered aluminium eartags. Overt external symptoms of myxomatosis were recorded for all captured rabbits. The dates of birth of animals caught during the year of their birth were estimated using a growth curve of body weight against age (Cowan 1983). Animals less than one year old are referred to here as juveniles and all others as adults.

Each year observations were made using a 15–60 by 50 zoom telescope during the four hours prior to dusk between March and November. Between 4 and 40 h of observation were carried out during each 28-day period between these dates. Repeated scans of the entire study site were made during observation periods. During a scan the identity was determined of each individually recognizable rabbit present above ground. The survival of these individuals could thus be followed without the need for recapture.

Juvenile survival was analysed by creating age-specific life-tables consisting of 28-day age intervals from 4–8 to 48–52 weeks of age. Although 33 individuals were captured when less than four weeks of age (the lightest being 150 g and, hence, perhaps no more than 15 days old), only 11 of these survived to enter the 4–8 week age interval. This particularly high mortality rate amongst very young animals suggests that capture may have reduced their subsequent survival. For this reason, only animals known to have survived to at least four weeks of age (i.e. fully weaned individuals) were included in the life-tables. Individuals were entered into the tables at the ages at which they were first captured (the 'fusion' method) rather than at age zero (the 'pseudo-cohort' method) because the latter technique tends to underestimate the mortality rate (Cowan & Roman 1985). Animals were removed from the tables at the ages when they were last known to be alive. Mortality rates and survival probabilities were estimated as described by Cowan & Roman (1985).

Adult survivorship was estimated from the individuals tagged before February 25 each year and known to be alive on that date (i.e. in the third 28-day period of the year, towards the beginning of the breeding season (Lloyd 1970a)), that were known to survive and enter each succeeding 28-day period until the following breeding season.

The estimates of survival given here are minima, as they do not distinguish dispersal off the site from mortality. Nor do they account for individuals which survived undetected after the last date on which they were known to be alive. The log-rank test was used to compare cohorts (Peto *et al.* 1977; Healy & Osborn 1981); hence, no assumptions were made concerning the independence of survival from age.

The minimum number known to be alive in each 28-day period of the study was calculated by summing the number of juveniles alive in each period, based on capture records, subsequent sightings and estimated birth dates, with the number of adults known to have been alive in each period. These adult population estimates included both marked and unmarked individuals seen during scans, some of which were individually recognizable from distinctive characteristics such as torn ears, eye loss or unusual pelage. More than 75% of the adult population was marked in all years except 1977, when the figure was 31%.

No multivariate techniques are available for analysing life-tables constructed by the fusion method. Hence, data were pooled with respect to the appropriate independent variables such as sex, time of birth and year. The potential biases in the analyses arising from this practice are considered where possible.

Results

The breeding season and changes in population size

In order to provide a framework within which to consider patterns of mortality it is first necessary to consider how annual changes in population size reflect the seasonal nature of rabbit reproduction. The numbers of young caught each year (i.e. the numbers known to have emerged above ground from their nests) and ascribed to particular 28-day birth periods are shown in Fig. 1. Data for the 44 young caught in 1977 are excluded, as early-born

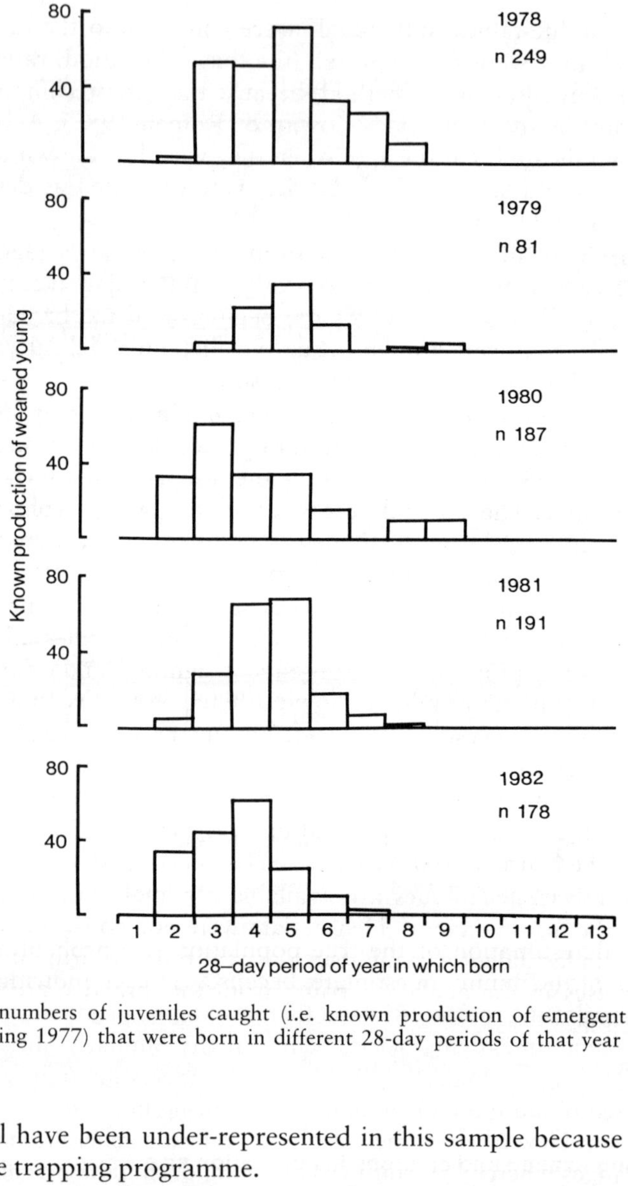

Fig. 1. The numbers of juveniles caught (i.e. known production of emergent young) each year (excluding 1977) that were born in different 28-day periods of that year starting from January 1.

young will have been under-represented in this sample because of the late start of the trapping programme.

The changes in the minimum population estimates made during the course of the study are given in Fig. 2. The agreement between these estimates and those from a capture-mark-recapture study, based on data from ferreted population samples (Cowan 1984), suggests that, with the possible exception of winter 1980/1981, the overwinter and adult breeding population estimates approached full enumerations. During the breeding seasons,

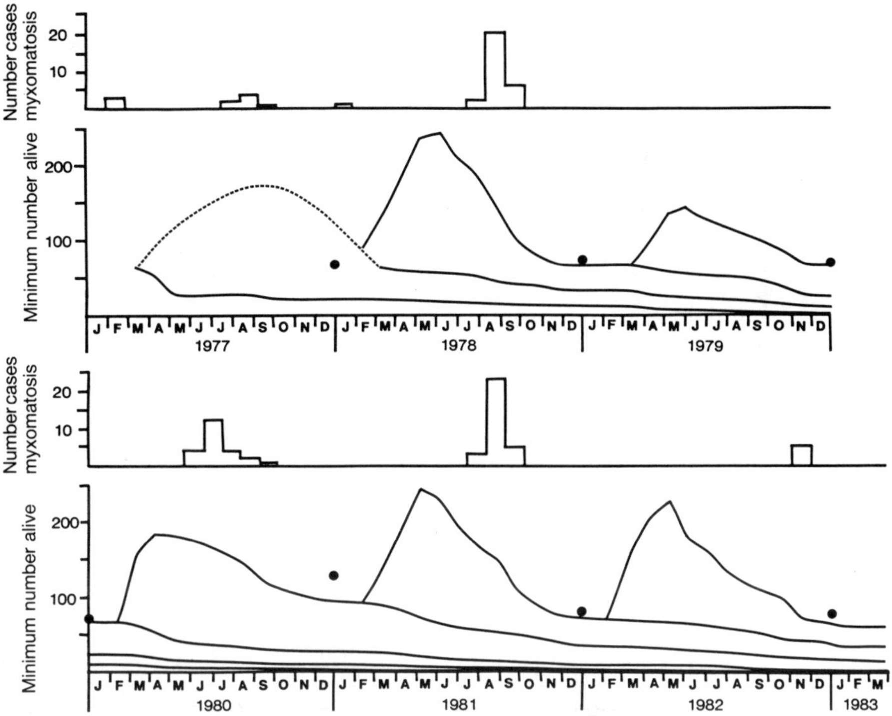

Fig. 2. Changes in the minimum numbers of rabbits known to be alive during the course of the study and the number of individuals captured when showing overt symptoms of myxomatosis in each 28-day period. Cohorts born in different years are separated (dotted line for 1977 cohort indicates period for which no realistic estimates are available owing to late start of trapping programme). Filled circles show population estimates from capture-mark-recapture data based on ferreted samples (Cowan 1984).

however, underestimation of the true population size probably increased with the size of the minimum estimate, because, for each individual caught and thus contributing to this estimate, there was some probability of others, born at the same time, dying before being caught. Similarly, in years when juvenile mortality was high, the likelihood of underestimation would have been increased because the probability of capturing individuals before they died was lower. Nevertheless, these estimates reflect a generalized pattern of seasonal changes whereby recruitment of young during the breeding season is matched and then superseded by mortality from May/June until the beginning of the following season. With the exception of 1981 (n = 92) the breeding population was relatively stable (n = 62–72). There appeared to be more variation in peak population size (n = 140–245), although the maximum/minimum ratio of 1.7 for these estimates was only slightly higher than that of 1.5 for the breeding populations. There was no correlation

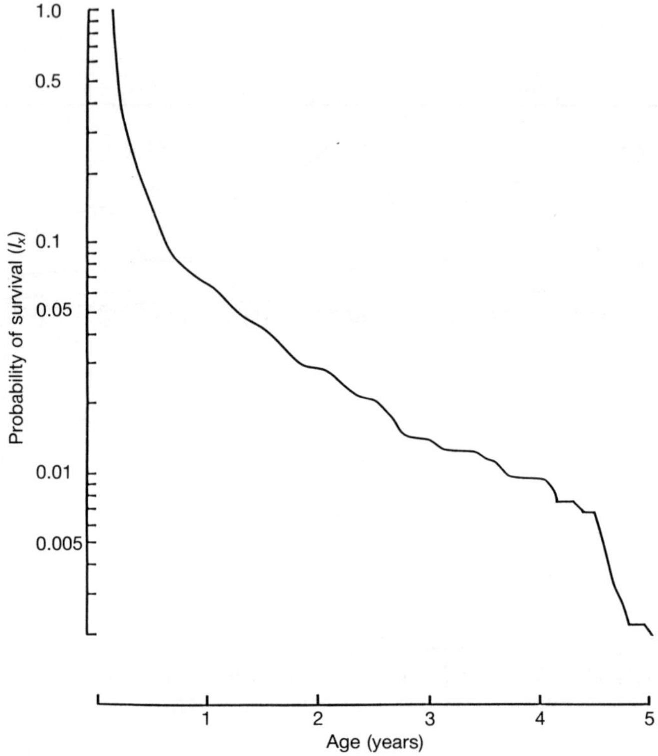

Fig. 3. Survival probabilities of rabbits from 4–8 week-old age interval to 5 years old. Life-table constructed by pooling data between sexes and amongst years but only includes animals whose age was known precisely. ($n < 1$ year old = 908, n 1-year-olds = 124, n 2-year-olds = 49, n 3-year-olds = 14, n >3 years old = 13).

between the peak estimates and the sizes of subsequent breeding populations ($r = -0.1$, $n = 5$, N.S.).

Variation in mortality with age and sex

The probability of survival from 4–8 weeks to five years of age is shown in Fig. 3. Mortality rates were high during the first year of life (the probability of survival from 4–8 weeks of age to more than 52 weeks of age was 0.065), became relatively low in one-, two- and three-year-olds, and rose again for animals four or more years old. The annual survival probability of 0.23 for animals more than three years old was significantly lower than that of 0.71 for three-year-olds (*S.N.D.* 2.46, $P < 0.05$), but not significantly different from the 0.44 for one-year-olds and 0.47 for two-year-olds. Life expectancy, calculated using the method given by Southwood (1966), rose from

Table 1. Variations with sex in the numbers of juveniles at risk of dying during different age intervals ($n_x = c_x + n_{x-1} - d_{x-1}$), their mortality rates ($q_x = d_x/n_x$), and their survival probabilities ($l_x = l_{x-1}\ (l - q_x)$).

Age interval (weeks)	*n* entering table (c_x)		*n* dying (d_x)		*n* at risk (n_x)		Mortality rate (q_x)		Survival probability (l_x)	
	♂♂	♀♀	♂♂	♀♀	♂♂	♀♀	♂♂	♀♀	♂♂	♀♀
4–8	242	313	153	162	242	313	0.63	0.52	1.000	1.000
8–12	110	109	85	75	199	260	0.43	0.29	0.368	0.482
12–16	25	41	41	47	139	226	0.30	0.21	0.211	0.343
16–20	12	31	23	45	110	210	0.21	0.21	0.149	0.272
20–24	9	14	16	39	96	179	0.17	0.22	0.117	0.214
24–28	1	0	16	25	81	140	0.20	0.18	0.098	0.167
28–32	0	1	11	22	65	116	0.17	0.19	0.079	0.137
32–36	0	0	4	10	54	94	0.07	0.11	0.065	0.111
36–40	0	0	2	6	50	84	0.04	0.07	0.060	0.099
40–44	0	0	1	8	48	78	0.02	0.10	0.058	0.092
44–48	0	0	3	1	47	70	0.06	0.01	0.057	0.083
48–52	0	0	4	1	44	69	0.09	0.01	0.053	0.082
>52	—	—	—	—	—		—	—	0.048	0.080

only 13 weeks at 4–8 weeks of age to 77 weeks in the middle of the third year of life.

Juvenile survival was significantly higher for females than for males (Table 1, *S.N.D.* 3.10, $P < 0.01$). This difference was predominantly due to higher male mortality rates in the 4–8, 8–12, and 12–16 week age-groups, as survival probabilities from 16–20 weeks to more than 52 weeks of age were similar for both sexes ($l_{16-20,\ >52}$ = 0.33 for males and 0.30 for females, *S.N.D.* 0.68, N.S.).

There was no difference in the annual survival probabilities of adult males (0.36) and females (0.47) pooled with respect to age (Fig. 4, *S.N.D.* 1.51, N.S.). However, survival of males outside the main breeding season, i.e. from Period 8 through to Period 2 of the following year, was significantly lower than that of females (*S.N.D.* 2.30, $P < 0.05$).

An inevitable consequence of these sex-specific differences in survival was a female-biased sex ratio in the adult breeding population in all years from 1977 through to 1983 (range = 1.14–1.63 females per male, $\bar{x} \pm S.D.$ = 1.35 ± 0.17).

Variation in juvenile mortality with time of birth

In order to compare the survival probabilities of young born at different times of the year, data were pooled over the whole study period, and with respect to sex, to create cohorts born in different 28-day periods of the year. Individuals born in Periods 8 and 9 were also pooled to provide an adequate sample size. There was no difference with sex in the relative numbers born during different periods of the year (χ^2 = 3.22, *d.f.* = 6, N.S.). However,

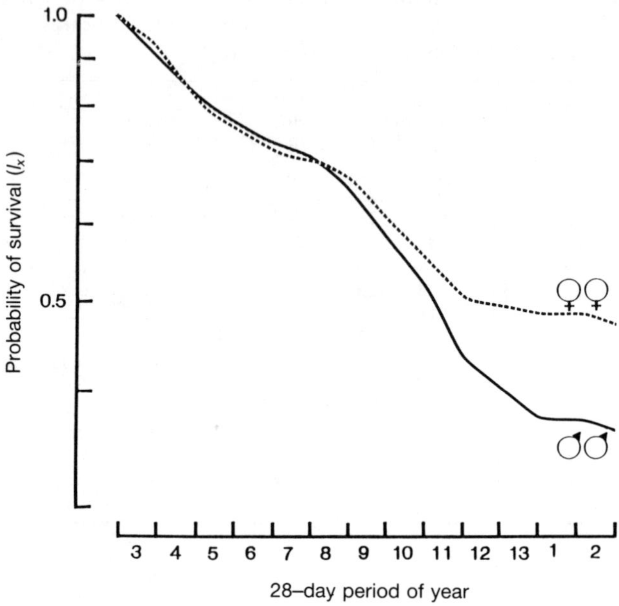

Fig. 4. Survival probabilities of adult males (n = 121) and adult females (n = 182) for 28-day periods from the start of one breeding season to the beginning of the next. Data were pooled both amongst years and with respect to age.

because of differences amongst years in distributions of young amongst birth periods (Fig. 1) this analysis needs to be interpreted with care. Nevertheless, most of the variation according to time of birth could not be explained by a confounding effect of this type. The highly significant decline in survival probabilities amongst successive cohorts born during the course of the breeding season is striking (Tables 2 and 3). The effect of period of birth on survival was particularly marked for the 4–8 and 8–12 week-old age intervals with the mortality rates (q_{4-8} and q_{8-12}, respectively) being significantly and positively correlated with period of birth ($r^2 = 0.76$, $n = 7$, $P < 0.02$, and $r^2 = 0.74$, $n = 7$, $P < 0.02$, respectively).

Variation in juvenile mortality between years

Cohorts of juveniles born in different years were created by pooling data with respect to sex and period of birth. The data from 1977 were excluded from the analysis because of the late start of the trapping programme. There were no differences between years in the relative numbers of juvenile males and females caught ($\chi^2 = 2.54$, $d.f. = 4$, N.S.). However, some bias may have arisen through the variation between years in the distributions of

Table 2. Variations in the mortality rates (q_x) and survival probabilities of juveniles (l_x), at different age intervals, with the 28-day period of the year in which they are born.

Age Intervals (weeks)	Mortality rate (*qx*)							Survival probability (*lx*)						
	Birth period							Birth period						
	2	3	4	5	6	7	8/9	2	3	4	5	6	7	8/9
n	76	190	234	228	99	57	24	76	190	234	228	99	57	24
4–8	0.47	0.42	0.48	0.70	0.57	0.69	0.88	1.000	1.000	1.000	1.000	1.000	1.000	1.000
8–12	0.20	0.30	0.39	0.29	0.33	0.72	0.78	0.526	0.582	0.503	0.299	0.431	0.311	0.118
12–16	0.26	0.20	0.17	0.21	0.48	0.50	0.00	0.421	0.408	0.310	0.212	0.290	0.087	0.026
16–20	0.08	0.10	0.24	0.34	0.31	0.00	0.00	0.313	0.328	0.256	0.168	0.151	0.044	0.026
20–24	0.06	0.12	0.32	0.24	0.22	0.25	0.00	0.287	0.297	0.195	0.111	0.105	0.044	0.026
24–28	0.16	0.22	0.26	0.16	0.00	0.00	0.00	0.270	0.261	0.132	0.085	0.081	0.033	0.026
28–32	0.18	0.31	0.16	0.08	0.07	0.00	0.00	0.228	0.205	0.099	0.071	0.081	0.033	0.026
32–36	0.17	0.10	0.06	0.06	0.15	0.00	0.00	0.187	0.141	0.083	0.066	0.076	0.033	0.026
36–40	0.11	0.03	0.07	0.03	0.09	0.00	0.05	0.154	0.127	0.078	0.062	0.064	0.033	0.026
40–44	0.12	0.06	0.07	0.06	0.10	0.00	0.00	0.138	0.124	0.073	0.060	0.058	0.033	0.013
44–48	0.00	0.06	0.04	0.03	0.00	0.00	0.00	0.122	0.117	0.068	0.056	0.052	0.033	0.013
48–52	0.00	0.03	0.04	0.07	0.11	0.00	0.00	0.122	0.110	0.065	0.054	0.052	0.033	0.013
>52	—	—	—	—	—	—	—	0.122	0.106	0.062	0.051	0.047	0.033	0.013

Table 3. Log-rank tests of differences between survival probabilities, from the 4–8 week age interval to >52 weeks of age, of juveniles born in different 28-day periods of the year. Only significant results are shown, with the cohort of lower survivorship given in the column on left of Table (* $=P < 0.05$, ** $=P < 0.01$, *** $=P < 0.001$).

28-day period of year in which born	28-day period of year in which born						
	2	3	4	5	6	7	8/9
2							
3							
4	*	**					
5	*	**					
6	**	**					
7	***	***	**		*		
8/9	**	***	**	*	*		

young born during different periods. Many of the differences in survivorship between years were, however, highly significant and inconsistent with differences arising from distributions of birth dates alone (Fig. 5 and Table 4). Survival was highest in 1980 and poorest in 1978. Although some of the differences were already apparent by the 12–16 week age interval, others were not. For instance, that 1979 cohorts had higher survival than 1981 and 1982 cohorts was attributable to lower mortality amongst animals more than 16 weeks old in 1979.

Causes of mortality

The cause of death was definitely established for only 73 tagged individuals. Death occurred from shooting (n = 35), predation (n = 14), myxomatosis (n = 11), coccidiosis (n = 7), trapping/ferreting accident (n = 5), and road accident (n = 1). These figures give no indication of the relative rates of death from each cause, because the great majority of all rabbits shot will be represented here, but only a small proportion of those that died from predation or disease.

The seasonal nature of myxomatosis epizootics can be seen in Fig. 2, with summer outbreaks in 1977, 1978, 1980 and 1981. Further outbreaks occurred in the winters of 1976/1977 and 1977/78, and in the autumn of 1982. No cases of myxomatosis were recorded between September 1978 and June 1980. The epizootics of summer 1978, summer 1981 and autumn 1982 all coincided with marked reductions in the minimum population estimates, although that of summer 1980 did not. The 1980 summer epizootic

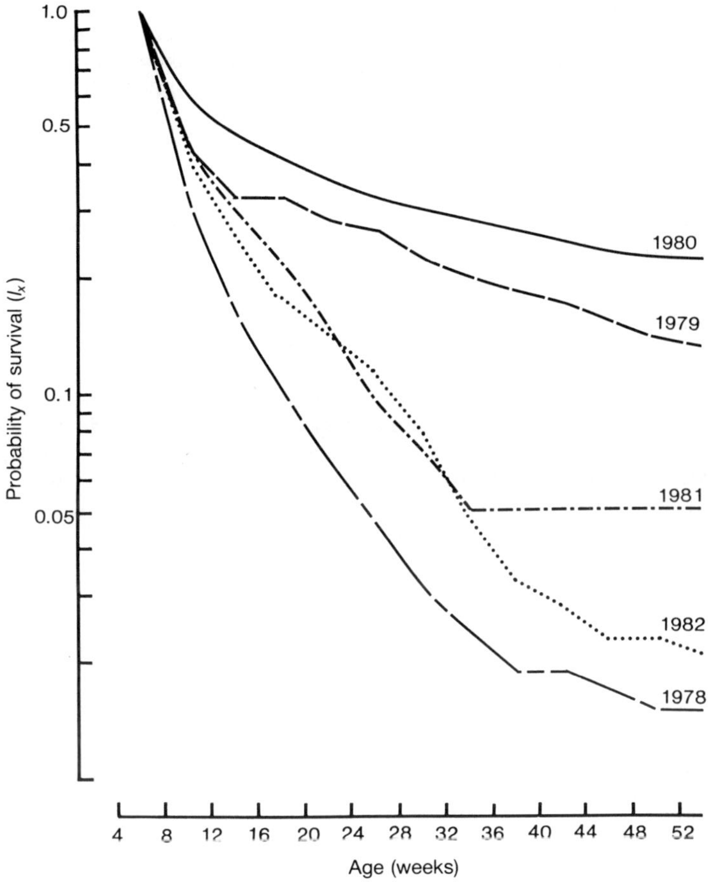

Fig. 5. Variations in survival probabilities of juveniles, in 28-day periods from 4–8 week age interval to >52 weeks of age, with the year in which they were born (*n* 1978 = 237, *n* 1979 = 81, *n* 1980 = 182, *n* 1981 = 191, *n* 1982 = 173). Data pooled with respect to sex and time of year when born.

was the only one during which any adults were known to be infected (as well as the adult caught and included in Table 5, two others were seen with overt symptoms but not captured). In addition, because the 1980 epizootic occurred earlier in the year, the mean age at first diagnosis amongst juveniles was lower than in 1978 and 1981 (Table 5). The proportion of infected animals that were known to recover from the disease was also higher in 1980 than in any other year. Overall, 31 of the 98 marked individuals (31.6%) infected with myxomatosis were known to recover from the disease.

Table 4. Log-rank tests of differences between years in the survival probabilities of juveniles from the 4–8 week age interval to >52 weeks of age. Only significant results are shown, with the cohort of lower survivorship given in the column on left of Table (* $=P < 0.05$, ** $=P < 0.01$, *** $=P < 0.001$).

Year of birth	Year of birth 1978	1979	1980	1981	1982
1978		***	***	**	*
1979					
1980					
1981		**	***		
1982		***	***		

Table 5. Numbers and mean age at first diagnosis of tagged individuals <9 months old caught when infected with myxomatosis, numbers of all tagged individuals caught when infected with myxomatosis and percentages of them that were known to recover from the disease during four summer epizootics and at other times during the course of the study (no meaningful data available for summer 1977 on age at diagnosis because of late start of trapping programme).

	Mean age at diagnosis (weeks) n	$\bar{x} \pm$ S.D.	All cases n	% recovering
Summer 1977	—	—	6	0
Summer 1978	28	16.4 ± 5.7 [a]	28	17.9 [c, d]
Summer 1980	22	11.8 ± 4.8 [a, b]	23	47.8 [d]
Summer 1981	31	16.9 ± 4.0 [b]	31	38.7
Others	6	33.5 ± 5.2	10	30.0

[a] Mann-Whitney U, $Z_{corr.} = 2.8$, $n_1 = 22$, $n_2 = 28$, $P < 0.01$
[b] Mann-Whitney U, $Z_{corr.} = 3.97$, $n_1 = 22$, $n_2 = 31$, $P < 0.001$
[c] $\chi^2_{corr.} = 2.82$, $d.f. = 1$, $P < 0.01$
[d] $\chi^2_{corr.} = 3.97$, $d.f. = 1$, $P < 0.05$

Discussion

Variation in mortality with age

The variation in mortality rates with age at Aston Rowant followed the typical mammalian pattern (Caughley 1966). The mortality rate for adults was consistent with the 50% to 80% annual loss estimated for free-living populations in Australia (Dunsmore 1974; Parer 1977; Wood 1980), New Zealand (Tyndale-Biscoe & Williams 1955) and France (Arthur 1980), and in enclosed populations generally (Lockley 1961; Myers & Poole 1963; Gibb, Ward & Ward 1978).

The mortality rate of juveniles was higher than in the two other studies where life-table analyses have been presented, particularly in the 4–8 and

8–12 week age intervals (Parer 1977; Wood 1980). This may, however, have been due to the method of life-table construction used in those studies, which tends to underestimate mortality in the younger age intervals (Cowan & Roman 1985). Juvenile mortality rates were almost certainly over-estimated in the current study owing to individuals disappearing, either dispersing undetected or dying without being captured. Nevertheless, the life-expectancy at 4–8 weeks of age calculated here is close to the 3.5 months at emergence estimated by Parer (1977), and the overall estimates of juvenile survival are similar to those of Southern (1940), Tyndale-Biscoe & Williams (1955), Lloyd (1970b), Parer (1977), Gibb *et al.* (1978), Arthur (1980), Wood (1980) and Richardson & Wood (1982).

Variation in mortality with sex

Neither Parer (1977) nor Wood (1980) observed either the differences in adult mortality between the sexes or the female-biased adult sex ratio reported here. However, Dunsmore (1974) reported higher mortality in males and a female-biased adult sex ratio during all five years of his study. His study population, like that at Aston Rowant (Cowan in press a), consisted of relatively large breeding groups in comparison with those observed by Parer (1977) and Wood (1980). Large groups offer greater opportunities for males to effectively monopolize matings with a large number of females (Cowan & Garson 1985). Higher mortality amongst males might be expected in such circumstances because of the costs of more intense competition for mates (Clutton-Brock, Guinness & Albon 1982). Survival of males was poorer than that of females only outside the breeding season at Aston Rowant, which suggests that competition associated with the reorganization of the breeding groups at this time was particularly important (Mykytowycz 1960; Myers & Poole 1961).

Except for one of the two populations studied by Wheeler & King (1985), higher male juvenile mortality has not been reported previously (Parer 1977; Gibb *et al.* 1978; Wood 1980). That differential mortality between the sexes was apparent from 4–8 weeks of age suggests that it was not simply a reflection of behavioral changes associated with attaining sexual maturity, for instance higher rates of male dispersal (Parer 1982; Cowan 1983).

Variation in juvenile mortality with time of birth

The increased mortality amongst successive cohorts born during the breeding season reported here is common to other rabbit populations (Mykytowycz 1961; Myers & Poole 1963; Parer 1977; Gibb *et al.* 1978; Wood 1980). Two exceptions to this pattern have been reported (Southern 1940;

Wheeler & King 1985). Southern did not take into account age-specific mortality amongst juveniles, hence the apparent improved survival during the course of the season probably reflected changes in the age structure of the juvenile population. Wheeler & King attributed the increase in age-specific survival during the course of the breeding season, in one of their two study populations, to an increase in ground cover affording greater protection against avian predators. In contrast, Parer (1977) considered that the decline in juvenile survival in his population was due to an increase in the rate of predation, particularly by cats, operating in a density-dependent manner. Gibb *et al.* (1978), Mykytowycz (1960) and Myers & Poole (1963) found that decreases in juvenile survival were associated with declines in the availability of food and reduced growth rates of rabbits. Late-born young have been found to show signs of protein deficiency (Richardson & Wood 1982). Growth rates increased during the course of the breeding season at Aston Rowant (Cowan 1983).

Richardson & Wood (1982) found raised serum alkaline phosphate levels and proportions of lymphocytes in the white cell counts in late-born young suggesting higher levels of infection. The densities of infective sporocysts of coccidia (*Eimeria* spp.) on the vegetation around burrow systems are likely to increase as life-cycles of these parasites are completed in successive cohorts of rabbits born during the course of the breeding season (Bull 1958). Coccidial oocyst counts, especially of one of the more pathogenic species *Eimeria steidia*, were highest in the faeces of late-born young at Aston Rowant and lowest amongst those born towards the beginning of the breeding season (Cowan 1985). The disease is most prevalent amongst individuals between six and 16 weeks of age (Mykytowycz 1962; Stodart 1968). These are the ages at which the decline in survival with date of birth was most marked. While such observations do not establish cause and effect, it is possible that coccidial infections contributed directly to the decline in juvenile survival.

Variation in juvenile mortality between years

The marked variation between years in juvenile survival at Aston Rowant was partly related to population density. Juvenile mortality was very high in 1978, the year of probably the highest peak in population size, given the biases inherent in the method of obtaining minimum population estimates. The lowest peak minimum estimate was recorded in 1979; mortality of the 1979 cohort was also low. The coccidial oocyst counts were much higher in 1978 than in 1979 (Cowan 1985). Similarly, coccidial oocyst counts were lower in young born into breeding groups containing a single adult female than in those born into groups containing more than one adult female (Cowan 1985), and these lower counts were associated with higher survival

(Cowan in press b). Parer (1977) emphasized differential rates of predation in relation to density-dependent patterns of juvenile mortality. He also showed that the number of young recruited into a given breeding population was not correlated with the number that emerged during the previous breeding season. Density-dependent variation in juvenile mortality between years has been recognized elsewhere in Britain (Cowan & Garson 1985; Garson 1986) and in New Zealand (Gibb *et al.* 1978; Gibb 1979), but the proximate cause was attributed to changes in food availability in those cases.

Myxomatosis may also operate in a density-dependent manner, since it is probable that a minimum density of susceptible individuals is required before an epizootic occurs (Ross & Tittensor 1986). The apparent absence of the disease in the summer of 1979 supports this view, given the low peak minimum estimate of population density in this year. The low number of emergent young in 1979 meant that relatively high juvenile survival did not lead to an increase in the size of the subsequent breeding population.

Three differences in the nature of the myxomatosis epizootic in summer 1980 were apparent. First, it was the only summer epizootic during which adults were known to have been infected. The outbreak followed a period of nearly two years when myxomatosis was apparently absent from the site. This suggests that in other years most juveniles were infected during the summer epizootics so that most animals reached adulthood having survived the disease and were immune.

Secondly, the mean age at diagnosis was relatively low in 1980. This was because the epizootic occurred earlier than in other years. The earlier onset of the disease could be attributed to the relatively high density of susceptible individuals in this year due to the presence of a high proportion of susceptible adults. Earlier and more intense summer/autumn epizootics have been associated with high densities of susceptible individuals elsewhere (J. Ross, pers. comm.).

Thirdly, the proportion of individuals known to survive the disease was relatively high. There is, however, an element of circularity in the argument that higher survivorship from myxomatosis during this epizootic (47.8%) contributed to the low overall mortality rate in this year. Nevertheless, the overall case mortality rate (68.4%) at Aston Rowant was in line with the maximum estimates for other populations in Britain (Ross & Tittensor 1986).

Perhaps the most important demographic aspect of the 1980 epizootic was its earlier timing and, hence, the relatively young ages of animals diagnosed as infected. Mortality due to myxomatosis thus occurred when other agents, such as coccidiosis and predation, were probably operating in a density-dependent manner. Animals were thus dying of myxomatosis when their probability of dying from other causes was already high. Hence, the contribution of myxomatosis to the overall rate of juvenile mortality in the 1980 cohort will have been relatively low. This might explain the increase in the size of the subsequent breeding population. In other years, the later occurrence of the disease meant that its effects were additive to those of more age-specific agents in terms of the numbers of juveniles surviving to enter the subsequent breeding season. This was particularly evident in 1982 when a late autumn epizootic caused relatively high mortality amongst juveniles more than 28 weeks old which otherwise would have had a reasonable probability of being recruited into the subsequent breeding population.

Conclusions

During the course of this study patterns of mortality varied with age, sex, time of birth and population density. Both coccidiosis and myxomatosis were implicated as influences on the way the summer peak populations declined and, hence, on the sizes of the subsequent breeding populations. The timing and density-dependent qualities of these influences were particularly important. Such observations suggest that the gradual increase in national rabbit abundance (Trout *et al.* 1986) might reflect the frequency of sudden, and sometimes transient, local increases in the size of breeding populations, as seen here in 1981.

Currently, only very general predictions can be made concerning the consequences for rabbit populations of further changes in the epidemiology of myxomatosis (Ross 1982; Ross & Sanders 1984), or of different rabbit management strategies. In order to devise effective strategies, the observations reported here need to be described more formally, perhaps through mathematical simulation, and used to design experimental manipulations of specific parameters. Nevertheless, it seems likely that the size of the subsequent breeding population would be reduced most efficiently if control was carried out after late autumn (Rees *et al.* 1985). At this time natural agents of mortality, particularly those exhibiting density-dependent characteristics, would already have exerted their main effects.

Acknowledgements

This study was partly funded by a N.E.R.C. studentship, and was initially undertaken at Royal Holloway College, University of London. The kind supervision and encouragement of Professor P. A. Jewell and Dr G. I. Twigg were gratefully received together with the support of the M.A.F.F. in completing the study. I should also like to thank Mr. W. G. Christer for assistance in the field from 1980 onwards, the N.C.C. for permission to work on their land, and Dr P. J. Garson, Dr. A. R. Hardy, Dr W. A. Rees, Dr J. Ross and Dr R. C. Trout for helpful comments on the manuscript.

References

Arthur, C.P. (1980). Démographie du lapin de Garenne *Oryctolagus cuniculus* (L.) 1978 en région Parisienne. *Bull. mens. Off. natn. Chasse No. spec. scient. techn.* **1980**: 127–62.

Bull, P.C. (1958). Incidence of coccidia (Sporozoa) in wild rabbits *Oryctolagus cuniculus* (L.) in Hawke's Bay, New Zealand. *N.Z. Jl Sci.* **1**: 289–329.

Caughley, G. (1966). Mortality patterns in mammals. *Ecology* **47**: 906–18.

Clutton-Brock, T.H., Guinness, F.E. & Albon, S.D. (1982). *Red Deer. Behavior and ecology of two sexes.* Edinburgh University Press, Edinburgh.

Cowan, D.P. (1983). *Aspects of the behavioural ecology of a free-living population of the European wild rabbit,* Oryctolagus cuniculus *L., in southern England.* Ph.D.thesis: University of London.

Cowan, D.P. (1984). The use of ferrets (*Mustela furo*) in the study and management of the European wild rabbit (*Oryctolagus cuniculus*). *J. Zool., Lond.* **204**: 570–4.

Cowan, D.P. (1985). Coccidiosis in rabbits. In *Population dynamics and epidemiology of territorial animals:* 25–7. (Eds Mollison, D. & Bacon, P.). (*I.T.E. Merlewood Res. Dev. Pap.* No. 106.). I.T.E. Merlewood, Cumbria.

Cowan, D.P. (In press a). Aspects of the social organisation of the European wild rabbit *Oryctolagus cuniculus. Ethology.*

Cowan, D.P. (In press b). Group living in the European wild rabbit (*Oryctolagus cuniculus* L.): mutual benefit or resource localisation? *J. anim. Ecol.*

Cowan, D.P. & Garson, P.J. (1985). Variations in the social structure of rabbit populations: causes and demographic consequences. In *Behavioural ecology: the ecological consequences of adaptive behaviour:* 537–55. (Eds Sibly, R.M. & Smith, R.H.). Blackwell Scientific Publications, Oxford.

Cowan, D.P. & Roman, E.A. (1985). The construction of life tables with special reference to the European wild rabbit (*Oryctolagus cuniculus*). *J. Zool., Lond. (A)* **207**: 607–9.

Dunsmore, J.D. (1974). The rabbit in subalpine south-eastern Australia. I. Population structure and productivity. *Aust. Wildl. Res.* **1**: 1–16.

Garson, P.J. (1986). Food limitation of a rabbit population: observational and experimental evidence from a study in coastal duneland habitat. *Mammal Rev.* **16**: 195–6.

Gibb, J.A. (1979). Factors affecting population density in the wild rabbit, *Oryctolagus cuniculus* (L.) and their relevance to small mammals. In *Evolutionary ecology*: 33–46. (Eds Stonehouse, B. & Perrins, C.). Macmillan, London.

Gibb, J.A., Ward, C.P. & Ward, G.D. (1978). Natural control of a population of rabbits, *Oryctolagus cuniculus* (L.), for ten years in the Kourarau enclosure. *Bull. Dep. scient. ind. Res. N.Z.* No. 223: 1–89.

Healy, M.J.R. & Osborn, J.F. (1981). *Manual of medical statistics* **2**. London School of Hygiene and Tropical Medicine, London.

Henderson, B.A. (1979). Regulation of the size of the breeding population of the European rabbit, *Oryctolagus cuniculus*, by social behaviour. *J. appl. Ecol.* **16**: 383–92.

Lloyd, H.G. (1970a). Variation and adaptation in reproductive performance. *Symp. zool. Soc. Lond.* No. 26: 165–88.

Lloyd, H.G. (1970b). Post-myxomatosis rabbit populations in England and Wales. *EPPO Pubs* (A) No 58: 197–215.

Lockley, R.M. (1961). Social structure and stress in the rabbit warren. *J. anim. Ecol.* **30**: 385–423.

Myers, K. & Poole, W.E. (1961). A study of the biology of the wild rabbit, *Oryctolagus cuniculus* (L.) in confined populations. II. The effects of season and population increase on behaviour. *CSIRO Wildl. Res.* **6**: 1–41.

Myers, K. & Poole, W.E. (1963). A study of the biology of the wild rabbit, *Oryctolagus cuniculus* (L.) in confined populations. V. Population dynamics. *CSIRO Wildl. Res.* **8**: 166–203.

Mykytowycz, R. (1960). Social behaviour of an experimental colony of wild rabbits, *Oryctolagus cuniculus* (L.). III. Second breeding season. *CSIRO Wildl. Res.* **5**: 1–20.

Mykytowycz, R. (1961). Social behaviour of an experimental colony of wild rabbits *Oryctolagus cuniculus* (L.). IV. Conclusion: outbreak of myxomatosis, third breeding season, and starvation. *CSIRO Wildl. Res.* **6**: 142–55.

Mykytowycz, R. (1962). Epidemiology of coccidiosis (*Eimeria* spp.) in an experimental population of the Australian wild rabbit, *Oryctolagus cuniculus* (L.). *Parasitology* **52**: 375–95.

Parer, I. (1977). The population ecology of the wild rabbit *Oryctolagus cuniculus* (L.) in a Mediterranean-type climate in New South Wales. *Aust. Wildl. Res.* **4**: 171–205.

Parer, I. (1982). Dispersal of wild rabbit, *Oryctolagus cuniculus*, at Urana in New South Wales. *Aust. Wildl. Res.* **9**: 427–41.

Peto, R., Pike, M.C., Armitage, P., Breslon, N.E., Cox, D.R., Howard, S.V., Mantel, N., MacPherson, K., Peto, K. & Smith, P.G. (1977). Design and analysis of randomised clinical trials requiring prolonged observation of each patient. II. Analysis and examples. *Br. J. Cancer Res.* **35**: 1–39.

Rees, W.A., Ross, J., Cowan, D.P., Tittensor, A.M. & Trout, R.C. (1985). Humane control of rabbits. In *Humane control of land mammals and birds*: 96–104. (Ed. Britt, D.P.). Universities Federation for Animal Welfare, London.

Richardson, B.J. & Wood D.H. (1982). Experimental ecological studies on a subalpine rabbit population. I. Mortality factors acting on emergent kittens. *Aust. Wildl. Res.* **9**: 443–50.

Ross, J. (1982). Myxomatosis: the natural evolution of the disease. *Symp. zool. Soc. Lond.* No. 50: 77–95.

Ross, J. & Sanders, M.F. (1984). The development of genetic resistance to myxomatosis in wild rabbits in Britain. *J. Hyg., Camb.* **92**: 255–61.

Ross, J. & Tittensor, A.M. (1986). Influence of myxomatosis in regulating rabbit numbers. *Mammal Rev.* **16**: 163–8.

Southern, H.E.N. (1940). The ecology and population dynamics of the wild rabbit (*Oryctolagus cuniculus*). *Ann. appl. Biol.* **27**: 509–26.

Southwood, T.R.E. (1966). *Ecological methods*. Methuen, London.

Stodart, E. (1968). Coccidiosis in wild rabbits, *Oryctolagus cuniculus* (L.), at four sites in different climatic regions in eastern Australia. I. Relationship with the age of rabbit. *Aust. J. Zool.* **16**: 69–85.

Trout, R.C., Tapper, S.C. & Harradine, J. (1986). Recent changes in rabbit populations in Britain. *Mammal Rev.* **16**: 117–23.

Tyndale-Biscoe, C.H. & Williams, R.M. (1955). A study of natural mortality in a wild population of the rabbit, *Oryctolagus cuniculus* (L.). *N.Z. J. Sci. Technol.* **36B**: 561–80.

Watson, J.S. & Tyndale-Biscoe, C.H. (1953). The apophyseal line as an age indicator for the wild rabbit *Oryctolagus cuniculus* (L.). *N.Z. J. Sci. Technol.* **34B**: 427–35.

Wheeler, S.H. & King, D.R. (1985). The European rabbit in south-western Australia III. Survival. *Aust. Wildl. Res.* **12**: 213–25.

Wood, D.H. (1980). The demography of a rabbit population in an arid region of New South Wales, Australia. *J. anim. Ecol.* **49**: 55–79.

Symp. zool. Soc. Lond. (1987) No. 58: 79–98

Cycles in game-bag records of hares and rabbits in Britain

Stephen TAPPER

The Game Conservancy
Fordingbridge
Hampshire

Synopsis

Archive records, in the form of fur returns and game shooting bags, have been used in the past to explore population fluctuations in animals over long periods of time. Although the principles of time-series analysis were known in these early studies, they were not widely applied in ecology papers. In many cases it was difficult to know whether variations were cyclic or random. However, with the introduction of computers many of these old data sets are being re-examined.

Game-bag records of the three British lagomorphs are examined with time-series analysis to test whether or not any show anything other than random variation.

Some populations of all three species are found to show quasi-cyclic fluctuations—different from the cyclic fluctuations found in North American and Scandinavian species.

Historical background

The foundation of many ecological studies of mammals in Britain can be traced back to a period from the 1920s to the 1940s, when a small team of researchers at the University of Oxford, led by Charles Elton and called The Bureau of Animal Population, began to investigate factors which limited animal abundance. A key question was the cause of fluctuations in numbers of some species over long periods of time; in particular it was proposed that yearly variations in number were not random but followed well developed cycles (Elton 1924). However, population studies were a new phenomenon, and there were no census data to test this idea.

Without question the best data to emerge were found in the archives of the Hudson's Bay Company which had been trading in North American furs since the mid 18th century. In a remarkable paper (Elton & Nicholson 1942a), the records of lynx (*Lynx canadensis*) taken in the Northern Department of the Company provided dramatic evidence for the now famous ten-year wildlife cycle in boreal North America. Later evidence

ZOOLOGICAL SYMPOSIUM No. 58
ISBN 0-19-854006-X

showed that a whole complex of mammals were caught up in this phenomenon (Elton & Nicholson 1942b; Chitty 1950a, b; Butler 1942).

This work paralleled a hunt for other kinds of data which would provide clues to cycles in other species in other areas. One such source was found to be the game books kept on sporting estates in England and Scotland. Driven game shooting developed as a fashion in the early 19th century, and by Edwardian times had reached an apogee of development. One aspect of the sport was a fascination with the numbers shot, so that daily and annual totals were often recorded in leather-bound ledgers, completed meticulously at the end of each day and season. Workers at the Bureau were quick to realize their potential for investigating fluctuations. Douglas Middleton (1934) analysed data from selected game books and concluded that there was evidence for a cycle of about eight years in rabbits (*Oryctolagus cuniculus*), brown hares (*Lepus europaeus*) and partridges (*Perdix perdix*); a six-year cycle in red grouse (*Lagopus lagopus*); and other fluctuations 'of a periodic nature' in mountain hares (*Lepus timidus*); weasels (*Mustela nivalis*), woodcock (*Scolopax rusticola*) and blackgame (*Tetrao tetrix*).

In 1938, in collaboration with a Major Eley who ran a cartridge manufacturing company, Douglas Middleton and Helen Chitty organized a game survey to draw together the data in these game books. Unfortunately the Second World War intervened and much of the data remained unanalysed.

After the war further investigation of British bag records of red grouse and other Tetraonidae seemed to suggest that they did indeed peak at approximately six-year intervals (Mackenzie 1952; G. R. Williams 1954). However, Cole (1951) had pointed out that a series drawn directly from a book of random numbers looked remarkably like some of the 'cyclic' series, and seriously questioned their reality. Later (Cole 1954), he even showed that the smoothing procedure adopted by some workers (including Middleton and Elton) was capable of producing cycles in otherwise random data. As a direct result, analysis of archival material largely foundered, and attention turned to field studies which would provide accurate census data of living animals, as well as measures of mortality and fecundity. This was a pity, since appropriate methods of analysis were already available and their use on such data had been established (Moran 1949, 1952, 1953). Although these methods were straightforward in a statistical sense, they were extremely arduous to perform without a computer, and few biologists had access to these early machines. However, a few statisticians pursued these ideas and the subject of 'time-series analysis' was well established by the mid 1970s (Box & Jenkins 1970; Kendall 1973). Williamson (1975) pointed out that these methods could easily be applied to biological series.

The result has been a renewed interest in analysing archival material with time-series techniques. Bulmer (1974, 1975) did an analysis and reappraisal of many of the Hudson's Bay records; he showed not only the range of species that cycled, but also that they were not all in synchrony but differed significantly in phase. He proposed that these differences in phase suggested ecological predator–prey relationships. More comprehensively, Finerty (1980) undertook a wide-ranging time-series analysis of fur-bearing mammals from Scandinavia and North America. Because he used the same methods throughout, he was able to compare cycles from different regions. For instance, red foxes (*Vulpes*) show a four-year cycle in northern Labrador but a ten-year cycle in the boreal forest—in Scandinavia the species shows no cycle in northern regions, a clear four-year cycle in mid-Norway, but none in south Norway.

In this reappraisal of cycles using time-series analysis very little more has been done with the game-bag records which continue to be kept on many estates in Britain. Apart from our own work on red grouse (Potts, Tapper & Hudson 1984), there has been one other study using time-series analysis on British data (J. Williams 1985) and this was a re-analysis of the same data published by Middleton (1934) and Mackenzie (1952) several decades previously.

The purpose of this paper is to determine whether or not any of the British lagomorphs for which bag records exist show any cyclic, as opposed to random, fluctuations in number.

Sources of data and methods

The Bureau of Animal Population game records, kept since the Second World War by the Game Conservancy, consist of completed questionnaires from about 460 estates who were asked for total numbers of game shot each year, and particularly during the periods 1888–1897 and 1928–1937. In 1961 the Game Research Association started the National Game Census which circulated over 500 shoots and estates and asked for annual returns of all game shot, including, where possible, predators taken by gamekeepers as well. This scheme, now run by the Game Conservancy, has gradually expanded since the late 1970s both by adding new contributors and by completing historical records; in particular, moorland estates in the north of England and, later, in Scotland were the subject of special surveys which increased the information from these districts considerably. These surveys have been sorted and aggregated into a database which consists of over 30,000 annual game bag returns from the early 19th century to the present day. Throughout this period changes in ownership and fragmentation of estates mean there are only a few places where there are continuous runs of data for more than 50 years. Also, gamebird bags are more likely to be completed than the mammalian series we are considering here. I have selected only the longest running series for analysis.

Time-series analysis has two primary tools to determine periodicity in a data set. The first is the correlogram, which is a graph relating correlation coefficients calculated from sets of paired records to the time intervals separating them. The second is spectral analysis, which defines the frequency and its variance for any periodic components in the data. Correlograms are easy to calculate with a computer, but spectrograms are mathematically more complex, and for the non-mathematician more difficult to interpret. For this reason I have not calculated spectra for any of the bag records, but have constructed correlograms only—Poole (1974) gives a straightforward account of how to calculate these.

It is common practice to treat data with various statistical techniques prior to calculating correlograms, first to remove trend which will normally ruin the correlogram, and second to reduce variance. Some workers prefer to use neither of these techniques (e.g. Finerty 1980), on the basis that cycles could be produced into series where none had existed before. However, with most game bag series some 'detrending' is usually necessary, since most series come from managed habitats such as farmland or grouse moor where changes in land management have led to huge changes in abundance—quite unrelated to the cycles under investigation. I have found that detrending was essential to most series, so all were detrended before calculating the correlogram. I have followed the method we have used before (Potts *et al.* 1984) and found to be successful for red grouse series. This is a simple unweighted moving average of five elements which is repeatedly run through the data, so becoming smoother at each pass. After ten runs this produces a curve which can be deducted from the original and which removes any trend. For details of this method see Potts *et al.* (1984).

Nisbet & Gurney (1982: 9) classify time series and show typical correlograms for each type. I shall follow their classification:

1. True cycles. The amplitude and period of these cycles is constant, resulting in a correlogram which shows coefficients that attain $+1$ and -1 for each multiple of the wave length and half wave length. These series are not found in animal populations and represent pure sine waves.

2. Phase-remembering quasi-cycles. These series show more noise than the above but the correlogram has significant coefficients for time lags corresponding to the half cycle and full cycle. Importantly, these significant coefficients are sustained and do not get weaker throughout the correlogram. Trapping returns from lynx, red fox and other fur-bearers produce time series of this type and are referred to as cyclic (Fig. 1.)

3. Phase-forgetting quasi-cycles. These series not only are noisy, but exhibit variation in cycle length which gradually draws the cycles out of line with sine-wave series of the same wavelength. Coefficients show a progressive reduction towards zero with increasing time intervals. Correlograms of this type are said to be damped. Some Canadian fur-bearers have slightly

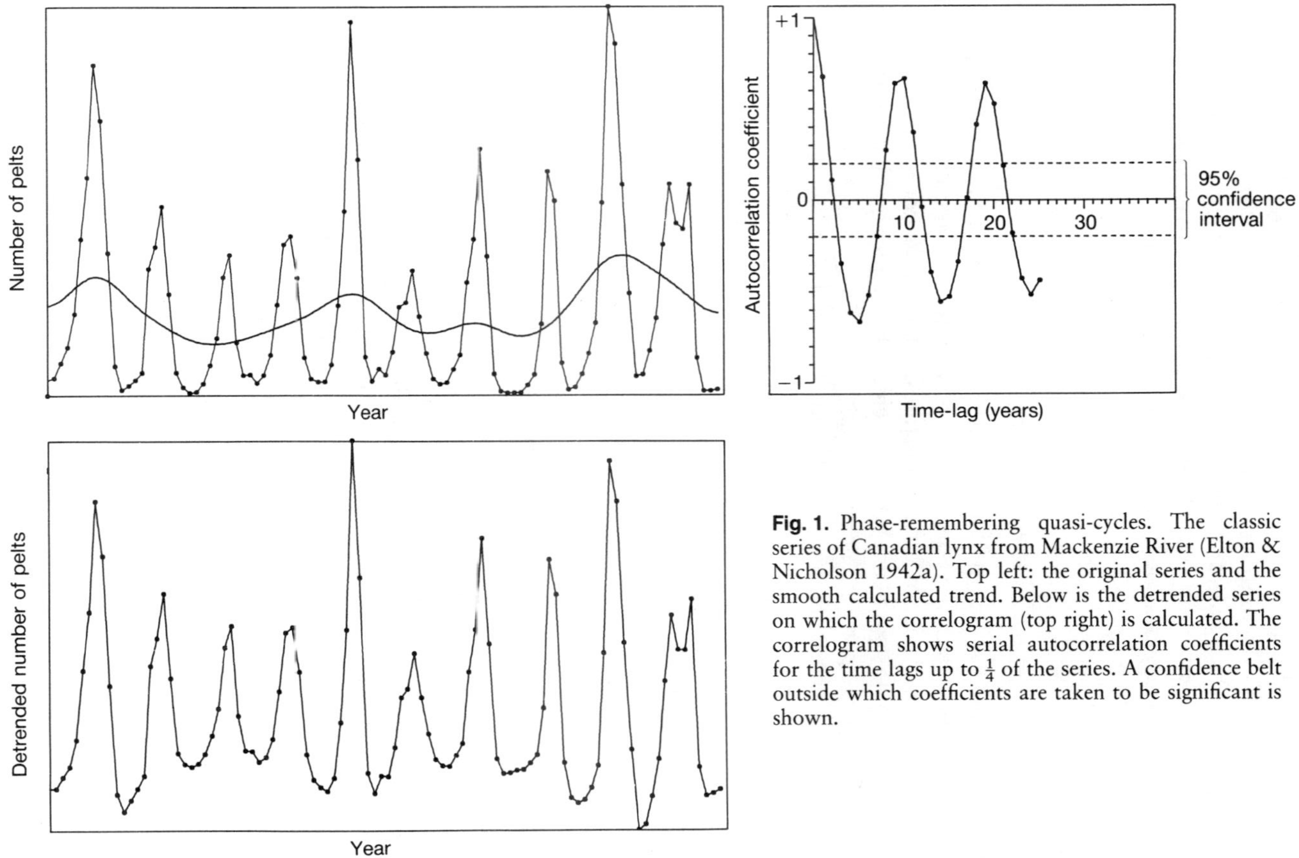

Fig. 1. Phase-remembering quasi-cycles. The classic series of Canadian lynx from Mackenzie River (Elton & Nicholson 1942a). Top left: the original series and the smooth calculated trend. Below is the detrended series on which the correlogram (top right) is calculated. The correlogram shows serial autocorrelation coefficients for the time lags up to $\frac{1}{4}$ of the series. A confidence belt outside which coefficients are taken to be significant is shown.

damped correlograms (see Finerty 1980). By contrast, red grouse bag records are usually heavily damped (Potts *et al.* 1984) (Fig. 2).

4. Non-cyclic fluctuations. These series can be divided into two subtypes: those that are basically random, where the correlogram shows no significant coefficients, and those which have some positive association with the previous years, and coefficients for the first one or two terms may be significantly positive (Fig. 3).

Results

Interpreting the results of correlograms is not easy if the data are only weakly quasi-cyclic. I have used the following terms to describe the results of time-series analysis for different estates: 'chaos', the correlogram shows no pattern and no significant terms; 'unclear', has some significant terms but no obvious cyclic pattern; 'not significant', shows a cyclic pattern at least for the first half cycle with none of the coefficients being significant; 'quasi-cyclic', a series of cyclic terms which have significant coefficients in at least the first half cycle.

Mountain hare

Twenty series from Scottish moors were examined, ranging in length from 37 to 102 years (Table 1). None of these series showed phase-remembering quasi-cycles, and 11 (55%) were non-cyclic in that either the variation was chaotic or autocorrelation coefficients were not significant (Fig. 4b). Nine series did have coefficients indicating phase-forgetting quasi-cycles; of these one had a period of between 13 and 15 years but most had a period of between six and ten years (Fig. 4a). The cyclic series were typically very damped, usually only coefficients for the first half cycle being significant.

Table 2 shows the average of all coefficients and demonstrates that the average population has at best a very weak quasi-cycle. To determine the average cycle length Fig. 5 shows a frequency distribution of only significant coefficients—this has the effect of cutting out chaotic variation and averaging the significantly quasi-cycle series. These data fit a cyclic period of 9.5 years, suggesting that this is the average wavelength for quasi-cyclic Scottish moors.

Brown hares

Fourteen series of brown hares from English lowland estates were examined and only four (28%) showed evidence of quasi-cycles (Table 3). Figure 6a shows a series that had a similar period to the average mountain hare series:

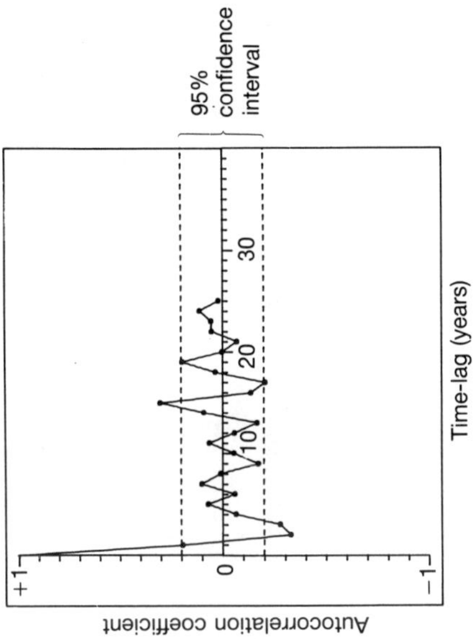

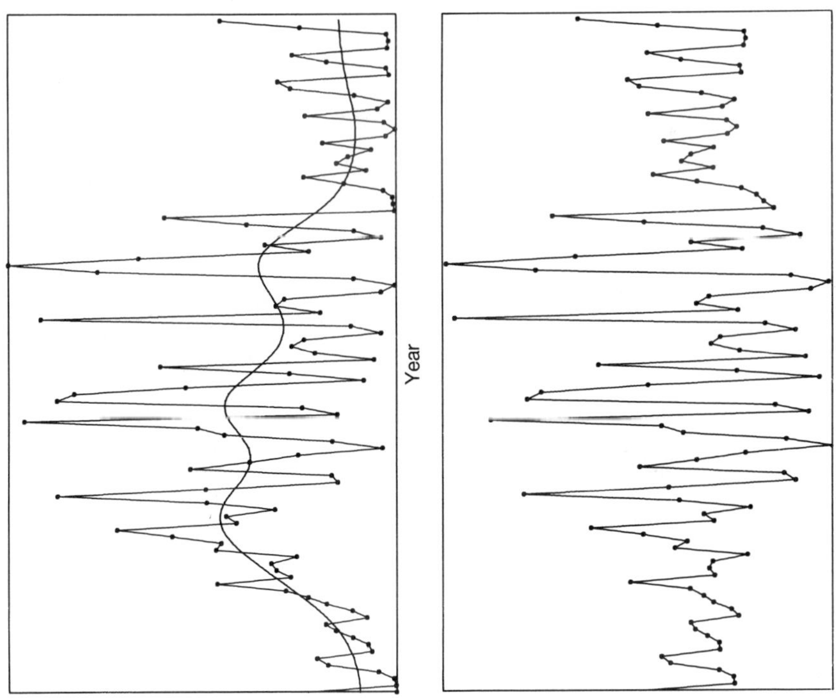

Fig. 2. Phase-forgetting quasi-cycles. Number of grouse shot on a Pennine moor. Layout and calculation as for Fig. 1. Notice the significant coefficients at time lags of 2 and 3 years.

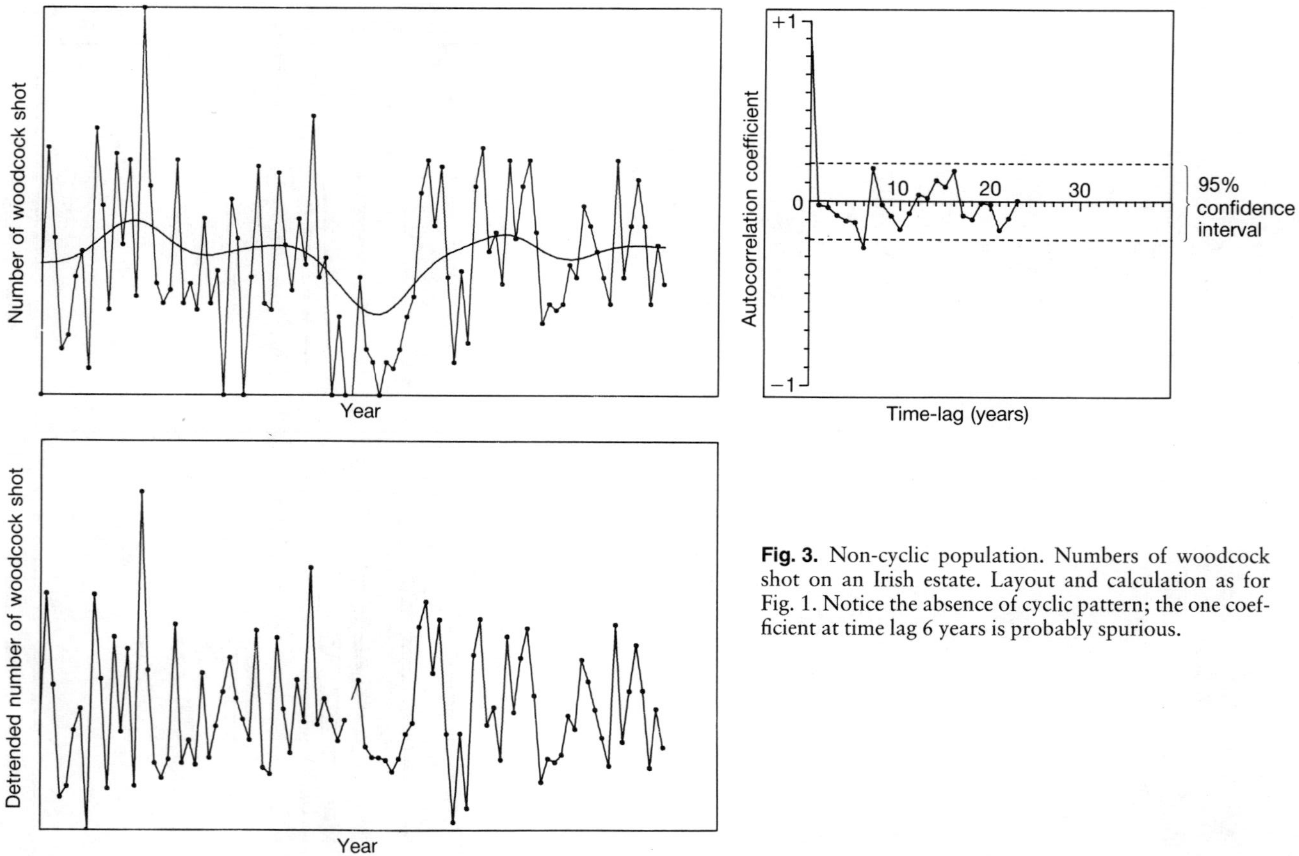

Fig. 3. Non-cyclic population. Numbers of woodcock shot on an Irish estate. Layout and calculation as for Fig. 1. Notice the absence of cyclic pattern; the one coefficient at time lag 6 years is probably spurious.

Table 1. Time series analysis of shooting records of mountain hares for 20 Scottish moors.

Moor	County	Cycle		Length of series (years)
Tulchan	Highland	Chaos		102
Fairburn	Highland	Quasi-cycle	8–10 years	56
Achentoul	Highland	Quasi-cycle	8–10 years	45
Moyhall	Highland	Chaos		95
Dalmgavie	Highland	Unclear		62
Carrbridge	Highland	Quasi-cycle	8–10 years	89
Altnaharra	Highland	Quasi-cycle	8–12 years	54
Cawdor	Highland	Chaos		70
Langwell	Highland	Unclear		67
Dinnet	Grampian	Not significant		37
Ballindalloch	Grampian	Chaos		99
Auchnafree	Tayside	Quasi-cycle	6–7 years	65
Drumour	Tayside	Unclear		73
Ardvorlich	Tayside	Quasi-cycle	8–12 years	90
Glenisla	Tayside	Not significant		84
Culter All	Strathclyde	Quasi-cycle	6–8 years	67
Newholm	Strathclyde	Quasi-cycle	6–8 years	37
Loganhouse	Lothian	Chaos		92
Philiphaugh	Borders	Unclear		81
Raeshaw	Borders	Quasi-cycle	4–6 years	58

however, two others had longer cycles of some 12 years. The two which showed cycles of eight to ten years were pre-war series extending back into the early 19th century. Figure 6b shows a non-cyclic series.

The average of coefficients for brown hares indicates only a very weak tendency to cycle (Table 2). Separating out the significant coefficients produces a frequency distribution less clear than for the mountain hare and can best be interpreted as comprising two types of population, one which shows a 9.5-year cycle, and another with a longer cycle of about 15 years (Fig. 7).

Rabbit

This was the most difficult of the three species to interpret, as the data were least reliable and myxomatosis has substantially altered abundance over the last 25 years. Figures 8a and 8b show two series, one with a long quasi-cycle, the other with no obvious pattern. Table 4 shows data from 11 series; only one pre-war series has an eight- to ten-year cycle like some of the hares, but three did show evidence of a long quasi-cycle of 13–15 years (Fig. 9). On average, however, autocorrelation coefficents appeared to be weaker than for either of the hares (Table 2).

Discussion

In North America the ten-year cycle of *Lepus americanus* is clear throughout the boreal region of Canada, but it is not present throughout its range;

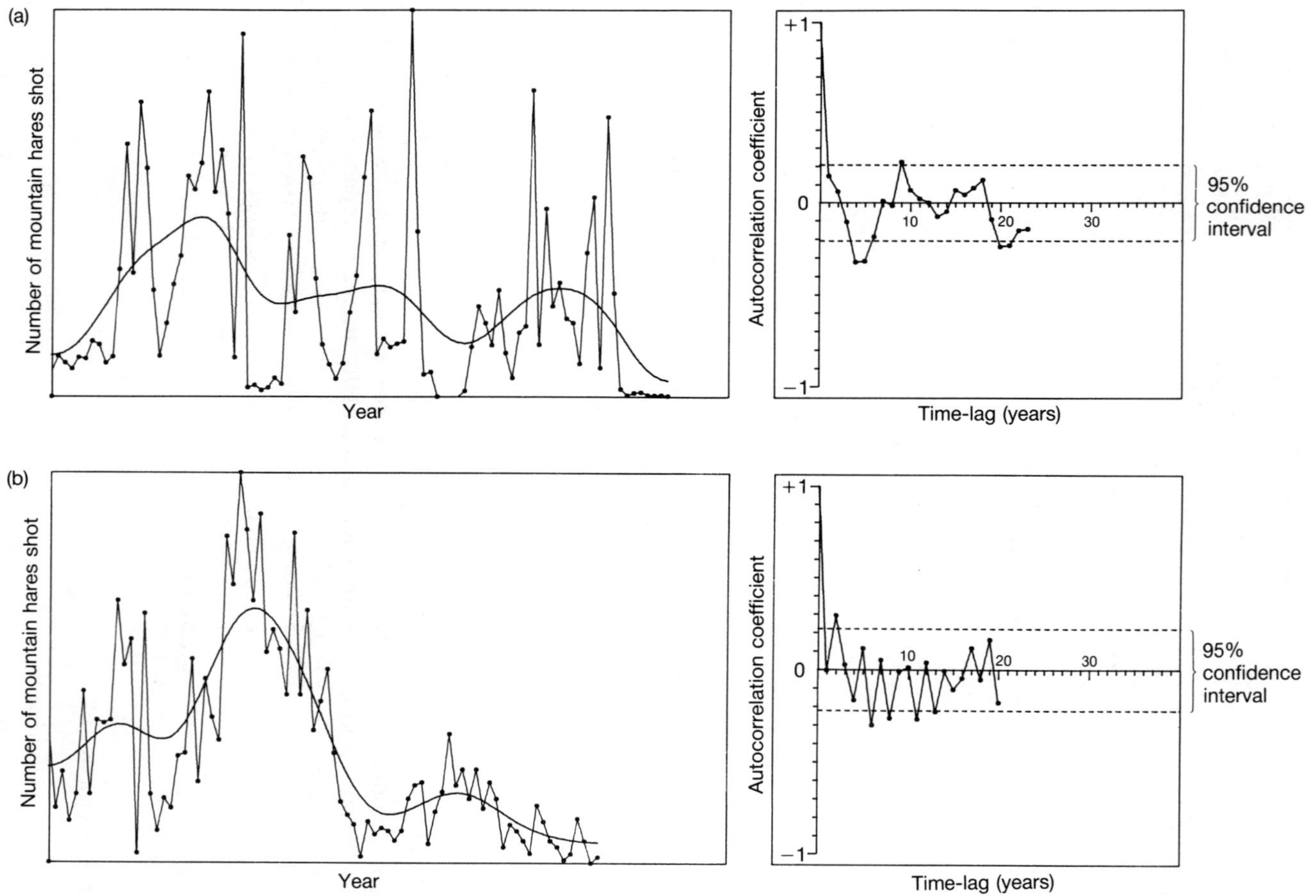

Fig. 4. Two series and correlograms of Scottish mountain hares. (a) A series with phase-forgetting quasi-cycles of 8–10 years. (b) An unclear series with chaotic or random fluctuations.

Table 2. Mean autocorrelation coefficients for three lagomorphs from British game-bag records. Numbers of estates in brackets.

Time-lag (years)	Mountain hare (20)	Brown hare (14)	Rabbit (13)
1	0.213	0.247	0.305
2	– 0.006	0.002	– 0.014
3	– 0.161	– 0.112	– 0.091
4	– 0.192	– 0.194	– 0.133
5	– 0.147	– 0.160	– 0.138
6	– 0.132	– 0.130	– 0.141
7	– 0.045	– 0.152	– 0.159
8	– 0.071	– 0.060	– 0.115
9	– 0.021	– 0.024	– 0.020
10	0.000	0.002	– 0.037

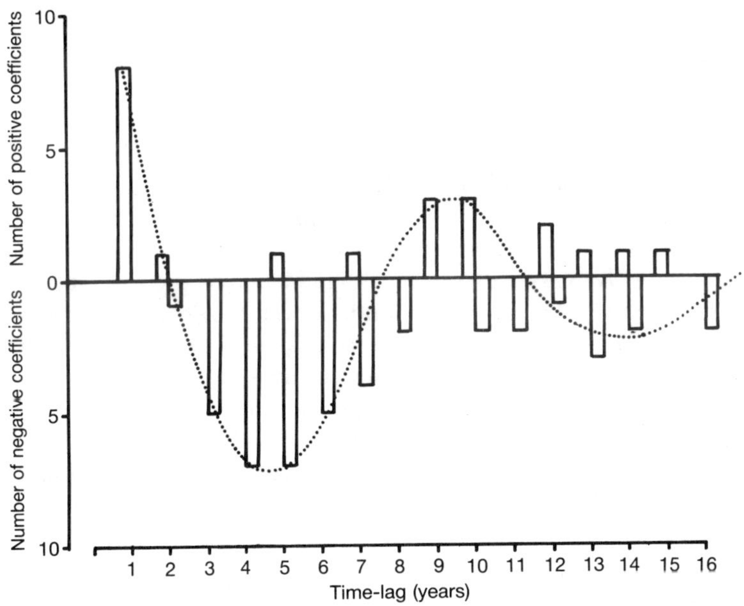

Fig. 5. Frequency distribution of significant coefficients from 20 mountain hare series (Table 1). The dotted line indicates an expected curve for a 9.5-year cycle.

in the fragmented habitats of the aspen parkland or the Rockies cycles are not evident (Keith 1981). *Lepus timidus* appears to have an approximate ten-year cycle in boreal Russia (Keith 1981) but in Scandinavia it shows a three- to four-year period, as do other species associated with a microtine cycle (Angelstam, Lindstrom & Widen 1985). *Lepus europaeus* is not regarded as cyclic in western Europe and the three- and nine-year periods

Table 3. Time-series analysis of shooting records of brown hares from 14 English estates.

Estate	County	Cycle		Length of series (years)
Middleton (1934)	Yorkshire	Quasi-cycle	8–10 years	89
Nocton	Lincolnshire	Unclear		72
Holkham	Norfolk	Unclear		158
Middleton (1934)	Norfolk	Unclear		70
Ryston	Norfolk	Unclear		100
Sall Moor Hall	Norfolk	Quasi-cycle	11–15 years	82
Woodbastwick	Norfolk	Unclear		66
Lexham Hall	Norfolk	Not significant		73
Elveden	Suffolk	Quasi-cycle	12–15 years	92
Chillesford	Suffolk	Unclear		63
Hendred	Berkshire	Quasi-cycle	8–10 years	100
Codford	Wiltshire	Chaos		66
Great Durnford	Wiltshire	Unclear		66
Mersham Hatch	Kent	Unclear		43

noted (but not analysed) by Andersen (1957) were taken to be correlates with weather.

The cycles shown here for British lagomorphs, although of similar period to those of North American species, appear to be quite different in nature. Time-series analysis of *L. americanus* trapped around the James Bay area of northern Canada shows them to undergo *phase-remembering quasi-cycles* (Finerty 1980)—the coefficients are, however, much smaller than for the Mackenzie River lynx. None of the cycles indicated by the game-bag records given here appears to be this strong, and, at best, they can only be described as *phase-forgetting quasi-cycles*, since usually only coefficients for the first half cycle are clearly marked and significant—at longer time lags they become weak or chaotic. Of the three species examined, the Scottish mountain hare, with an average period of 9.5 years, is the most cyclic, although not all series showed cycles. In the brown hare and the rabbit smaller proportions of the series showed evidence of cycling and these had periods of 9.5 or 15 years.

The first question is whether these data reflect quasi-cyclic fluctuations in animal abundance, or simply quasi-cycles in the records themselves. The time-series analysis has demonstrated that they are not random, but they may be spurious. The long period of 15 years is the most suspect and may have to be discounted; the Elveden rabbit and hare series show well-marked lows in 1914 and 1943 coincident with two world wars when management of the estate was considerably disrupted (Turner 1954), and in 1954 when myxomatosis drastically reduced the numbers of rabbits. However, the Elveden rabbits do show two distinct cyclical peaks in abundance in the 32 years since myxomatosis—so final judgement should perhaps be deferred. The 9.5-year cycle present in all species is not so easily dismissed. Both the

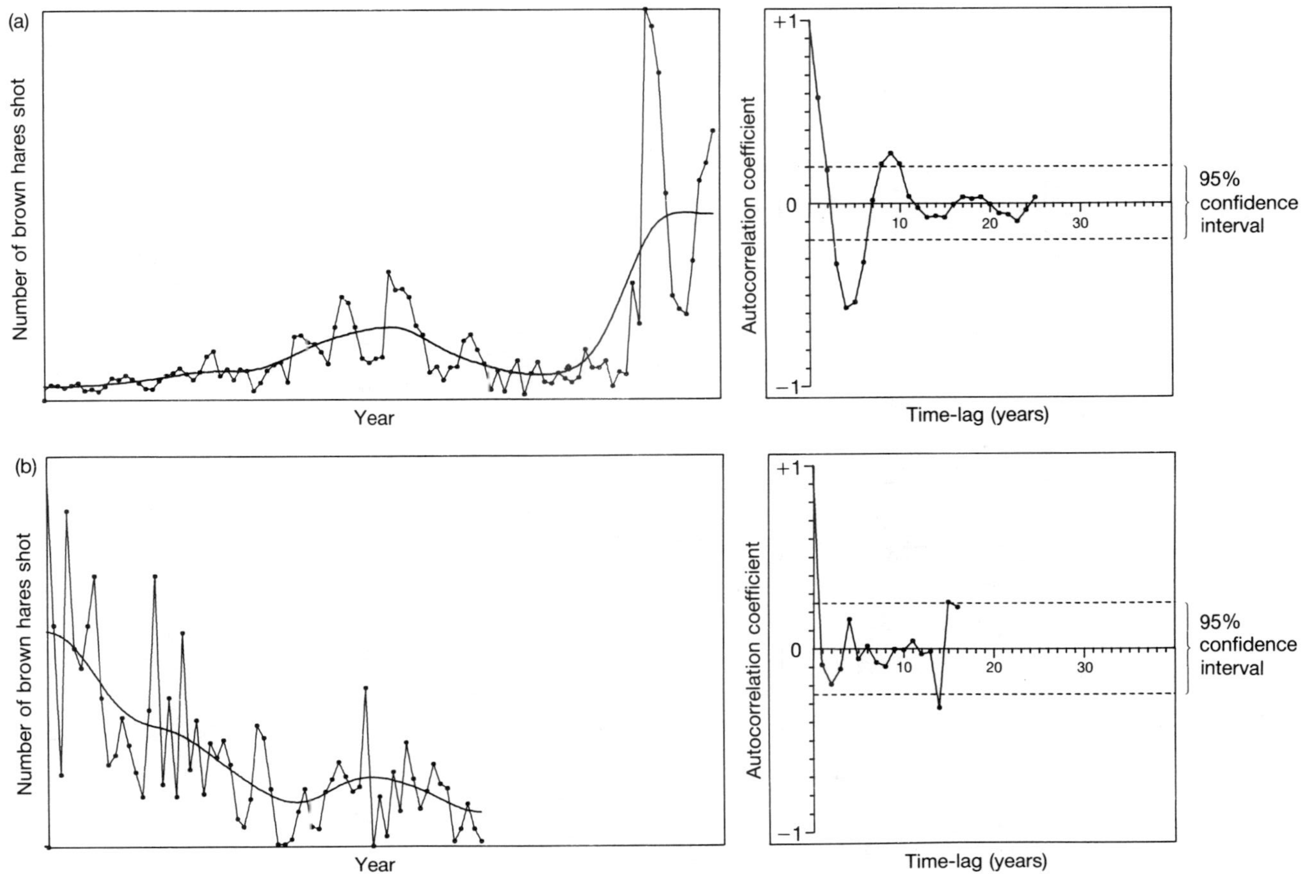

Fig. 6. Two series and correlograms of lowland brown hare populations. (a) A quasi-cyclic series of a period of about 9 years. (b) A random or chaotic series.

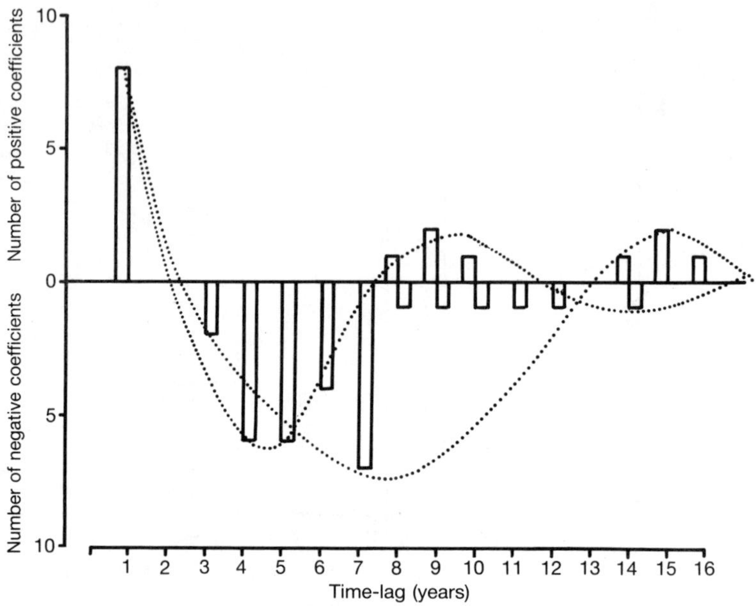

Fig. 7. Frequency distribution of significant coefficients from 14 brown hare series (Table 3). The two dotted lines indicate expected curves for 9.5- and 15-year cycles.

brown hare series which show the 9.5-year cycle are pre-war, and extend well into the 19th century.

Shooting records are not always good indicators of game abundance, since weather and other variables may affect the bag. Also, and more importantly, they usually tend to accentuate population changes. In grey partridge there is a logistic relationship between the bag and autumn numbers (Potts 1980); this is also true for red grouse (Hudson 1986). For hares this relationship is less well defined but appears to be similar in the two species. In mountain hares the proportion shot normally varies between 10 and 50% (Flux 1970), and in brown hares between 5 and 60% (R. Barnes & S. Tapper, unpublished). Higher proportions shot are associated with higher populations, since gamekeepers and farmers organize large shoots when they see lots of hares around but often do not bother if there are few. This exaggeration produced in the game bag can be seen in Hewson's (1976) study of mountain hares (Fig. 10); however, Hewson noted that his trap–night index may underestimate changes. Thus the real population change may have fallen somewhere between these two curves. The evidence for a relationship between game-bags and densities in rabbits is less clear, although gross changes in abundance caused by myxomatosis are well marked in game-bag records (Barnes & Tapper 1986).

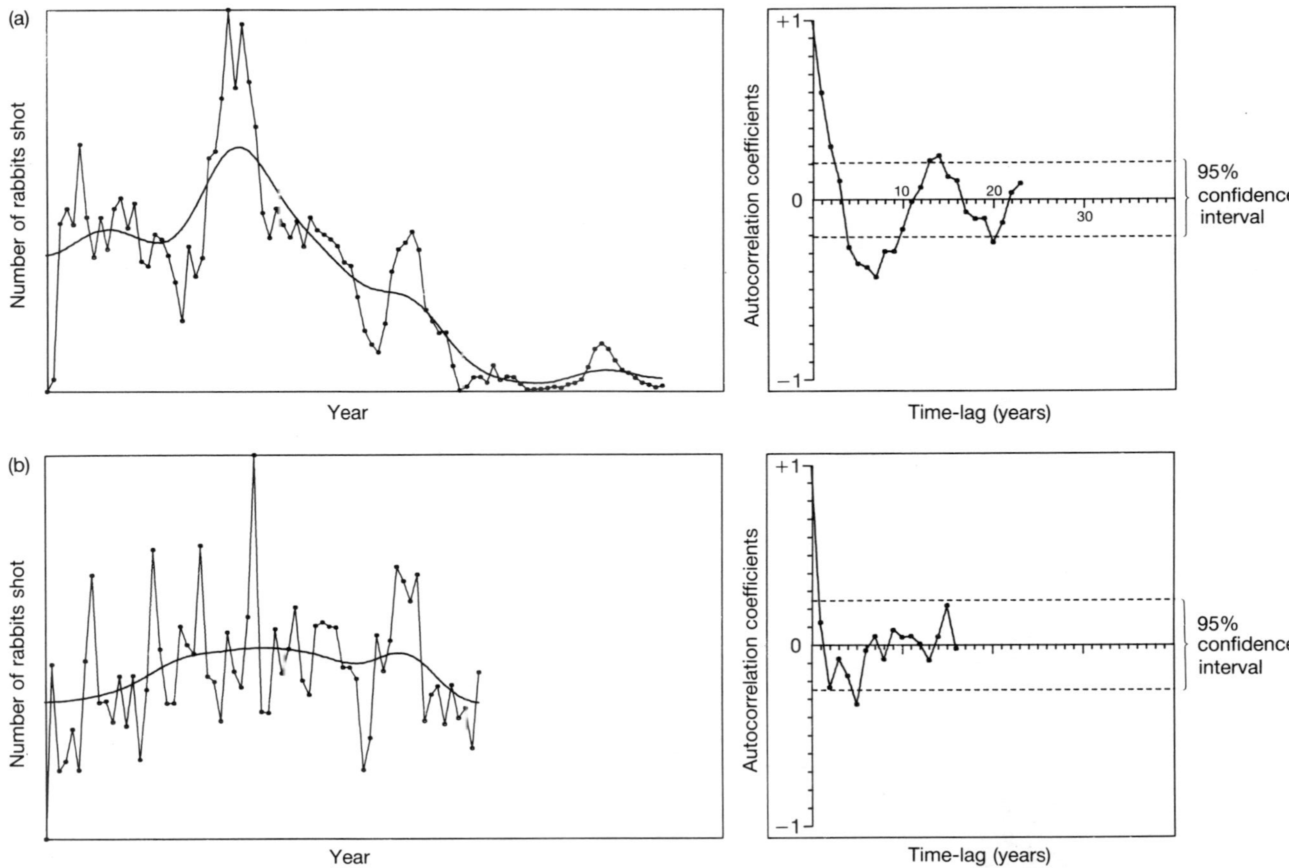

Fig. 8. Two series and correlograms of rabbit bags. (a) A series with an uneven quasi-cycle of 13–15 years. (b) An unclear series with no obvious pattern.

Table 4. Time series analysis of shooting records of rabbits from 11 English estates.

Estate	County	Cycle		Length of series (years)
Middleton (1934)	Yorkshire	Unclear		89
Nocton	Lincolnshire	Quasi-cycle	13–14 years	72
Middleton (1934)	Norfolk	Unclear		70
Middleton (1934)	Norfolk	Quasi-cycle	8–9 years	70
Sall Moor Hall	Norfolk	Unclear		82
Lexham Hall	Norfolk	Chaos		73
Elveden	Suffolk	Quasi-cycle	13–15 years	92
Chillesford	Suffolk	Chaos		63
Hendred	Berkshire	Unclear		100
Great Dunford	Wiltshire	Not significant		66
Mersham Hatch	Kent	Quasi-cycle	14–15 years	43

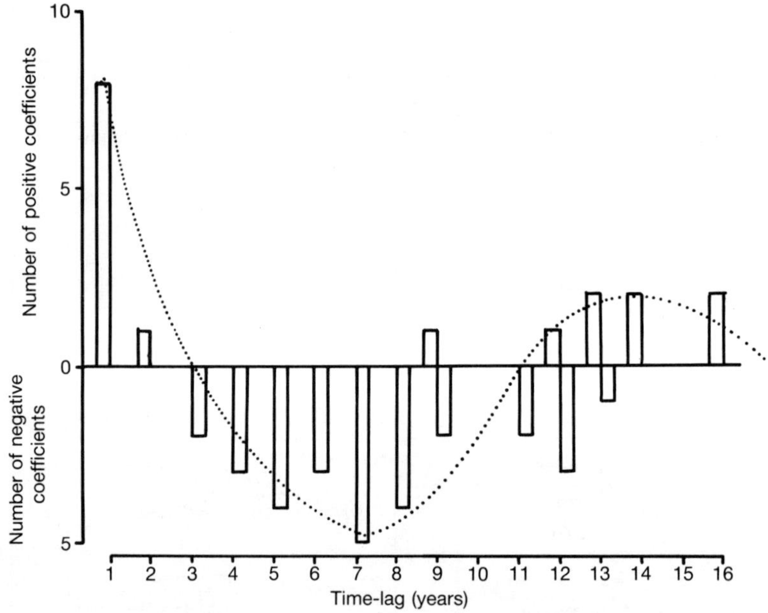

Fig. 9. Frequency distribution of significant coefficients from 11 rabbit series (Table 4). The dotted line shows an expected curve for a 14-year cycle.

Additional evidence of the reality of these cycles can be seen in Hewson's (1976) study of mountain hares in Aberdeenshire. His population at Jock's Hill peaked in 1958, declined the following year and then remained low until 1966 after which numbers increased until at least 1969—making a cycle of at least 11 years. In brown hares perhaps the best study is that of Abildgård, Anderson & Barndorff-Nielsen (1972) on Illumø Island where the entire population was netted and released each year. This shows a very

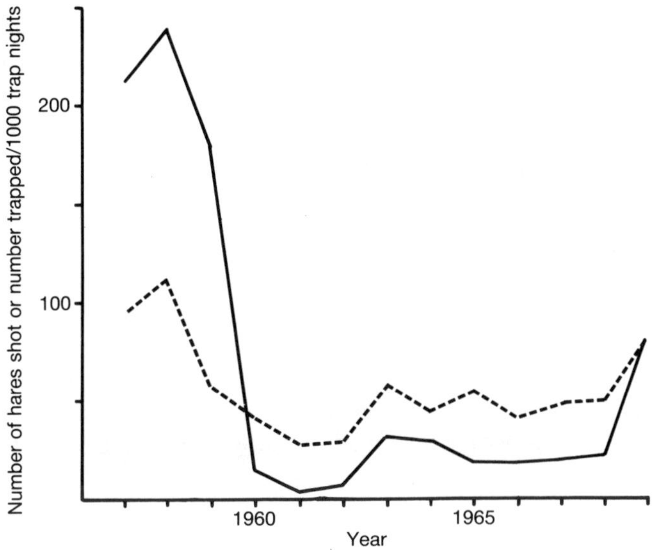

Fig. 10. The relationship between the bag of mountain hares (solid line) and an index of abundance–numbers trapped per 1000 trapnights (broken line). Data from Hewson (1976).

marked eight-year cycle in a comparatively dense population. The studies of Hewson (1976) and Abilgård *et al.* (1972) show that variable breeding success is probably insufficient to have caused these changes and that survival of adults and leverets is likely to be more important. Their studies also suggest that these cycles are quite different from those found in *L. americanus*, and *L. timidus* in Sweden where predation appears to be a significant component in the cycle (Keith, Todd, Brand, Adamcik & Rusch 1977; Lindstrom, Andren, Angelstam & Widen 1986). Hewson (1976) observed that predators were greatly reduced on his study area, and the population studied by Abildgård *et al.* (1972) was an island population with no mammalian predators.

There is further evidence that the quasi-cycles of British lagomorphs are fundamentally different from those observed in boreal north America and Scandinavia. In these areas other species, particularly the Tetraonidae, are clearly caught up in the changes. Ruffed grouse (*Bonasa umbellus*) numbers track snowshoe hare densities within one or two years (Keith 1963) and willow grouse (*Lagopus lagopus*) appear to do the same in Scandinavia (Angelstam, Lindstrom & Widen 1985). Red grouse populations in Scotland are not in phase with the hares and indeed overall tend to have shorter quasi-cycles, of about six years in Scotland (J. Williams 1985; R. Barnes, unpublished) and four to five years in England (Potts *et al.* 1984). British grouse studies do not give us a useful lead since the work of Watson and others suggests a behavioural mechanism as the cause of the fluctuations in Scotland (Moss & Watson 1985), whereas parasitism is suggested as the cause of cycles in the Pennines (Hudson, Dobson & Newborn 1985).

An explanation of these cycles may not be complex—May (1976) has demonstrated that density-dependent mechanisms acting with a time-delay can generate cycles. Looking for delayed density-dependent relationships is likely to provide the best chance of an explanation. However, just because some of the populations appear not to be cyclic it does not necessarily mean that delayed density-dependence is not operational in non-cyclic populations. May (1976) also showed that by altering the values in a difference equation (e.g. intrinsic rate of growth), model populations would switch from cyclic to chaotic. This is unfortunate since we do not know whether or not the failure of some of the populations we have observed here to show cycles results from normal stochasticity or is the product of chaos generated by the same mechanism which produced the quasi-cycle in others.

Acknowledgements

Many people have helped gather game bag records over the years and this work would not have been possible without their efforts. In particular Richard Barnes gathered the bulk of the records on Scottish mountain hares and Mike Rands collected some useful series on brown hares. Dick Potts kindly read and commented on the manuscript.

References

Abildgård, F., Anderson, J. & Barndorff-Nielsen, O. (1972). The hare population (*Lepus europaeus* Pallas) of Illumø Island, Denmark. A report on the analysis of the data from 1957–1970. *Dan. Rev. Game Biol.* **6** (5): 1–32.

Andersen, J. (1957). Studies in Danish hare-populations. I. Population fluctuations. *Dan. Rev. Game Biol.* **3**: 85–131.

Angelstam, P., Lindstrom, E. & Widen, P. (1985). Synchronous short-term population fluctuations of some birds and mammals in Fennoscandia—occurrence and distribution. *Holarctic Ecol.* **8**: 285–98.

Barnes, R.F.W. & Tapper, S.C. (1986). Consequences of the myxomatosis epidemic in Britain's Rabbit (*Oryctolagus cuniculus* L.) population on the numbers of Brown Hares (*Lepus europaeus* Pallas). *Mammal Rev.* **16**: 111–6.

Box, G.E. & Jenkins, G.H. (1970). *Time-series analysis. Forecasting and control.* Holden-Day, San Francisco.

Bulmer, M.G. (1974). A statistical analysis of the 10-year cycle in Canada. *J. Anim. Ecol.* **43**: 701–18.

Bulmer, M.G. (1975). Phase relations in the ten-year cycle. *J. Anim. Ecol.* **44**: 609–21.

Butler, L. (1942). Fur cycles and conservation. *Trans. N. Am. Wildl. Conf.* 7: 463–72.

Chitty, H. (1950). The snowshoe rabbit enquiry, 1946–48. *J. Anim. Ecol.* **19**: 15–20.

Chitty, H. (1950). Canadian Arctic wild life enquiry, 1943–49: with a summary of results since 1933. *J. Anim. Ecol.* **19**: 180–93.

Cole, L.C. (1951). Population cycles and random oscillations. *J. Wildl. Mgmt* **15**: 233–52.

Cole, L.C. (1954). Some features of random population cycles. *J. Wildl. Mgmt* **18**: 2–24.

Elton, C.S. (1924). Periodic fluctuations in the numbers of animals: their causes and effects. *Br. J. exp. Biol.* **2**: 119–63.

Elton, C. & Nicholson, M. (1942a), The ten-year cycle in numbers of the lynx in Canada. *J. Anim. Ecol.* **11**: 215–44.

Elton, C. & Nicholson, M. (1942b). Fluctuations in numbers of the Muskrat (*Ondatra zibethica*) in Canada. *J. Anim. Ecol.* **11**: 96–126.

Finerty, J.P. (1980), *The population ecology of cycles in small mammals.* Yale University Press, New Haven & London.

Flux, J.E.C. (1970). Life history of the Mountain hare (*Lepus timidus scoticus*) in north-east Scotland. *J. Zool., Lond.* **161**: 75–123.

Hewson, R. (1976). A population study of mountain hares (*Lepus timidus*) in north-east Scotland from 1956–1969. *J. Anim. Ecol.* **45**: 395–414.

Hudson, P. (1986). *Red Grouse, the biology and management of a wild gamebird.* The Game Conservancy Trust, Fordingbridge.

Hudson, P.J., Dobson, A.P. & Newborn, D. (1985). Cyclic and non-cyclic populations of red grouse: a role for parasitism? In *Ecology and genetics of host-parasite interactions*: 79–86. (Eds Rollinson, D. & Anderson, R.M.) Academic Press, London & New York.

Keith, L.B. (1963). *Wildlife's ten-year cycle.* The University of Wisconsin Press, Madison.

Keith, L.B. (1981). Population dynamics of hares. In *Proceedings of the World Lagomorph Conference*: 395–440. (Eds Myers, K. & MacInnes, C.D.). University of Guelph, Canada.

Keith, L.B., Todd, A.W., Brand, C.J., Adamcik, R.S & Rusch, D.H. (1977). An analysis of predation during a cyclic fluctuation of Snowshoe hares. In *XIIIth International Congress of Game Biologists*: 151–75. (*Int. Un. Game Biol.* **13**.) (Ed. Peterle, T.J.). The Wildlife Society, Washington.

Kendall, M. (1973). *Time-series.* Charles Griffen, London.

Lindstrom, E., Andren, H., Angelstam, P. & Widen, P. (1986). Influence of predators on hare populations in Sweden. *Mammal Rev.* **16**: 151–6.

Mackenzie, J.M.D. (1952). Fluctuations in the numbers of British tetraonids. *J. Anim. Ecol.* **21**: 128–53.

May, R.M. (1976). Models for single populations. In *Theoretical ecology. Principles and applications*: 4–25. (Ed. May, R.M.). Blackwell Scientific Publications, Oxford.

Middleton, A.D. (1934). Periodic fluctuations in British game populations. *J. Anim. Ecol.* 3: 231–49.

Moran, P.A.P. (1949). The statistical analysis of the sunspot and lynx cycles. *J. Anim. Ecol.* **18**: 115–6.

Moran, P.A.P. (1952). The statistical analysis of game-bird records. *J. Anim. Ecol.* **21**: 154–8.

Moran, P.A.P. (1953). The statistical analysis of the Canadian lynx cycle. I. Structure and prediction. *Aust. J. Zool.* **1**: 163–73.

Moss, R. & Watson, A. (1985). Adaptive value of spacing behaviour in population cycles of red grouse and other animals. In *Behavioural ecology—ecological consequences of adaptive behaviour*: 275–94. (Eds Sibly, R.M. & Smith, R.H.). Blackwells, Oxford.

Nisbet, R.M. & Gurney, W.S.C. (1982). *Modelling fluctuating populations.* John Wiley & Sons, Chichester, New York etc.

Poole, R.W. (1974). *An introduction to quantitative ecology.* McGraw-Hill, Kogakusha Ltd, Tokyo.

Potts, G.R. (1980). The effects of modern agriculture, nest predation and game management on the population ecology of partridges (*Perdix perdix* and *Alectoris rufa*). *Adv. ecol. Res.* **11**: 1–79.

Potts, G.R., Tapper, S.C. & Hudson, P.J. (1984). Population fluctuations in Red grouse: analysis of bag records and a simulation model. *J. Anim. Ecol.* **53**: 21–36.

Turner, T.W. (1954). *Memoirs of a gamekeeper.* Geoffrey Bles, London.

Williams, G.R. (1954). Population fluctuations in some northern hemisphere game birds (Tetraonidae). *J. Anim. Ecol.* **23**: 1–34.

Williams, J. (1985). The statistical analysis of fluctuations in red grouse bag data. *Oecologia* **65**: 269–72.

Williamson, M. (1975). The biological interpretation of time series analysis. *Bull. Inst. Math. Appl.* **11**: 67–9.

Symp. zool. Soc. Lond. (1987) No. 58: 99–113

Planning and monitoring an attempt to eradicate coypus from Britain

L. M. GOSLING and
S. J. BAKER

*Coypu Research Laboratory,
M.A.F.F.,
Jupiter Road,
Norwich NR6 6SP,
U.K.*

Synopsis

Population simulations based on data from post-mortem examinations of animals killed during control trapping were used to help plan a campaign to eradicate feral coypus, *Myocastor coypus*, in East Anglia. The simulations allowed estimates of the number of trappers needed to reduce coypu numbers under any climatic conditions. Other elements of the eradication plan were to increase the numbers of traps, improve equipment, especially vehicles, and provide an incentive bonus scheme to overcome the problem that the trappers might be reluctant to work themselves out of a job. The campaign started in April 1981 with a force of 24 trappers; between 1981 and 1986 annual trapping effort averaged 216,000 trap nights. Estimated numbers of adult females declined from about 3,000 in April 1981 to about 20 in April 1986. Average patterns of mortality and reproduction during the campaign were estimated from trapping results and post-mortem examinations. This information was incorporated into the simulation model and the results obtained were then used to test whether the measured population variables were sufficient to explain the observed reduction in population size. The pattern of decline indicated by the simulations was similar to the actual decline estimated from population reconstruction except in 1985-86 when the number of adult females had fallen below 200. During these years the estimated decline in numbers was more rapid than that calculated in the simulations. Two unusually cold winters may have been partly responsible but a significant fall in the proportion of pregnant females during summer months suggests that there may also have been density-dependent reduction in reproductive success.

Introduction

Coypus are hystricomorph rodents native to South America which weigh between 6 and 7 kg when fully grown. They were introduced into Britain in the 1930s for fur farming. Following escapes, a feral population became established in the Norfolk wetlands and eventually expanded to surround-

ZOOLOGICAL SYMPOSIUM No. 58
ISBN 0-19-854006-X

ing counties (Laurie 1946; Davis 1963; Norris 1967a; Gosling, Watt & Baker 1981). Like most mammals coypus have a polygynous mating system; the ranges of dominant males overlap those of several females (Gosling 1977; unpubl. data). Females breed throughout the year and the mean litter size at birth is 5.3 (Newson 1966). The young are suckled for about eight weeks and, although in captivity they can live for up to six years, in the wild fewer than 2% survive to three years old because of intensive trapping (Gosling & Baker 1981). Mortality due to trapping and the effects of cold winters are the two most important factors in coypu population dynamics.

Coypus dig extensive burrow systems in the banks of rivers and ditches (de Soriano 1960) and feed on a wide range of agricultural crops and native plants. As a result they damage drainage systems (Cotton 1963) and crops (Norris 1967a) and have a devastating impact on wetland plant communities (Ellis 1963; Boorman & Fuller 1981; Moss 1983). In 1962 they were added to the list of animals subject to the provisions of the Destructive Imported Animals Act. Organized control of their numbers also started in 1962 and, since 1971, the organization responsible has been called Coypu Control.

In 1977 an independent committee called the Coypu Strategy Group was set up to consider long-term options for the control of coypus in Great Britain (Anon. 1978). The group took evidence from a wide range of opinion, as well as considering the biological information then available. It recommended that an attempt should be made to eradicate the population. These recommendations were accepted in 1980 and an eradication campaign started in 1981. By this time further research by the Coypu Research Laboratory had clarified some remaining problems and it was decided to attempt eradication within 10 years. A reconstituted Coypu Control Organization employed 24 trappers and was funded by the Ministry of Agriculture, Fisheries and Food (50%), Anglian Water Authority (40%) and Association of Drainage Authorities (10%). At the time of writing the expected eventual cost is about £2.5 million.

This paper discusses the information used to plan the eradication campaign, most of which was available to the Coypu Strategy Group in 1977. Next, it outlines events during the campaign, including the response of the population to intensified trapping. Finally, population simulations are used to test whether the known mortality and the average rates of reproduction were sufficient to explain the observed population decline.

Materials and methods

The sample

All control is by cage-trapping and coypus are shot when the traps are inspected each day (Norris 1967b). A sample of the animals killed each month were dissected to record such variables as the litter size of pregnant females, and to estimate projected dates of parturition from mean embryo weight. The animals were aged using eye-lens weight (Gosling, Huson & Addison 1980) and classified as adult or juvenile using criteria described by Gosling, Watt *et al.* (1981). The numbers sampled averaged 136±54 (*S.D.*) per month between February 1970 and April 1986, the period used for most of the analyses which follow. The total numbers killed per month over the same period were 597±399.

Population reconstruction and simulation

Monthly estimates were made of the entire British coypu population between 1970 and 1986 by reconstructing the population as described by Gosling, Watt *et al.* (1981). The procedure assumes that the number of adults that die each month is known. This is possible for coypus because of the continuous control trapping over the whole of the population range; trappers report coypus killed by gamekeepers, wildfowlers etc. as well as their own results. Unrecorded deaths of adults are believed to have been rare since 1970. The ages of all animals that died each month were estimated from the large dissected sample. Their lives were extrapolated back from the month of death and the population was reconstructed by summing the animals then known to have been alive each month in the past. The most recent 36 months were adjusted for animals still alive by using known relationships between the total number already accrued, with the final value. The reconstructed population and simulations presented here are restricted to the adult female population.

The simulation model is constructed as a time-series of monthly units. The reproductive inputs are the proportion of females littering each month, numbers of females born per litter and the proportion of these that reach maturity in each month after their birth. Death rates were calculated from the results of the Coypu Control trappers and from all other mortality that could be quantified. The model is 'seeded' with the number of adult females alive each month for 12 months, derived from the population reconstruction. These values are used to create the recruits (animals born some months previously that become adult in a particular month) for the first few months after the end of the seed data. The simulation model calculates the population size thereafter, using the reproductive and mortality parameters given. Further details of the inputs and the structure of the simulation are given in an Appendix.

Coypus are vulnerable to cold weather and all population parameters differ in years following mild and cold winters (Gosling, Baker & Skinner 1983). Sufficient data to calculate average years of both types were eventually collected, but, when the Coypu Strategy Group was discussing its report in 1977, only one cold winter (1975–76) had occurred during our study. However, the information collected during the unusual run of five mild winters between 1970 and 1975 was valuable because 'worst case' simulations (i.e., simulations of the behaviour of a population when environmental conditions are favourable for the animal) are most important when planning population reduction.

A number of elements were unknown when the advice was given to the Strategy Group. It was not possible to predict how many more coypus would be trapped either by deploying greater numbers of traps (although we knew that trappers could manage to deploy considerably more) or as a result of technical improvements: in practice, the use of traps set on rafts gave a large increase in efficiency (Baker & Clarke in prep.). Similarly, it was not known to what extent juvenile mortality would increase with increased trapping and the simulations made the conservative assumption that it was independent of trapping intensity. Finally, there was a delay of three years between the recommendations of the Strategy Group and the implementation of the eradication programme: during this time the population's range expanded, which increased the difficulty of comprehensive trapping (Fig. 1).

Planning the eradication campaign

Trapper numbers

Simulations assuming continuous mild weather were run with various numbers of trappers to provide the Coypu Strategy Group with a number of strategic options (Anon. 1978; Gosling 1981a). These simulations showed that a force of 20 trappers would keep the population declining by about 20% per year and this was suggested as a trapper force which, giving some margin for error, could probably reduce the coypu population under any climatic circumstances. Since some cold winters would almost certainly occur it was assumed that the actual fall in numbers would be greater in practice. At the time of these recommendations parts of the area were trapped by other agencies whose effort was more difficult to quantify than that of the Coypu Control trappers. Using past experience we estimated that about four more trappers would be required for these additional areas, making a total recommended force of 24 trappers. Other simulations showed the effect of larger and smaller forces and the very rapid increase in coypu numbers likely if control was given up altogether. The 24-trapper option was eventually recommended by the Strategy Group (Anon. 1978). It was consistent with their conclusion that an attempt should be made to eradicate coypus and, unlike the option of employing a larger force, it gave time for a number of potential obstacles to eradication to be overcome.

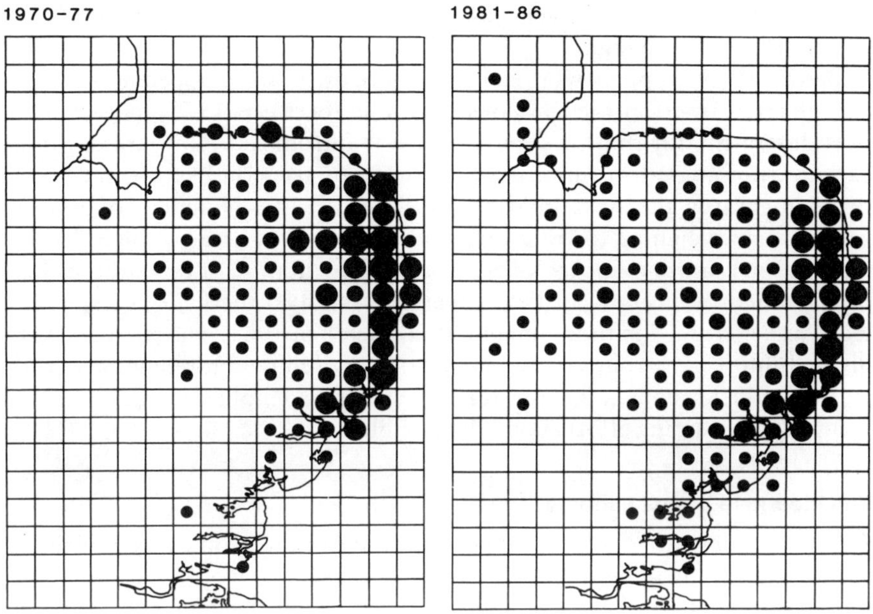

Fig. 1. The known distribution of coypus at the time of the Coypu Strategy Group's report (data from 1970–1977) and the distribution during the eradication campaign (1981–1986). The population probably expanded between 1977 and 1981 although part of the change is due to additional records obtained by more extensive trapping. In descending order of size, the solid circles represent >5.0%, 1.0–4.9%, 0.5–0.9% and <0.5% of the totals killed in each period per 10 km^2. Numbers killed in the two periods were respectively 59,290 and 34,838.

Equipment and strategic planning

The mortality sections of the simulations considered by the Strategy Group used historical information about the behaviour of trappers. The unit of trapping intensity was a night of trapping (a night when traps were set). The number of trap-nights per Coypu Control trapper per night between 1973 (when records started) and 1977 averaged 32±14. However, between 1975 and 1981 field staff from the Coypu Research Laboratory were engaged on a pilot eradication scheme along a 30 km stretch of the River Yare valley (Gosling, Baker & Clarke in prep.) which showed that larger numbers of traps were practical and effective: by 1980, three trappers, setting an average of 41±15 traps per night, had removed coypus from most of the study area (this average, and the previous one for Coypu Control trappers, is restricted to trappers working alone, since the proportion of trappers that worked in pairs—mainly from boats—differed in the two samples and this would bias the comparison). Other technical limitations in the past included a lack of four-wheel-drive vehicles and these were provided, where necessary, in the new campaign.

Before 1975, trappers worked in exclusive areas and trapped at an intensity dictated by local tactical considerations. Between 1975 and 1980 an overall strategy for trapper deployment was introduced. This strategy used capture:trapping-effort ratios over the previous six months to plan trapper deployment for the next quarter (Gosling 1981a). Numbers of coypus caught were weighted to concentrate future effort in areas of high density and thus maximize numbers killed. The implementation of this strategy had been limited by insufficient field supervision and so the field management team was increased in the campaign that started in 1981. The trappers were divided into three groups of eight, each supervised by a foreman, and with an overall field manager. From the outset, it was emphasized that deployment of the trapping effort should be flexible.

The incentive bonus scheme

A potential obstacle to eradication is that some of the people involved in the work may be unwilling to kill all coypus because they would thus lose their jobs. Trappers may also become bored when coypus become rare, or when they are required to continue trapping to confirm the absence of coypus. An incentive bonus scheme was built into the eradication scheme to help overcome these problems. The contracts of employment of the trappers are limited to 10 years (up to 1991) and trappers can earn a bonus of three times their annual salary if coypus are eradicated by April 1987. The bonus declines *pro rata* up to the end of the campaign with none payable if coypus are not eradicated. The attraction is not the amount eventually earned but the prospect of earning a capital sum at the end of a limited period of employment.

Mortality and recruitment during the eradication campaign

Trapping intensity and adult mortality rates

The average number of traps set per trapper per day increased from 32±15 in 1973–80 to 48±20 in 1981–86. The annual trapping effort by the 24 Coypu Control trappers, plus some trapping by MAFF research staff, averaged 216,000 trap-nights in 1981–86. This was an increase of 96% over the total effort in the two years immediately before trap numbers were increased for the start of the campaign. As trappers set more traps there could be an interference effect between traps, with a corresponding decline in the probability that each trap will catch an animal. The number of animals killed, taking into account the number available for capture, should thus reach an asymptote as the number of traps set per trapper increases. In

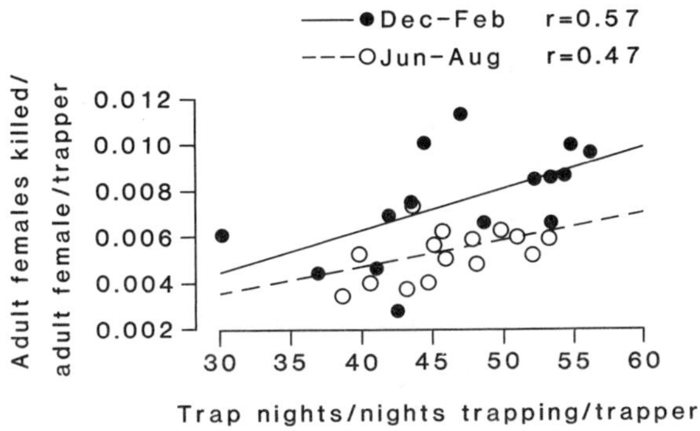

Fig. 2. The capture rate of adult female coypus in relation to increasing trapping effort. The dependent variable is the number of adult females killed per number of live adult females (from the population reconstruction; see text, and Gosling, Watt *et al.* 1981) per trapper, in the seasons shown. The independent variable is the mean number of traps set per trapper per night of trapping (i.e., a night when one or more traps was set). Surprisingly, there is no tendency for the capture rate to level off at high trapping intensities.

practice this did not occur. Numbers killed continued to increase in a linear fashion over the range of traps set per trapper in the new campaign (Fig. 2). This was probably because trappers avoided setting too many traps close together and instead increased the area covered as trap numbers increased. The slope of the increasing trend could have been reduced by an interference effect, but if so, this was a constant effect throughout the range of trap numbers, rather than one which appeared when large numbers of traps were set. As a result, the average number of adult females trapped per female alive per month increased by 77% in the eradication campaign (from 0.092 in 1970–80 to 0.163 in 1981–86).

Recruitment rates

Recruitment was estimated directly from the number of adults determined, by reconstruction, to be entering the population (see Gosling, Watt *et al.* 1981 for details). Rates were expressed in relation to the mean number of adult females in the reconstructed population. Because the recruitment of coypus born in late summer and autumn before a given winter, as well as those born during and just after it, will be affected by its severity (Gosling, Baker & Skinner 1983) the year classes considered ran from August to July.

Recruitment would be expected to decline as more juveniles are killed at higher trapping intensity. However, cold winter is also known to affect juvenile survival, partly because young animals may find it difficult to excavate roots and rhizomes in frozen ground. Over the period 1970 to 1986 recruitment proved to be negatively correlated with both trapping intensity and an

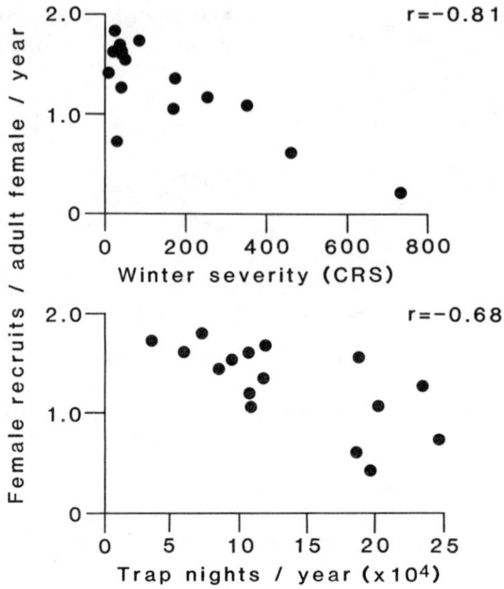

Fig. 3. The effect of winter severity and trapping intensity on female recruitment rate. The dependent variable is the number of female recruits per adult female in the population, pooled for each of the 16 years (August–July) between 1970 and 1986. Values were taken directly from the population reconstruction (see text, and Gosling, Watt *et al.* 1981, for details). Winter severity is expressed using an index (CRS) that weights and accumulates runs of freezing days each winter (see Gosling, Baker & Skinner 1983); trapping intensity is the total number of trap nights each year.

Table 1. Multiple regression analysis of the factors affecting the recruitment rate of females. The variables are as described in the legend of Fig. 3.

Independent variable	*t* value	*p*
Winter severity (days^2)	−5.19	<0.001
Trapping intensity (trap-nights)	−3.43	<0.01
Coefficient of determination = 0.80; F = 30.3 (p<0.001).		

index of winter severity (Fig. 3) and, together, these variables accounted for 80% of its variation (Table 1).

Population changes during the campaign and the effect of density-dependent change in reproductive success

The estimated number of adult females declined from 2,975 in April 1981 to about 12 in April 1986, the lowest level since accurate monitoring started in 1970 (Fig. 4). Annual reductions (April to April) ranged from 28% to 86% with an average of 41%.

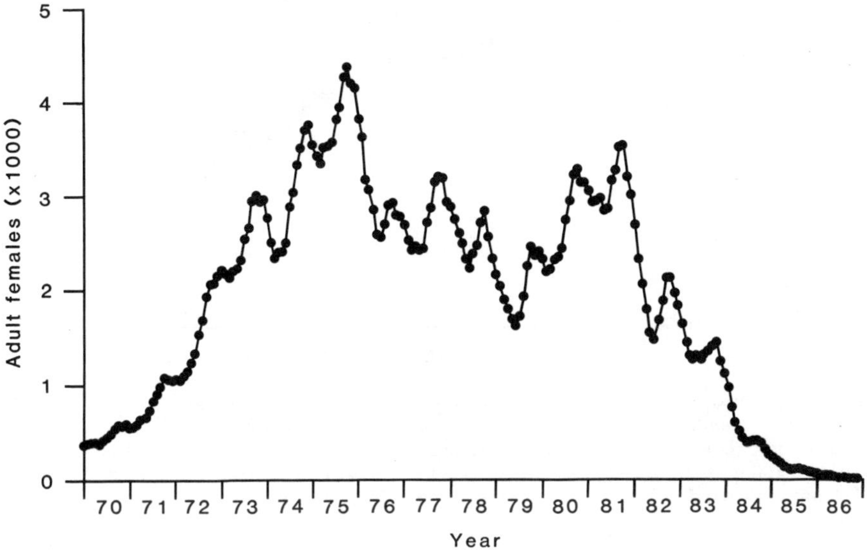

Fig. 4. Reconstructed numbers of adult females between 1970 and 1986. The rapid increase in the early 1970s occurred in a run of mild winters when trapping intensity was low. Numbers fluctuated between 1975 and 1980 with a higher trapping intensity and occasional cold winters. Numbers fell sharply after the start of the eradication campaign in 1981.

Are the mortality and reproduction variables measured during the campaign sufficient to account for these changes? To test this, the inputs to the population simulation (see Appendix) were first modified to the values which occurred on average between 1981 and 1986. The simulated curve of adult females showed quite close agreement with reconstructed numbers during the first four years of the campaign (Fig. 5). In the last two years, when the number of adult females had declined to less than about 200, the simulated curve continued to decline at a similar rate but the reconstructed population started to decline more quickly (Fig. 5). Two factors could have caused this effect. The first is the unusually cold weather in the winters of 1984–85 and 1985–86. These winters were the coldest since the exceptionally cold winter of 1962–63. Cold weather has a well-documented impact on coypu dynamics (Norris 1967b; Gosling, Watt *et al.* 1981; and others) and it seems certain that it accounted for at least part of the decline.

In addition, there could have been a density-dependent reduction in reproductive success at low population density, perhaps because animals were failing to find mates. We tested this by comparing the proportion of adult females that were not visibly pregnant (i.e., not more than 10 days pregnant) in post-mortem examinations (Fig. 6). The comparison was restricted to summer months, when reproductive failure immediately after cold winters would have no effect, and was between data pooled in two

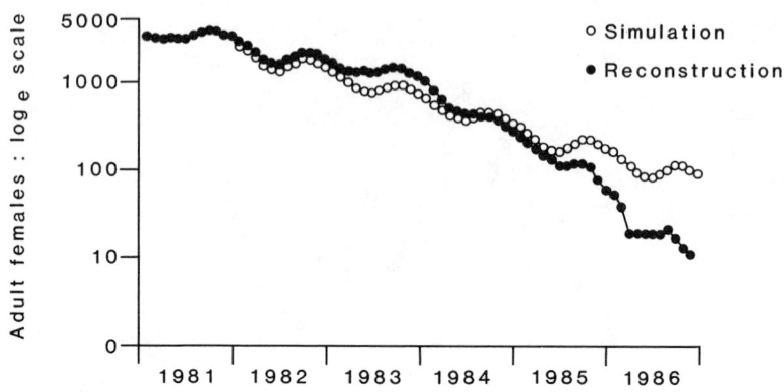

Fig. 5. A comparison of the reconstructed population and the output of a simulation model which uses population variables averaged from the whole eradication period. The reconstructed population starts to decline more rapidly than the simulation after about 1984–85. The possibility that the increased rate of decline is density-dependent is discussed in the text. The unusually cold winters of 1984–85 and 1985–86 probably contributed to the accelerated decline.

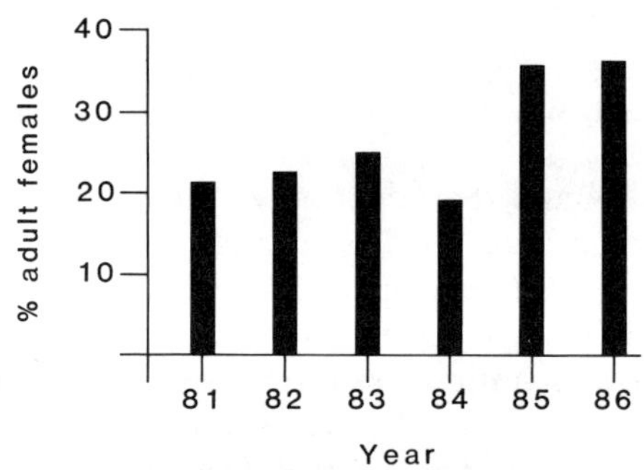

Fig. 6. The proportion of adult females that were not visibly pregnant (i.e., not more than 10 days pregnant) amongst those caught during the eradication campaign. The period considered each year is restricted to the summer and autumn (April–November) to avoid the effect of variation in winter severity. The increase after 1984 may be because some animals fail to find mates at low population density. See text for analysis.

time-classes; the classes were separated at the point when the simulated population curve (using inputs which were averages from the whole campaign) diverged from the reconstructed curve. The analysis showed that the proportion of females which were not visibly pregnant increased from 22% in the first four years of the campaign to 36% in the last two ($\chi^2 = 17.6$, $p<0.001$).

When the summer (i.e., density-dependent) decline in recruitment was incorporated into the fecundity section of the simulation, the discrepancy between simulated and reconstructed numbers of adult females in April

1986 was reduced by 33%. Most of the remaining discrepancy is probably due to unusually cold weather in the winters of 1984–85 and 1985–86.

Discussion

As expected, the large increase in trapping intensity that was achieved in the current campaign has resulted in a sharp decline in the coypu population. The decline is somewhat faster than that in the simulation models that were used to plan trapping intensity in the campaign. This is partly because we used 'worst case' simulations to ensure that trapping effort was not underestimated. The incentive bonus scheme designed to overcome the problem that the trappers might be reluctant to work themselves out of a job has also had a noticeable effect on the attitude of the workforce. Some results of their increased motivation may be reflected in larger numbers of traps but others, such as a willingness to search for better trap positions, are less easy to quantify. In spite of these minor qualifications the results of the campaign confirm the value of using fundamental investigations of population dynamics to plan a practical control operation.

Although population reconstruction suggests that the number of adult females had been reduced to about 20 by April 1986 it is still far from certain that eradication will be achieved. At low population density the practical problem shifts from reducing the population to finding and catching individual animals. Captures in 1986 are thinly scattered over the entire known range so that trappers must continue to search and trap over very wide areas. It also becomes difficult to predict population trends since chance events are more important in small populations.

Conversely, there are no *a priori* reasons why coypus cannot be eradicated from Britain. They breed slowly (Newson 1966; Gosling 1981b, 1986) and, in spite of some problems at low population density, are easier to trap than most mammals (Norris 1967b). Some wetland areas are difficult to penetrate but none is completely inaccessible. Finally, coypus were successfully removed from a representative area of wetland habitat after six years of continuous trapping in a pilot eradication trial (Gosling, Baker & Clarke in prep.). There is also a precedent for the removal of an introduced semi-aquatic rodent: muskrats, *Ondatra zibethecus*, which were established in several areas of the British Isles, were eradicated by trapping in the 1930s (Warwick 1940).

References

Anon. (1978). *Coypu: Report of the Coypu Strategy Group*. Ministry of Agriculture, Fisheries and Food, HMSO, Chesham.

Baker, S.J. & Clarke, C.N. (In prep.). *Cage trapping coypus on baited rafts*.

Boorman, L.A. & Fuller, R.M. (1981). The changing status of reedswamp in the Norfolk Broads. *J. appl. Ecol.* **18**: 241–69.

Cotton, K.E. (1963). The coypu. *River Bd Assoc. Yb.* **11**: 31–9.

Davis, R.A. (1963). Feral coypus in Britain. *Ann. appl. Biol.* **51**: 345–8.

Ellis, E.A. (1963). Some effects of selective feeding by the coypu (*Myocastor coypus*) on the vegetation of Broadland. *Trans. Norfolk Norwich Nat. Soc.* **20**: 32–5.

Gosling, L.M. (1977). Coypu, *Myocastor coypus*. In *The handbook of British mammals*: 256–65. (Eds Corbet, G.B. & Southern, H.N.). Blackwell Scientific Publications, Oxford.

Gosling, L.M. (1980). The duration of lactation in feral coypus (*Myocastor coypus*). *J. Zool., Lond.* **191**: 461–74.

Gosling, L.M. (1981a). The dynamics and control of a feral coypu population. In *The worldwide furbearer conference proceedings*: 1806–25. (Eds Chapman, J.A. & Pursley, D.). Worldwide Furbearer Conference Inc., Frostburg, Maryland.

Gosling, L.M. (1981b). Climatic determinants of spring littering by feral coypus, *Myocastor coypus*. *J. Zool., Lond.* **195**: 281–8.

Gosling, L.M. (1986). Selective abortion of entire litters in the coypu: adaptive control of offspring production in relation to quality and sex. *Am. Nat.* **127**: 772–95.

Gosling, L.M. & Baker, S.J. (1981). Coypu (*Myocastor coypus*) potential longevity. *J. Zool., Lond.* **197**: 285–9.

Gosling, L.M., Baker, S.J. & Clarke, C.N. (In prep.). *An attempt to remove coypus* (Myocastor coypus) *from a representative area of wetland habitat in East Anglia.*

Gosling, L.M., Baker, S.J. & Skinner, J.R. (1983). A simulation approach to investigating the response of a coypu population to climatic variation. *EPPO Bull.* **13**: 183–92.

Gosling, L.M., Huson, L.W. & Addison, G.C. (1980). Age estimation of coypus (*Myocastor coypus*) from eye lens weight. *J. appl. Ecol.* **17**: 641–7.

Gosling, L.M., Watt, A.D. & Baker, S.J. (1981). Continuous retrospective census of the East Anglian coypu population between 1970 and 1979. *J. Anim. Ecol.* **50**: 885–901.

Laurie, E.M.O. (1946). The coypu (*Myocastor coypus*) in Great Britain. *J. Anim. Ecol.* **15**: 22–34.

Moss, B. (1983). The Norfolk Broadland: experiments in the restoration of a complex wetland. *Biol. Rev.* **58**: 521–61.

Newson, R.M. (1966). Reproduction in the feral coypu (*Myocastor coypus*). *Symp. zool. Soc. Lond.* No. 15: 323–34.

Norris, J.D. (1967a). A campaign against feral coypus (*Myocastor coypus* Molina) in Great Britain. *J. appl. Ecol.* **4**: 191–9.

Norris, J.D. (1967b). The control of coypus (*Myocastor coypus* Molina) by cage trapping. *J. appl. Ecol.* **4**: 167–89.

Soriano, B.S. de (1960). La habitación de *Myocastor coypus bonariensis* Geoffroy, 'nutria'. *Actas Trab. Congr. Sud-am. Zool.* **1**: 145–52. (In Spanish).

Warwick, T. (1940). A contribution to the ecology of the Musk-rat (*Ondatra zibethica*) in the British Isles. *Proc. zool. Soc. Lond.* (*A*) **110**: 165–201.

Appendix: Inputs to the simulation

Proportion of females littering

The proportion of females giving birth was estimated for each month from the animals sampled using calculations of the month of birth (from mean embryo weight) and from the proportion of females lactating (Gosling 1981b). Monthly estimates from the two techniques were significantly correlated over the study period ($r_{182} = 0.47$, $p<0.001$), but early estimates were relatively unreliable because of small sample sizes. Consequently, the data from the entire study period were pooled to obtain weighted monthly estimates from each technique and these were averaged for mild and cold years.

The percentage of females littering varies seasonally and between cold and mild years. However, there is little difference in the proportion littering in cold and mild winters: the cumulative percentages between December and February averaged 27% and 34% respectively. Spring values following cold winters are relatively low, probably because of abortions in cold weather (Gosling 1981b). The proportion littering in the four months between March and June after mild winters averaged 56%. The summer peak of littering following cold winters is later and better defined, with an average of 56% littering in the three months between June and August. Smaller peaks of littering from reconceptions at a post-partum oestrus follow these summer peaks. Differences due to the previous winter disappear after October.

Litter size

Coypus frequently resorb individual embryos and abort entire litters (Newson 1966; Gosling 1986). To reduce any consequent bias, only viable embryos in the last six weeks of the 19-week gestation period were considered. The birth date of each litter was calculated after estimating gestation stage from mean embryo weight (Newson 1966). The numbers of female embryos per litter were pooled by month of parturition to obtain average months for cold and mild years ($n = 413$ and 699 litters respectively).

Litter size and composition varies seasonally and between cold and mild years. In general, litters are large in winter (mean females per litter for December–February = 2.77) and spring, and smaller in summer (mean for June–August = 2.41) and early autumn. This effect is more pronounced in cold than mild years and may be due to the effects of nutritional constraints at conception and slight differences in the age distribution of females that litter in different seasons.

Recruitment

The recruitment curves used in the simulations consist of a series of estimates of the proportion of the young born in a particular month that in each successive month have survived to become adult. These curves are derived from the proportions of adults in the animals sampled in successive months after their estimated month of birth (this proportion rises to 1.0 when recruitment is complete), estimates of the absolute numbers born per month and estimates of the absolute number that survive to become adult in successive months. The derivation of these variables and their use in estimating recruitment frequencies are described by Gosling, Watt *et al.* (1981).

The timing of recruitment is similar in cold and mild years, perhaps because only juveniles that maintain a minimum level of condition will survive a cold winter, and those that do are therefore likely to be as advanced in their development as the survivors of a mild winter. In contrast, there are large differences in timing after births in different months of the year. Recruitment is slowest after births in late summer when offspring wean in the winter after an eight-week lactation (Gosling 1980). Recruitment is rapid after late winter births, presumably because the young can wean on spring plant growth. When recruits are expressed in the form required for the simulation, as proportions of the estimated number born (that is, taking into account juvenile survival), there are clear differences between cold and mild winters. These are most easily envisaged by comparing the total proportion that is eventually recruited from each month of birth: from births in July and August the proportion recruited is similar before cold and mild winters but from births in September and October it is much lower in cold than mild years (on average 22% compared to 49%). These differences persist for birth dates through the winter up to January. For young born in February the average values are similar (at about 53%), perhaps because they can survive on their mother's milk until spring growth commences in March–April.

Adult mortality

The mean number of adult females killed per trap-night/trapper/month since 1981 was regressed against the estimated number of adult females alive in that month, and the relationships obtained were used to simulate kill rates.

Adult female mortality rate, measured as the number killed/trap-night/month/10,000 live adults, is high in the winter and spring and low in the summer and autumn. These values (which reflect the ease of capture in different months of the year) reach a peak of 7.2 in March and a minimum of 3.0 in July. Possibly coypus are easier to catch in winter because the baited traps are more attractive to them then and because it is easier to detect their presence and position traps accordingly when summer vegetation has died down.

It is difficult to test whether or not the mortality rate varies between cold and mild years, because numbers killed are the primary input to the population reconstruction. Since the reconstruction is used to obtain the above values such estimates would be biased. However, there is no independent indication (such as large differences in simple kill rate) between cold and mild years and so the same mortality relationships were used in both types of year.

A few coypus (about 4%) were killed by agencies other than official trappers (gamekeepers, cars, etc.). These show a similar pattern of more kills in the winter and spring months than in the summer and autumn. The problems of potential bias do not apply to these data but there were no significant differences between either winter or summer values in cold and mild years ($t = -0.04$, $d.f. = 33$, $p = 0.48$ and $t = -0.7$, $d.f. = 42$, $p = 0.32$ respectively).

Population size

Population sizes were needed to calculate recruitment frequencies and as 'seed data' for the simulations. All estimates were from the population reconstruction (Gosling, Watt *et al.* 1981): no estimates of error are available for this method.

Structure of the simulation

The simulation is constructed as a time-series of monthly units. It is 'seeded' with a 12-month series of adult female numbers. There are three main sections in the model: calculation of the number of females born each month; calculation of the number of recruits each month from earlier birth months and addition of these to surviving adults from the previous month; and calculation of the number of adults killed each month by a given force of trappers.

The number of females born each month is expressed as $x = ny$, when n = the number of adult females littering, and y = the mean number of female offspring per litter. In this expression, $n = Nl$, where N = the number of adult females alive, and l = the proportion of adult females littering.

As described above, all these values are available as monthly estimates for cold and mild years.

The number of females that reach maturity in a particular month is determined by summing the contributions from births in previous months (see Gosling, Baker & Skinner 1983 for further details). This monthly total is then added to the adults surviving from the previous month. Mortality is calculated from the relationships between kill rate and population size that have been described, and converted to the numbers that would be killed by whole trapper units.

Symp. zool. Soc. Lond. (1987) No. 58: 115–130

Group stability and the structure of a capybara population

Emilio A. HERRERA* and
David W. MACDONALD

Animal Behaviour Research Group
Department of Zoology
University of Oxford
South Parks Road
Oxford OX1 3PS

Synopsis

A capybara population in the seasonally flooded savannas of Venezuela was found to be structured in groups of five to 14 adult individuals of both sexes, with a socionomic sex ratio biased towards females (♀:♂ = 1.7:1.0, N = 16), and a variable number of young (range 0–68% of group membership). Although such groups amalgamated during the dry season, they appeared to be of closed membership, and remained stable in size and composition for two or more years. Social integrity of groups was maintained by territorial aggressive behaviour exhibited by all adult group members against intruders.

Although some groups produced many young, these were rarely recruited as adults into their parental group. There was no support for the hypothesis that groups develop from a breeding pair. However, we present circumstantial evidence that yearling subadults disperse together from their natal group in the company of a subordinate adult male. Favourable habitat was saturated with capybara groups, therefore such dispersing, incipient groups probably had little chance of successful establishment, with the probable result that subadult mortality was high.

The importance of group living in capybaras is discussed with reference to the distribution of essential resources and the reproductive success of females.

Introduction

The social behaviour and grouping patterns of individuals of any species have a strong influence on the structure and dynamics of their population (e.g. Dunbar 1985: Łomnicki 1978). Thus, the mating system, mechanisms of dispersal and other elements of a species' social system will affect each individual's survival and reproduction. The aggregate outcomes of these

* Present address: Smithsonian Tropical Research Institute, Apdo 2072, Balboa, Panama.

ZOOLOGICAL SYMPOSIUM No. 58
ISBN 0-19-854006-X

individual life histories will determine population parameters such as density and growth rate. In this paper, we present data on the grouping patterns of capybaras (*Hydrochoerus hydrochaeris*) in the seasonally flooded savannas of Venezuela, and discuss the relevance of those patterns for the structure of the population at large. Details of the social relationships within groups, which may also be relevant to population structure, are discussed elsewhere (Herrera 1986).

Capybaras are caviomorph rodents of up to 50 kg in weight which feed mainly by grazing (Escobar & Gonzalez-Jimenez 1976). They are abundant in the northern neotropical savannas, where they live in groups, invariably associated with water (Macdonald 1981; Schaller & Crawshaw 1981). In the study area, most of the matings occur at the beginning of the wet season and the young—on average four per female—are born five months later. A general account of capybara biology is given in Herrera & Macdonald (1984), and the population dynamics are described by Ojasti (1973).

The data reported here come from our three-year study of capybara social behaviour and ecology (Herrera 1986). Our study has shown that the groups of capybaras in the Llanos are constructed of a dominant male, several females and young, and one or more subordinate males. Among the males there is a social hierarchy, one consequence of which is that each dominant male mates more often than any of the subordinates in his group. The relationships between females are less well understood since females interact very little with each other.

In our study area the mean size of groups of capybaras in the dry season was 10.9, but this figure embraces some significantly smaller groups in areas that were artificially irrigated (mean group size = 7.8) and larger groups in areas which dried up naturally (mean group size = 18.6) (Macdonald 1981). The same factors may underlie a similar seasonal difference: small groups are typical of the wet season, but some large groups of up to 60 members can be found in the dry season (Ojasti 1973). In addition to seasonal and local variations in mean group sizes, up to 7% of individuals appear to be solitary (Macdonald 1981).

Hitherto, nothing has been known of the stability or social dynamics of groups of capybara in either the short or the long term. In seeking to determine how stable and cohesive are groups of capybaras in the Llanos, we ask how groups relate to one another, how groups are formed and whether group size affects the reproductive success of females. We also discuss the ecological factors that may underlie the social organization observed and, more specifically, we ask whether the larger dry-season groups are actually aggregations of smaller wet-season groups and, if so, whether they fall into their original component units when they subsequently fragment again.

Study area

We carried out our study on a cattle ranch, Hato El Frio (7° 46′ N, 68°57′ W), located in the low Llanos region of Venezuela. This area consists of tropical savannas, i.e. vast flat plains dominated by grasses (Sarmiento 1984), and its climate is dominated by a marked wet/dry seasonality. The wet season lasts approximately from May to October, the dry season from November to April. In the late dry season (February and March), evaporation from surface water reaches a rate of 1 cm per day, with the result that most lakes and ponds dry up, leaving a parched landscape. The wet season receives 90% of the annual average rainfall of 1600 mm, leading to extensive flooding over most of the savanna. In this habitat, Ojasti (1978) found a high (up to 30%) density-independent adult mortality of capybaras due to starvation. Capybaras may also starve during the wet season if the flooding is unusually severe. In a later review, Ojasti (1983:435) concluded that 'the demographic strategy of a capybara in the Llanos region is to breed and grow in the rainy season and survive the drought'.

Methods

The behaviour of individual capybaras was studied from March to October in 1982, 1983 and 1984. To achieve this, it was necessary to capture and mark animals. This was done with the help of cowboys (llaneros) in February of each year. The most successful method of marking involved a combination of expanding Darvic PVC (ICI) collars (designed by Fiona Guinness, see Clutton-Brock, Guinness & Albon 1982), painted with unique combinations of two letters, and coloured plastic eartags (Dalton Supplies, Nettlebed).

Capybaras were observed by means of 10×22 binoculars or a 60mm telescope, and their behaviour was recorded according to both the scan (10 min intervals) and instantaneous sampling regimes (Altmann 1974). Animals were classified into four age groups (infants, juveniles, sub-adults and adults) on the basis of size, and each class was aged on the assumption that most individuals were born during September and October at the end of the wet season.

Results

Five groups were observed in detail during 1983 and 1984, and one of these (P5) was also observed in 1982. The numbers of animals present in each group were scored at each scan throughout every observation session. The analyses of group size and group stability were decoupled in so far as data on both marked and unmarked animals were used in the former, whereas only data on marked individuals could be used for the latter.

Group stability

Table 1 presents examples of the pattern of presence and absence of marked individuals across observation sessions for two groups in 1983. It can be seen that some animals were present repeatedly throughout the study period while others were seen only rarely (once or twice). The pattern observed is caused by factors of both biological and methodological origin. Capybara groups generally moved and rested as a cohesive unit; in these circumstances it was straightforward to note which animals were 'in' the group. It was not uncommon, however, for a group to be more scattered; for example, one or a few animals might lag behind, sometimes for several hours. There were also occasions when animals from neighbouring groups strayed so close together that members of one might mistakenly be recorded as members of the other. Despite these sources of confusion, the recurrent presence of a cadre of animals in each group (and the fact that these individuals were never seen in other groups) demonstrates the existence of a social cohesion among group members. The behaviour of animals towards each other provided a further clue to membership of groups. As will be shown below, group members aggressively reject intruders. So, individuals were recorded as group members (as indicated on Table 1) on the grounds that they were frequently present and generally tolerated.

It should be noted, nonetheless, that the presence of group members was less consistent in the dry season: until the rains started in mid-April, there appeared to be more members absent and more non-members present. This was due in part to the difficulty of defining groups when they had coalesced into the larger, and sometimes rather diffuse, aggregations typical of that season (see below). We conclude that capybara groups are stable social units during one year, but that at a given moment a minority of animals may be more or less loosely affiliated with the group.

Group size

Group size and composition appeared to vary, but prolonged observation revealed that this was largely caused by animals moving into and out of view (cf. Aldrich Blake 1970). Over several hours all of the marked animals present would be seen, even if they were never all in sight simultaneously. The number of unmarked animals was estimated from the proportion of those marked individuals known to be present that were in sight on average; assuming that marked and unmarked animals were equally likely to be in view, the total group size could be estimated from the number in sight. Since both the confounding variables (absentee members and interloping non-

Table 1. Presence of identified individuals in all systematic observation sessions in 1983 (dates across the top) of (a) group C2 and (b) group NC. An X means presence, blank absence. The last column indicates whether each individual is considered a member (Y) of the group or not (N).

(a) Date of observation session

Age/sex and status[a]	Identity code	Mar	Apr				May				June						July		Aug		Member of group
		29	6	14	20	25	7	10	17	31	3	16	20	21	24	28	9	10	11	19	
DM	OJA				X	X	X	X	X	X	X	X	X	X	X	X	X	X	X	X	Y
F	OXH	X			X	X	X	X	X	X	X	X	X	X	X	X	X	X	X	X	Y
F	OXB				X	X	X	X	X	X	X	X	X	X	X	X	X	X	X	X	Y
F	OVN	X	X		X	X	X	X	X	X	X	X	X	X	X	X	X	X	X	X	Y
M	OHX				X	X		X	X	X	X		X							X	Y
M	OHD				X	X	X	X	X	X			X							X	Y
M	M14							?	X	X		X	X	X	X	X					Y
F	YCZ							X													N
M	OJX	X	X	X						X											N
M	OEL					X															N
M	WNT		X																		N

(b) Date of observation session

Age/sex and status[a]	Identity code	Mar	Apr			May				June					July	Aug		Sept		Oct	Member of group
		23	4	8	21	6	10	18	24	1	2	17	23	27	1	9	25	9	23	7	
DM	R15	X	X	X	X	X	X	X	X		X	X	X	X		X	X	X		X	Y
F	YCC	X	X	X		X	X	X	X	X	X	X	X		X	X	X	X			Y
F	YBV				X	X	X	X	X		X	X	X	X	X	X	X	X	X	X	Y
F	YED	X	X			X	X	X	X		X	X	X	X		X		X			Y
F	YCH			X		X	X	X	X			X	X			X		X	X		Y
F	YVI					X	X	X	X		X	X	X	X		X	X	X	X		Y
M	WTE		X	X	X	X	X	X	X	X	X	X	X	X	X		X	X		X	Y
M	WNA	X	X		X	X	X	X	X	X	X	X	X	X	X	X	X	X		X	Y
M	1DG						X		X			X	X	X	X		X	X			Y
M	WPT	X	X	X																	N
M	WLH		X																		N
M	WNH		X																		N
F	YVA	X																			N
F	YDN				X																N
M	WNT			X																	N

[a] DM, dominant male; F, adult female; M, adult male.

members) proved to be rare events, we may expect the estimated group sizes to be accurate.

Figure 1 shows a histogram of the numbers of adults in 16 groups. In this histogram, for those groups that were seen for more than one field season, only the first year is used, since the values of one year and the next are not independent. This sample gives a mean group size of 9.6 (± 3.8), with an average composition of 3.6 males (± 1.8) and 6.1 females (± 2.4), giving a socionomic sex ratio of 1.7:1.0 (female:male). Including juveniles, the mean group size is 14.0 (± 9.9) with juveniles contributing 0–68% of group membership (0–27 individuals). These figures do not include solitary individuals or pairs and may therefore be biased towards the more stable social situations (see below).

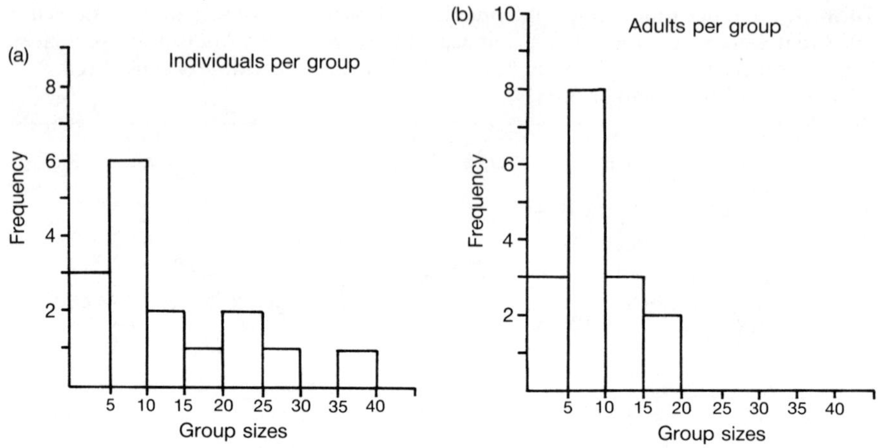

Fig. 1. Histograms showing the distributions of the sizes of (a) 16 capybara groups and (b) the numbers of adults in these groups.

Jarman (1982) has pointed out that the commonly used mean for group size may not represent the group size that the average animal experiences. He proposes the formula:

$$\bar{g} = \frac{\sum_{i=1}^{n} g_i^2}{\sum_{i=1}^{n} g_i}$$

where *g* represents group sizes and *n*, number of groups. This formula indicated that the average female capybara is found in a group of 7.0 females, the average male in a group of 4.4 males, and young among 17.8 of their contemporaries.

The histogram in Fig. 1 indicates that sizes of capybara groups vary from four to 16 individuals with a peak at eight and a long tail towards the larger groups. These features are consistent with the interpretation that there is a threshold size below which a group is not advantageous, while to be in a large group has fewer disadvantages than being in one that is smaller than this threshold or in no group at all (see Sibly 1983).

Interannual group stability

Because of the difficulties in determining group membership described above, and because collars were so often shed between field seasons, the structures of only five groups could be accurately determined in two successive years. Diagrams of these structures for all five groups in 1983 and 1984 are given in Tables 2a–e. These show that many, and probably most, adult capybaras remained in the same group from year to year. In the five groups shown in Table 2a–e, five out of 11 males marked in 1983 (excluding the dominants of each group) and nine out of 19 females were still present in 1984. There was no indication from these data of any difference between the sexes in terms of tendency to stay or disappear (χ^2 = 0.001, *d.f.* = 1, N.S.).

It is unlikely that all the adults which disappeared also died. Thorough searches of the study area, combined with field signs such as circling vultures, led to the discovery of many corpses but few of the missing marked animals were among them. The likely provenances of some of the adults first marked in 1984 (Table 2) were determined on the basis of their size and appearance. If an animal was a young adult, it was listed as a possible recruit from that group's young of the previous year. If its size and appearance indicated that it was older than 1.5 years, it could have been either an immigrant or an unmarked animal from the previous year. From Table 2a–e it is clear that there must be some immigration of adults, since some animals could not have been either recruits from the previous year's young or previously (i.e. 1983) unmarked members of that group. However, of 858 animals marked over the three years, we never saw one case of an adult member of one group transferring permanently to another group within the study area. These newcomers were, therefore, probably immigrants from outside the study area. One female (YCA: Table 2d) was seen some 5 km south of the study area two weeks before she was first seen in Group P5. Two other females were observed to leave their groups. Thereafter they were both found alone, or occasionally accompanied by a male and occupying small home ranges next to those of their former groups.

The few changes observed in the adult composition of groups (Tables 2a–e) imply that there was little recruitment of young into their parental groups and little change in group size from one year to the next. For instance, Group NC had 27 young in 1983 but only four of the newly marked adult members of 1984 could have been young recruited from the previous year. Similarly, the number of adults in Group P5 from 1983 to 1984 rose by only four, although this group had 12 young in 1983. In fact, the general trend was for a decrease in the number of adult members of groups. From the changes in group composition (Table 2), it can be seen that the net change in the number of male members is 0 while that of females is −2.

Relations between groups

All groups of capybaras observed in the study area occupied stable home ranges of 6–16 ha (Herrera & Macdonald in prep.). The overlap between

Table 2. Changes in group composition from 1983 to 1984 in five groups (a) C1, (b) C2, (c) NC, (d) P5 and (e) VQ. Identification codes (three characters) of animals unambiguously identified in both years appear joined by a solid line: unmarked animals in either year that could have been marked the previous or subsequent year are joined by a broken line. Total number of animals in each age/sex class is shown.

(a)

	1983		1984
Females	OXD	———	OXD
	YBH	- - -	YQK
	YCZ		
	Fem	- - -	YLL
			YL4 Newcomers
			Fem
N =	4		5
Males	WSC	———	WSC
	OJX	———	OJX
	WTC		
	Male	- - -	WHI
			Male Newcomer
N =	4		3
Young	0		0

(b)

	1983		1984
Females	OVN (Died)		
	OXH	———	OXH
	OXB		
	Fem1	- - -	YLU
N =	4		2
Males	OJA	———	OJA
	M14	———	M14
	OHD	———	OHD
	OHX		
N =	4		3
Young	0		2

(c)

	1983		1984
Females	YCC	———	YCC
	YVI	———	YVI
	YBV	———	YBV
	YED		
	YCH		
	Fem1	- - -	YQB
	Fem2	- - -	YT4
			YLQ
			YSB Young recruits
			WWD Newcomer?
N =	7		8

Males	R15		
	WTE	——	WTE
	WNA	——	WNA
	1DG		
	Male1	- - -	WHC
	Male2	- - -	Male
			W14 Young recruit
N =	6		5
Young	27		20

(d)

	1983		1984
Females	2GB	——	2GB
	YDS	——	YDS
	YVA	——	YVA
	YCA	——	YCA
	Fem1	- - -	2LW
	Fem2	- - -	YPU
			YST Newcomer
N =	6		8
Males	165	——	165
	D14	- - -	Male1
	WLJ		
			W14 Young recruit
			WW7 Newcomer
			Male2
			Male3 Newcomers?
N =	4		6
Young	12		7

(e)

	1983		1984
Females	OVP Died		
	YBI	- - -	Y76
	YBD	- - -	YTS
	YCD		
	Fem1		
	Fem2		
N =	6		2
Males	1BY	——	1BY
	WTN		Male1
			Male2
N =	2		3
Young	0		7

neighbouring ranges was minimal, so we regarded them as territories. During 729 hours of observation, a total of 31 group territorial border encounters involving 13 groups was observed. In general, these encounters were aggressive. The typical course of events was that the members of two groups would stray closer together as they grazed. Peripheral members of each group would gradually intermingle, and those that were closest would launch attacks on the individual nearest them. Other individuals of both groups, who happened not to meet a stranger, would show no interest in these individual squabbles. During territorial encounters, several ritualized behaviour patterns were noted. These included walking in parallel along the territorial border, with both capybaras co-ordinating their movements so that they started and stopped in synchrony. Very aggressive, all-out fights were also observed during inter-group encounters, involving animals of either sex.

In total, during 31 inter-group encounters, 67 such aggressive interactions between two males from different groups were recorded, 41 between a male and a female, 42 between two females, and 13 in which the sex of either or both of the participants could not be determined. It is noteworthy that intersexual aggressive interactions between groups involved a high proportion of female–female clashes. This is in marked contrast to the situation within groups, where female–female aggression constituted only 8.3% of all aggressive interactions (Herrera 1986). In three of the 31 cases, several members of both interacting groups were simultaneously involved in the rejection of intruders. In one of these cases 16 chases between individuals of both sexes were recorded in the 15 min for which the episode lasted. It is noteworthy that males which were low-ranking during interactions within their own group were actively involved in aggression towards other groups and, in at least one case, a subordinate from one group was able to drive away the dominant from another.

Despite the great variation in group size, there was no evidence of any group being consistently dominant over another. Rather, the outcome of an interaction between members of neighbouring groups was determined by its location: each individual acted as dominant over its neighbour when inside its own territory.

How does this general finding of antagonism between neighbours bear on the possibility of groups coalescing during the dry season? During observations at the end of the 1983–1984 dry season, three large aggregations were found of which members in each case included marked individuals known to have belonged to up to four separate groups from the preceding wet season. When the rains returned, these same individuals reverted to their original groupings. There was no evidence of any exchange of members between the groups composing each aggregation. These observations made clear that the recurrent reports of dry season groups being larger than wet season ones (Ojasti 1973; Azcarate 1980; Macdonald 1981) reflect the combination of several independent social units rather than the breakdown and amalgamation of such units.

The splitting and coalescing of groups occurred on a daily basis with groups forming aggregations while resting and wallowing in the mornings and separating to a greater or lesser extent to graze in the evenings. When separating from the large aggregations, members of each group would move as a unit. The spatial tolerance of the dry season does not destroy the group's integrity since neighbouring groups appear simply to rest side by side around receding pools. The observation of territorial behaviour in the dry season indicates that the groups can maintain their identities despite the apparent mingling.

Group formation

Ojasti (1973) postulated that new groups arise when newly formed pairs breed, the parents and offspring subsequently remaining together. In the present study, at least three possible pairs were recorded within the 300 ha study area. None of these animals that was relocated was still in the same company. No pair was seen to rear young (the smallest group to do so numbered five). Of course, in an area where suitable habitat is saturated with group territories the establishment of new groups (and therefore sightings of pairs) might be a rare event. Nonetheless, group sizes were fairly constant (Table 2) and numbers of young surviving to April averaged 7.2 ($\pm$ 7.8, $N = 6$) per group annually. What was the fate of this surplus annual production, and was there any possibility of some of them forming new groups?

On 20 occasions during the latter part of the wet seasons of 1983 and 1984 (September and October), marked subordinate males from nine different groups were seen grazing between 30 and 50 m from the core of their group accompanied by a number (ranging from one to 11) of that group's young. In the context of mean nearest-neighbour distances between grazing group members of 12.5 m (Macdonald 1981), the distances between these cliques and the remainder of the group were notable, and the fact that they invariably involved subordinate males was striking. In September and October, the young of the year are approaching sub-adult status and the next generation is about to be born. At this time of year systematic observation of behaviour is hampered by extensive flooding and thick vegetation, so it was not possible to monitor any change in adult–young aggression that might presage dispersal. These observations of subordinate males accompanied by sub-adult individuals of unknown sex suggested that they were in the process of splitting from the parent group. In this case the subordinate male would be expected to become dominant in the incipient group by virtue of being bigger and older.

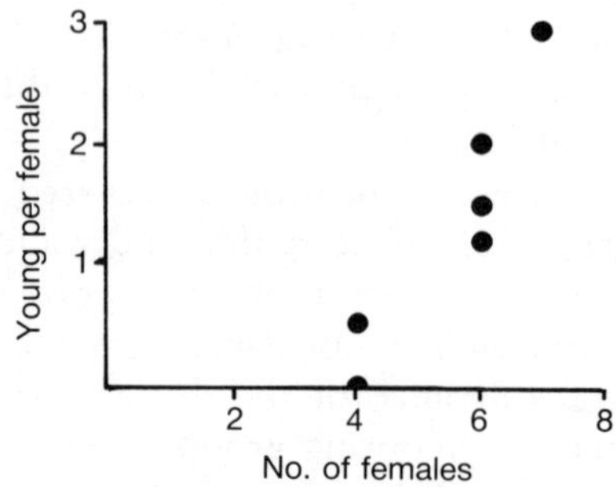

Fig. 2. Reproductive success of females and group size: number of young surviving to April per female plotted against number of females in the group the previous year.

Group size and the reproductive success of females

The observation that pairs failed to breed in our study area raises the question of which factors affect capybara reproductive success and, in particular, whether survival of offspring is related to group size. The reproductive success of an individual is best measured as the number of its offspring that survive to sexual maturity or to a certain critical age. In the case of capybaras, most young are born at the end of the wet season, a time of relatively abundant forage and cover. But soon after this, they face their first dry season, a time of high adult mortality (Ojasti 1978). Survival during this first dry period is crucial. Thus, it seems appropriate to use the number of young surviving to April as an indicator of the reproductive success of females.

The number of young born to any particular female is difficult to determine. This is because of the social system of the capybaras: female capybaras give birth away from the group and hidden under cover, and the mothers are wary and retiring when their offspring are very young; within the first week of life the young join creches; and females nurse each others' young communally (Macdonald 1981; Schaller & Crawshaw 1981). Hence only the total number of young born to a group could be observed reliably, and from this the average number of young per female was calculated. For five groups accurate estimates of adult membership in 1983 and the number of young in April 1984 are available, along with comparable data for one group from 1982 to 1983. Figure 2 shows that there is a significant positive correlation between the mean number of surviving young per female and the number of females in a group (Spearman rank correlation coefficient, $Rs = 0.926, p < 0.05, N = 6$).

Discussion

The niche occupied by the capybara has much in common with those of many species of tropical ungulates (see Jarman 1974; Leuthold 1977; see also Dubost 1968). Like them, the capybara is a relatively large grazer and subject to predation on open plains. Capybaras are both group-living and territorial, but the selective forces favouring these two facets of their social system may be separate. Capybaras are, or have been in the recent past, prey to puma, *Felis concolor*, jaguar, *Panthera onca*, feral dogs, *Canis familiaris*, caiman, *Cayman crocodilus*, and man. Macdonald (1981) described how a group of capybaras, under attack by feral dogs, bunched together with youngsters within a protective cordon. There are three reasons why capybaras might form groups in response to predation: 1. increased vigilance to forestall attack, 2. increased possibility of intimidating the predator through greater strength of numbers; 3. the 'selfish herd' effect (Hamilton 1971), in which individuals share the risk of capture, whether or not this also reduces the capture rate of the predator. Bertram (1978) reviews evidence for these commonly presumed advantages of grouping among some other species. Turner & Pitcher (1986) present a model in which the dilution effect (predator is less likely to find a group than an individual) and the avoidance effect (point 3 above) are combined to produce what they call 'attack abatement', so that animals living in groups may have greater fitness than scattered ones. A separate possible function of group living relates to McNaughton's (1984) suggestion that the stimulating effect of a group of grazers on a patch of grass is advantageous in maintaining a nutritious grazing lawn. However, the premise that grazing stimulates grass growth has recently been challenged (Belsky 1986).

The demands of predation and perhaps grazing may explain why capybaras are gregarious, but alone they do not explain either why they are territorial or why their groups sizes are not larger. The answers to these questions are probably to do with the dwindling of resources as the dry season progresses. In particular, capybaras are dependent upon water to sustain their food plants, and they also thermoregulate in water, mate in water, and seek refuge from predators in water (Ojasti 1973; Azcarate 1980; Macdonald 1981; Herrera & Macdonald 1984; Herrera 1986). As the dry season progresses, lakes that were previously almost contiguous over the landscape contract until they are little more than puddles surrounded by kilometres of dry, dusty and, to the capybaras, useless land. The remaining pools and the associated food become limiting, and highly patchy, resources. We suggest that it is the need to guarantee access to a permanent dry-season lagoon that favours territoriality. The presence of the water hole next to a grazing patch makes a piece of land defensible (Brown 1964). We observed that every territory did indeed contain both a waterhole and a grazing lawn (Herrera & Macdonald in prep.).

The foregoing discussion of predation, together with the fact that females nurse each other's infants (and occasionally carry them on their backs while swimming from danger) might suggest that individual capybaras should benefit from being members of groups that are as large as possible. This suggestion is compatible with the observed distribution of group sizes with its long tail towards large groups (Fig. 1). Within the range of group sizes that we found, any possible disadvantages of larger membership (e.g. greater discord with more males) were apparently overridden by the advantages, since females in larger groups reared more young. However, data on reproductive success are lacking for the largest groups observed (16 adults). It would be premature to conclude that there was a causal link between the postulated advantages of gregariousness and this increased reproductive success. Another possibility is that group size and reproductive success are both affected, partially or completely independently, by a third variable, such as the nutritional quality of the grassy patches in each territory.

The stability of groups observed in these rodents is comparable to that of several territorial species of primates and carnivores (e.g. Hrdy 1977; Rasa 1985). In the case of capybaras, the interannual group stability, particularly the fact that subordinate males remained in their groups for two years or more while being deprived of some resources (particularly access to sexually receptive females), demonstrates the importance of group living for these individuals.

The stability of group sizes between years was particularly notable in those groups in which a large number of young was not reflected in an increase in the number of adults the following year. Clearly, many young were not recruited into their parental groups, and their fate is crucially important to understanding the structure of the capybara population in the Llanos. If the young of a group do indeed disperse as a unit, as suggested above, they would find the area surrounding the water holes in our study area populated by contiguous capybara territories, and these were maintained over at least three years (Herrera 1986). In fact, between 1983 and 1984 virtually all groups using the study area were known, and there was no evidence of any new group becoming established. This suggests that incipient groups would have little chance to find a vacant territory adjoining that of their parents. The alternatives would be to colonize suboptimal habitat close at hand or to disperse over greater distances. The second option is likely to be hazardous since suitable blocks of habitat are widely dispersed and, anyway, likely to be fully colonized. Therefore, the circumstantial evidence is that dispersing sub-adults have poor prospects and probably suffer high mortality.

How do these patterns fit with Ojasti's (1973) apparently correct finding (Ojasti 1980) that there is a sustained yield of 30% of the population which can be harvested every year? This 30% is calculated on the basis of the total population, including all age and sex classes. However, the slaughtermen only kill adult individuals and they attempt to spare pregnant females. In practice, the cull moves from a selected point across favourable areas killing all adults found until the quota is reached. Field observations revealed large areas which were populated almost exclusively by young capybaras after the cull, thereby indicating that in some areas most of the adult population is wiped out. Having no adults with which to compete, these young are probably able to remain in their natal territories. Conditions in our study area were a marked contrast to this, since it was spared from the annual slaughter; in this case the majority of young faced dispersal at 1.5 years of age. Thus, the cull probably has the effect of reducing dispersal and mortality among sub-adults. These animals and the surviving adults will form the population to be harvested one year later.

Acknowledgements

E. A. H. was supported by a scholarship from the Consejo Nacional de Investigaciones Cientificas y Tecnologicas of Venezuela. D. W. M.'s visits to Venezuela were supported by the Royal Society and Universidad Simon Bolivar (Caracas), during a period when he was an Ernest Cook Fellow. We gratefully acknowledge the co-operation of the owners and workers of El Frio Ranch, and of Fundacion La Salle for permission to work at El Frio Biological Station. Our colleagues at the ABRG in the Zoology Department at Oxford were an important source of stimulation and support; in particular, G. M. Carr and A. B. Taber made helpful comments on this paper.

References

Aldrich-Blake, F.P.G. (1970). Problems of social structure in forest monkeys. In *Social behaviour in birds and mammals*: 79–101. (Ed. Crook, J.H.). Academic Press, London and New York.

Altmann, J. (1974). Observational study of behavior: sampling methods. *Behaviour* **49**: 227–67.

Azcarate, T. (1980). Sociobiologia y manejo del capibara. *Doñana Acta Vert.* (Spec. No.) **7–6**: 1–228.

Belsky, A.J. (1986). Does herbivory benefit plants? A review of the evidence. *Am. Nat.* **127**: 870–92.

Bertram, B.C.R. (1978). Living in groups: predators and prey. In *Behavioural ecology: an evolutionary approach*: 64–96. (Eds Krebs, J.R. & Davies, N.B.). Blackwell Scientific Publications, Oxford.

Brown, J.L. (1964). The evolution of diversity in avian territorial systems. *Wilson Bull.* **76**: 160–9.

Clutton-Brock, T.H., Guinness, F.E. & Albon, S.D. (1982). *Red deer. Behavior and ecology of two sexes.* Chicago University Press, Chicago.

Dubost, G. (1968). Les niches écologiques de forêts tropicales sud-américaines et africaines, sources de convergence remarquable entre rongeurs et artiodactyles. *Terre Vie* **1**: 3–28.

Dunbar, R.I.M. (1985). Population consequences of social structure. In *Behavioural ecology: ecological consequences of adaptive behaviour*: 507–20. (Eds Sibly, R.M. & Smith, R.). Blackwell Scientific Publications, Oxford.

Escobar, A. & Gonzalez-Jimenez, E. (1976). Estudio de la competencia alimenticia de los herbivoros mayores del llano inondable con referencia especial al chiguire (*Hydrochoerus hydrochaeris*). *Agron. trop.* **26:** 215–27.

Hamilton, W.D. (1971). Geometry for the selfish herd. *J. theor. Biol.* **31:** 295–311.

Herrera, E.A. (1986). *The behavioural ecology of the capybara.* D. Phil. Thesis: University of Oxford.

Herrera, E.A. & Macdonald, D.W. (1984). The capybara. In *Encyclopaedia of mammals*: 696–9. (Ed. Macdonald, D.W.). George Allen & Unwin, London.

Herrera, E.A. & Macdonald, D.W. (In preparation). *Ranging behaviour and spatial organisation of capybaras.*

Hrdy, S.B. (1977). *The langurs of Abu, female and male strategies of reproduction.* Harvard University Press, Cambridge, Mass.

Jarman, P.J. (1974). The social organisation of antelope in relation to their ecology. *Behaviour* **48:** 215–67.

Jarman, P.J. (1982). Prospects for intraspecific comparisons in sociobiology. In *Current problems in sociobiology*: 323–42. (Eds King's College Sociobiology Group). Cambridge University Press, Cambridge.

Leuthold, W. (1977). African ungulates: a comparative review of their ethology and behavioural ecology. *Zoophysiol. Ecol.* **8:** 1–307.

Łomnicki, A. (1978). Individual differences between animals and the natural regulation of their numbers. *J. Anim. Ecol.* **47:** 461–75.

Macdonald, D.W. (1981). Dwindling resources and the social behaviour of Capybaras, (*Hydrochoerus hydrochaeris*) (Mammalia). *J. Zool., Lond.* **194:** 371–91.

McNaughton, S.J. (1984). Grazing lawns: animals in herds, plant form, and coevolution. *Am. Nat.* **124:** 863–86.

Ojasti, J. (1973). *Estudio biologico del chiguire o capibara.* Fond. Nac. Invest. Agropec., Caracas.

Ojasti, J. (1978). *The relation between population and production of the capybara.* Ph.D. Thesis: University of Georgia.

Ojasti, J. (1980). Ecology of capybara raising on inundated savannas of Venezuela. In *Tropical ecology and development*: 287–93. (Ed. Furtado, J.I.). Int. Soc. Trop. Ecol., Kuala Lumpur.

Ojasti, J. (1983). Ungulates and large rodents of South America. In *Tropical savannas*: 427–39. (Ed. Bourliere, F.). Elsevier, Amsterdam.

Rasa, O.A.E. (1985). *Mongoose watch.* John Murray, London.

Sarmiento, G. (1984). *Ecology of neotropical savannas.* (Trans. by O. Solbrig.) Harvard University Press, Cambridge, Mass.

Schaller, G.B. & Crawshaw, P.G. (1981). Social organization in a capybara population. *Säugetierk. Mitt.* **29:** 3–16.

Sibly, R.M. (1983). Optimal group size is unstable. *Anim. Behav.* **31:** 947–8.

Turner, G.F. & Pitcher, T.J. (1986). Attack abatement: a model for group protection by combined avoidance and dilution. *Am. Nat.* **128:** 228–40.

Symp. zool. Soc. Lond. (1987) No. 58: 131–152

Estimating the abundance of red deer (*Cervus elaphus* L.) and roe deer (*Capreolus capreolus* L.) and their current status in Great Britain

Brian W. STAINES* and
Philip R. RATCLIFFE**

*Institute of Terrestrial Ecology
Hill of Brathens, Banchory
Scotland, AB3 4BY
**Forestry Commission
Wildlife and Conservation Research Branch
Alice Holt Lodge
Wrecclesham
Surrey, GU10 4LH*

Synopsis

Methods of estimating the absolute or relative abundance of deer in Britain are discussed. The techniques used will vary according to the level of precision required, habitat types, animal behaviour and logistics. Counts from direct observations are generally preferred, and are more reliable in open habitats, but indirect methods, particularly faecal counts, are more useful in concealing ones. Recent developments in the techniques for assessing the numbers of red and roe deer in commercial plantations are highlighted.

Red deer on open range in Scotland have increased from an estimated 180K in the 1960s to around 300K now. The increase in numbers is greater in the east and central Highlands than in the north and west. Low culling rates of females, and in some areas reduction in competition with domestic stock, are thought to have allowed densities of red deer to increase and may also have promoted extension in range. Densities vary from 0.5 to 31 per km^2. Red deer now resident in Scottish conifer plantations commonly achieve 5–15 deer per km^2, with greater densities reached in favourable habitats.

Estimated densities of roe deer in conifer plantations in the Scottish Borders vary from 8–25 deer per km^2 according to habitat type, being greatest in the younger forest stages before canopy closure. In small mixed woods in southern England densities of 30 or more roe deer per km^2 have been suggested.

Current culling levels, coupled with increasing afforestation, indicate that both deer species will continue to increase in numbers and range for the foreseeable future.

ZOOLOGICAL SYMPOSIUM No. 58
ISBN 0-19-854006-X

Introduction

This paper reviews the methods used for assessing the distribution and abundance of red deer (*Cervus elaphus* L.) and roe deer (*Capreolus capreolus* L.) in Great Britain, and gives information on recent developments in the study of these species in forestry plantations. Finally, it briefly reviews the recent history of these deer in Britain and their current status, and predicts the likely population trends for the future.

Habitats

Both red and roe deer are indigenous to Britain and occupy a variety of habitats. Red deer are especially abundant on the open hills of the Scottish Highlands and Islands and in Galloway (south-west Scotland), and are resident in most of the large conifer forests planted there since the 1920s. Sizeable populations associated with hill land also occur in Exmoor and the English Lake District, with scattered populations, usually in woodlands, elsewhere (Fig. 1).

Roe deer live mainly in woodlands, although in some areas a few will continuously occupy open moorland or agricultural land. They are found in most forests throughout northern and southern Britain (Fig. 2), and particularly favour mixed woodlands (Andersen 1953) or the younger growth-stages of conifer plantations (Loudon 1982; Staines & Welch 1984).

Because of their value for sport and because both conflict with forestry, agriculture and nature conservation, the distribution and abundance of both species have received much attention, particularly in recent years.

Methods

Methods for estimating the abundance of large ungulates have been much studied and there is an extensive literature (see Overton 1971; Seber 1973; Caughley 1977; Riney 1982). Table 1 summarizes the common techniques used for cervids.

The choice of method depends on the specific purposes of the enquiry and the accuracy required. For detailed studies of animal population dynamics, the level of precision needed, both in enumeration and animal classification, is greater than for many management purposes. For some investigations, it may be sufficient to know that there are more or fewer animals in one place than in another, or more or fewer animals in the same area over time. Studies of habitat utilization may require to know only relative differences in animal abundance, rather than absolute numbers. Choice of method is also constrained by costs and logistics. Although direct methods of population assessment are usually regarded as more reliable than indirect ones, these may not be feasible because of terrain, amount of cover or the animals'

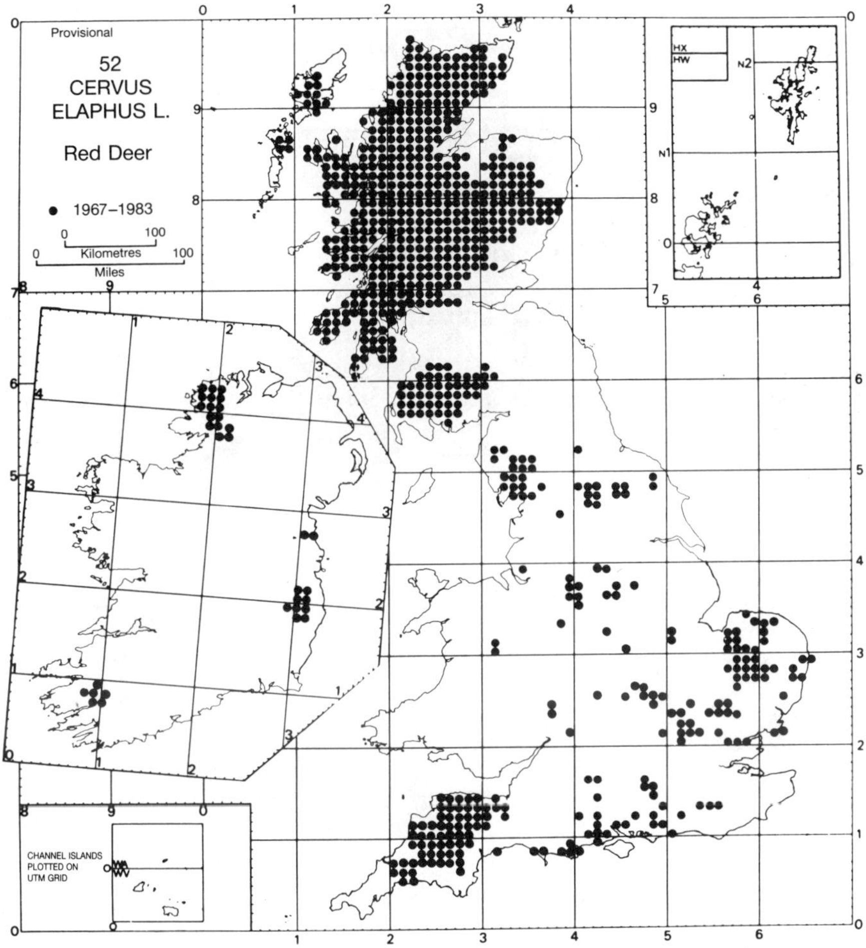

Fig. 1. The distribution of red deer in the British Isles. Source: Arnold (1984).

secretive behaviour. Different approaches may be needed also in highly dispersed or aggregated populations of the same species.

A further consideration is animal classification. It is easier to estimate total numbers than to classify the populations accurately into sexes or age classes. Lowe (1969), for example, showed that, whilst counts of red deer in open country (see below) were relatively reliable, errors in classification could be large. Experienced forest rangers may also seriously misclassify roe deer (Rowe 1982). Of seven kids (animals less than one year old) and seven female yearlings presented for age-testing, 43% and 100% respect-

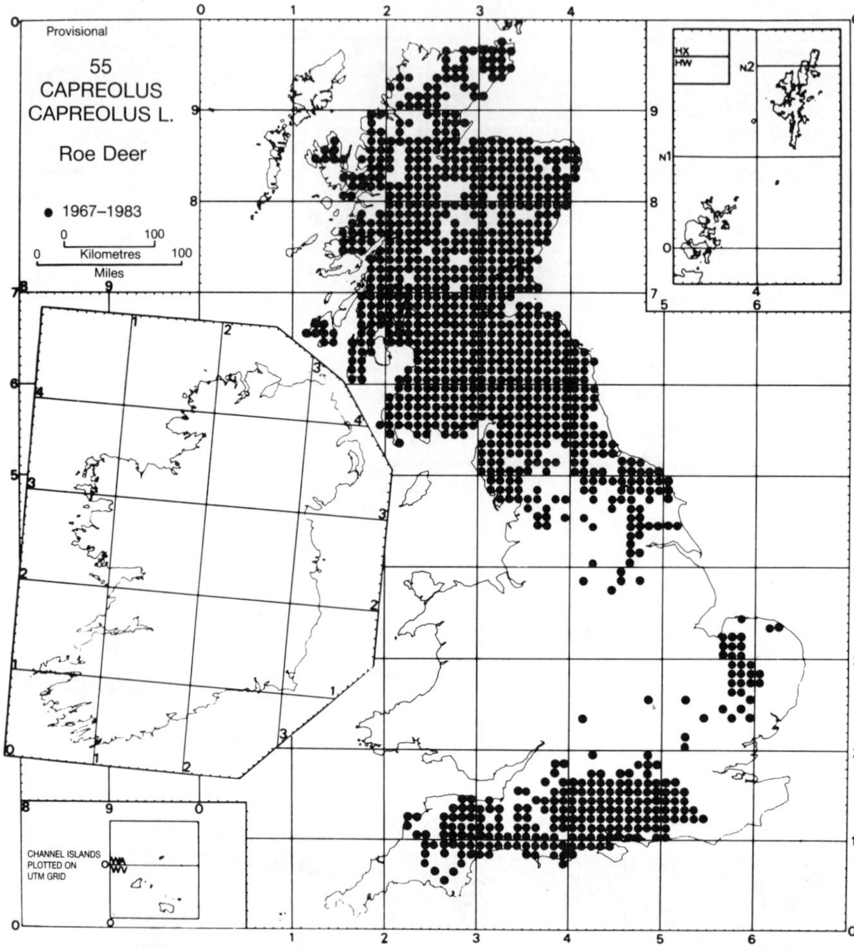

Fig. 2. The distribution of roe deer in the British Isles. Source: Arnold (1984).

ively were misclassified to year-classes. Of the adults, 91% were incorrectly aged ($n = 11$). Ageing deer is frequently influenced by the body size of the animal rather than any behavioural or other attributes (Rowe 1982; Ratcliffe 1984a).

Direct methods

Several techniques for population assessment are used which rely on animal sightings. The direct counting of red deer on open range is widely used, and the first serious attempt was made by Evans (1890) on the Island of Jura,

Table 1. Principal methods used for estimating deer abundance.

- Direct
  - Counting
    - Total
    - Sample
      - Areas—transects, plots or standard routes
      - Groups, e.g. at feeding sites
    - (Direct counting or use of photographs of areas or groups)
      - Ground surveys
      - Aerial surveys
        - Visual
        - Photographic
  - Marking
    - Capture–recapture
    - Saturation
      - Inventory
      - Cumulative catch
    - (Marking calves or adults by catching, trapping or immobilization; by self-marking; or by individual recognition)
- Using mortality data
  - Total population kill—as an experimental check procedure
  - Population reconstruction—from total deaths by sex and age over one max. life span period
  - Sex-ratio change in sample counts on killing known number of one sex
- Indirect assessments
  - Faeces (numbers, weight, volume/unit area/unit time)
    - Fresh pellet groups
    - Clearance counts
    - Standing crop
    - (Used mainly to give relative indices of animal abundance, but correction factors often applied to give estimates of population size or density)
  - Tracks or other signs

After Mitchell, Staines *et al.* (1977).

west Scotland. Over a period of 11 years, he made total counts of the adult males in the autumn and the females and immature animals in the spring. He used a 'balance-sheet' approach and by allowing for all known deaths (natural and culled) he was able to predict the adult population for the following year. His estimates and counts agreed reasonably well, but such an approach is a test of counting consistency rather than of accuracy, with possible compensatory errors operating in the same direction (Mitchell, Staines & Welch 1977).

The technique was developed further by Darling (1955) and Lowe (1961, 1969). Lowe studied the population dynamics of red deer on the Island of Rhum, west Scotland, from 1957 to 1966. He used a team of three to seven men and counted and classified the deer on the island over four to five consecutive days in the late winter and spring each year. He also used the balance-sheet approach to estimate counting consistency, and with intensive searching for natural deaths, together with the known culls, he was able to reconstruct the early populations using age-specific mortality data (Lowe 1969). The reconstructions agreed with the original counts within 2% over the first four years. However, he found that on average 11% of males and 19% of calves (young of the year) were misclassified as females during the counts. The misclassification of calves was variable and reached 30% in one year when body size was above average following a good plant growing season.

A similar census technique is used in Scotland by the Red Deer Commission (RDC) (Stewart 1976). The RDC use larger counting teams who are equipped with two-way radios to avoid double counting. Because of these factors, the accuracy of the counts is assumed to be as good, if not better, than Lowe's (1969) earlier counts, although specific tests of accuracy or consistency have not been made. The teams count deer within large tracts of open land which vary from 16–60 Kha on the islands and from 20–600 Kha on the mainland (Stewart 1985). These areas are thought to have relatively self-contained populations of deer, and most counts are made in late winter or early spring when the animals are less widely dispersed, are confined to lower ground, and are not inclined to move far. The deer are usually counted from higher ground and without disturbance unless this is necessary to flush them from cover. The results are regarded as minimum rather than absolute counts and are given later (p. 146). One drawback of the total census is that one is unable to attach confidence limits to the count even if the method is adequately tested against populations of known size. Such statistical considerations do not occur in sample counting as, for example, is widely used for large mammals inhabiting the open plains of Africa (e.g. see Norton-Griffiths 1978). Nevertheless, we suggest that the total-count approach as currently practised is probably the best one for red deer on open ground in Britain. Local populations in winter are highly aggregated both into groups and into relatively discrete areas, and a wide range of densities occur (p. 146). Sample counts would be highly variable and necessitate greater sampling intensity.

Aerial counts of red deer on open ground are seldom used in Britain. The nature of the terrain and the unpredictable weather in the Scottish uplands make low flying hazardous in hilly country (Lowe 1961). Herd composition is also more difficult to assess from aerial counts or photographs than with counts from the ground. Helicopters have been used successfully by the RDC over extensive flat areas with a highly dispersed and low density deer stock, as in Caithness and Sutherland. Helicopters alarm red deer and cause them to bunch and run off wildly; this makes both counting and classification difficult. Modern techniques of aerial surveillance such as thermal imagery have not so far provided reliable results for counting animals in woodland cover, and they are expensive to apply.

Deer 'drives' are frequently used in woodlands where direct counting is not possible (e.g. Holloway 1967; Pucek, Bobek, Labudzki, Milkowski, Morow & Tomek 1975; Dzieciolowski 1976a; Loudon 1980; Mitchell & McCowan 1980). An area is systematically traversed by a line of 'beaters' who should be close enough to see and count any deer breaking back through the line. Other deer leaving the forest cover are seen by observers strategically placed around the area being surveyed.

Neither red nor roe deer are easily flushed from cover. Our own experience, through observation and by radio-tracking deer during drives, has shown that many individuals of both species either remain stationary and hidden, or quietly work back through the line of beaters, remaining unseen by them. We have found red deer particularly difficult to flush from cover or to see during drives by people or with trained dogs. In addition, the young of both sexes are usually more likely to be seen than are adults (Strandgaard 1972; own data) although Mitchell & McCowan (1980) observed the opposite in red deer. It is also difficult to co-ordinate large numbers of people in dense thickets (see also p. 143).

Capture–mark–recapture (CMR) methods are frequently used for a variety of mammals, although with large species the recapture phase is usually replaced by observation. However, they are costly in terms of time and labour. They are more relevant to intensive research than to management. The basic assumptions are reviewed by Seber (1973) and Montgomery (this volume). However, the underlying assumptions rarely hold true for deer populations because of the non-random mixing of individuals within populations (Mitchell, Staines *et al.* 1977; Lowe 1979). This is particularly the case with roe deer. In a population of roe deer at Kalø, Denmark, it was suggested that between three-quarters and two-thirds of the stock needed to be marked before 'reliable estimates' of the known population could be achieved (Andersen 1962; Strandgaard 1972). Nevertheless, Strandgaard points out that 'Even samples comprising one-fifth to one-quarter of the estimated population size are preferred in relation to estimates made by guesswork.'

A new method for assessing roe deer numbers has been suggested which depends on a knowledge of the mean territory size of males (bucks) (Ratcliffe & Rowe 1985). Bucks are territorial from April to August, and the ratio of territorial males to adult females of approximately 1:2 is relatively constant throughout British and many continental populations. Summer density is estimated from the mean territory sizes related to fecundity, survival and male:female ratios in particular areas. Fecundity can be assessed from observation or from the examination of reproductive tracts of females shot in the annual culls coupled with estimates of survival (Ratcliffe & Rowe 1985). A more variable parameter is the number of non-territorial males, but this can be allowed for in the model.

In parts of continental Europe, deer are often counted at artificial feeding sites, especially during periods of heavy snowfall. The assumption is that all animals must emerge from woodland at some time to feed. However, both red and roe deer may remain in thick cover for several days, feeding on small open areas within or adjacent to the forest blocks (Staines, Welch, Catt, Scott & Hinge 1985; Hinge 1986; Catt & Staines 1987). The climate in Britain, relatively mild and unpredictable, particularly with regard to snow-fall, makes this a less useful approach except perhaps in small woodlands.

Vantage-point counts (Dasmann & Taber 1955) have proved successful for counting red (Ratcliffe 1984b) and also roe deer (P. R. Ratcliffe & B. A. Mayle unpublished) in conifer plantations in Scotland. The principle is to find an elevated position and to observe an area of forest and count deer seen feeding and moving within it. During the day, red deer are most abundant in the late pre-thicket and thicket stages, especially where these have open areas within the forest blocks (Ratcliffe 1984b; Staines & Welch 1984; Catt & Staines 1987). Feeding during the daytime mostly takes place within these open areas, but nevertheless the average time a red deer is seen under these circumstances is only 2–3 min. Studies in Scottish spruce forests have shown that both red and roe deer rarely rest for more than three hours at a time but feed periodically throughout the 24 hours (Hinge 1986; Catt & Staines 1987), although this may not always be the case (P. Forde & S. Harris, pers. comm.). Ratcliffe (1984b) used vantage-points and observed red deer in measured areas (range 80–140 ha) for a minimum of three hours in early morning or late evening. He systematically counted, classified and mapped all deer seen feeding and moving and noted the direction of their movements to avoid double counting. Two or three repeat counts were made over the next one to two days on the same area. The maximum count was the one used, and related to the known area surveyed so that deer density could be estimated. Repeat counts were made in other blocks of similar structure in the same forest so that mean densities for the forest could be derived. One test of the technique using data from a known cull gave good agreement. Initial counts made in a fenced area suggested a population of 61 deer, and subsequent counts, following a cull of 26 deer, showed 36 deer. The use of vantage-point counts is limited to hilly country and is not practicable on flat or rolling terrain.

Sample transects from roadsides are a popular technique for index counts of cervids in concealing habitats in North America (Overton 1971). They have not been widely used in Great Britain except for roe deer by Loudon (1980). One attraction of the method is that large areas can be covered, assuming that there is a good road system. However, there are many factors against the technique including differences in the area visible in various types of habitat and differences in the activities of animals according to season, weather conditions and time of day. Roe use roadsides relatively more than do red deer (Staines, Catt, Welch & Scott 1982). This multitude of factors affecting the index increases the required sampling intensity and therefore decreases its value. Nevertheless the technique may have promise as an indicator of population trends within forests with good road systems, provided that the sampling is adequate and statistically based and that the habitats are not rapidly changing in their cover characteristics.

Mortality data

Population reconstruction using known-age mortality data has been described already, and is a useful technique for retrospectively checking estimates made by other methods. Another method relying on known-age mortality data was proposed by Holgate (1973) for red deer. In a stable population, recruitment will equal deaths. Therefore, assuming that all dead animals can be found and correctly aged, the sum of ages in years of all the animals dying will equal the number of animals surviving. Allowances can be made in the model for populations with known, or estimated, trends in population size. With modifications it could be adapted for species with different population structures and dynamics. This seems a useful technique for testing other methods and particularly for monitoring populations of red deer in those woodlands where most mortality is thought to be from the cull (Ratcliffe in prep.). Unfortunately, this approach has not been developed further since Holgate's initial proposal.

Kelker (1940, 1944) used changes in the sex ratios observed before and after a known cull of males to estimate population size. This technique is not applicable to many red or roe deer populations. For example, the areas used by male red deer during summer and autumn (when most culling occurs) can be far removed from those used in winter. Counts of males at different seasons may therefore relate to different female populations. In addition, the behaviour of one sex may vary seasonally making them more or less visible. In roe deer, young males frequently move between local populations in summer.

A further method is the unit herd concept of Leopold (1933). Leopold studied white-tailed deer (*Odocoileus virginianus*) in California, and suggested that a herd of 24 deer was needed for every buck shot. This was subsequently used by Whitehead (1959) for red deer in Scotland, although he used a unit herd of 25 for every stag shot. Despite reasonable agreement with the results of the Nature Conservancy's estimates (Lowe 1961), the technique has no sound theoretical or statistical base and was adequately dismissed by Lowe (1961) and Mitchell (1966).

Indirect assessments

Indirect population assessments rely on the tracks, signs or products (e.g. faeces) of animals and are particularly useful in dense woodlands. Estimating numbers from tracks, especially in snow, is widely used in eastern Europe. The logic is that density indices are based on the difference between the numbers of deer tracks counted entering an area on one day and those leaving the next (Dzieciolowski 1976a). Dzieciolowski found significant correlations between the relative densities derived from such track counts and drive counts in a number of Polish ungulates, including roe deer, although the relationship for red deer was not statistically significant. The technique was criticized by Pucek *et al.* (1975) (also in Poland) as being unreliable. The method is more an index of animal activity than of relative density, and it is unlikely to be a valuable tool for estimating deer numbers in Great Britain because of the lack of reliable snow cover in most years.

Estimate of relative densities from signs of damaged trees (e.g. fraying or thrashing by males with their antlers, or browsing and bark-stripping damage) has sometimes been considered by deer managers but remains untested. It assumes a simple relationship between damaged trees and deer numbers, which appears not to be the case (Staines & Welch 1984; Welch, Staines, Scott & Catt in press).

A possible method, previously used on badgers (*Meles meles*) in this country (Kruuk, Gorman & Parish 1980), involves radio-active markers. A number of animals are caught and injected with a harmless radio-isotope, such as Zn^{65}. These markers are then detected in the pellet groups, and the ratio of marked to unmarked groups can be used to indicate population size. Kruuk and his colleagues used this technique to estimate the number of badgers within individual territories; it is difficult to see how the technique could be applied to free-ranging red and roe deer, when marked animals would be unlikely to mix randomly within a study population. It may, however, be of value in the study of restricted deer populations. For practical reasons, this technique is more attractive to research than to management.

The most widespread indirect techniques for assessing the abundance of ungulates involve the use of faeces. These techniques are attractive for studying deer in concealing habitats and may be quantified in several ways, for instance by the number of pellet groups, by weight or by volume. There are two basic approaches: the first relies on the standing crop of faeces (faecal standing crop or FSC) and the other on the rate of accumulation of faeces over a known time (faecal accumulation rate or FAR). The validity of the techniques, the statistical assumptions and the errors involved are reviewed by van Etten & Bennett (1965), Neff (1968), Mitchell, Staines *et al.* (1977), Mitchell & McCowan (1980) and Putman (1984).

The FSC method is frequently used to detect relative differences or trends in populations. It is based on the assumption that there must be some stable relationship between the amounts of dung present and the numbers of animals. The dung is measured on temporary or permanent plots or by using various nearest-neighbour techniques (e.g. Batcheler 1973, 1975). FSC has considerable merit in being easy to apply, requiring only one visit (whereas the FAR needs at least two) and being relatively quick to use. It is therefore particularly attractive for management. The disadvantages are that there are few tests of its validity under different circumstances, and usually the method is only related empirically to the results from other techniques (but see below, and Ratcliffe in prep.). In addition, rates of dung decomposition vary according to the richness of the site and its vegetation (Mitchell & McCowan 1984; B. W. Staines & D. Welch unpubl.), and this needs to be allowed for in the sampling scheme adopted. Nevertheless, because of its potential use in this country, we feel that more research should be made into validating the FSC technique. It is more likely to be valuable when used in late winter because of the greater variability in dunging and decomposition rates in summer.

The FAR method requires knowledge of:

1. Dunging rates related to season, habitat and forage quality.
2. Rates of disappearance per season and habitat type.
3. Accuracy of identification when more than one species is involved.
4. A measure of the findability of pellet groups by different observers in different habitats and seasons.

In addition, studies of habitat-use need information on when deer defaecate and during which activities; i.e., what are the relative differences in dunging rates during feeding and resting periods?

Both the FAR and FSC methods have been recently investigated for red and roe deer in Britain. Mitchell & McCowan (1980, 1984) measured the dunging rate for red deer in enclosures of varying habitat quality (Table 2).

Table 2. Recent estimates of defaecation rates (pellet groups/day) of red deer in different habitats in Scotland.

Habitat					
Parkland: good quality		Scots pine plantation: poor quality		Heather moorland: intermediate quality	
April–June	28.5				
		May–July	26.1		—
July–Sept.	26.6				
		July–Aug.	23.9	Aug.–Sept.	25.0
		Sept.–Oct.	24.7		
Oct.–Nov.	23.5				
		Nov.–Jan.	18.9		—
Dec.–Mar.	14.9*				—
		Jan.–Feb.	24.7		
		Mar.–April	20.8		

Source: Mitchell & McCowan (1984).
* Thought to be an underestimate because of heavy snow followed by flooding during the sampling period.

Table 3. Days taken for different proportions of red deer faecal pellet groups to disappear in contrasting habitats.

		25%	50%	100%
Lush pasture	Summer	15– 40	30– 65	95–155
	Winter	45–120	60–165	90–220
Woodland heath	Summer	70–115	95–180	190–340
	Winter	150–230	180–270	370–450

Data from Mitchell & McCowan (1984).

The results were variable, but showed seasonal trends, with highest rates being found in summer and the lowest in winter. These rates were greater than those previously described for red deer (e.g. 10 pellet groups/day: Riney 1957) or for the related North American elk, *Cervus elaphus canadensis* (e.g. 11 pellet groups/day: Julander, Ferguson & Dealy 1963; 12 pellet groups/day: Neff, Wallmo & Morrison 1965). Rates were only slightly higher in the lusher habitats. Rates of decomposition and other losses were also greater in summer and on richer soil types (Table 3). Other studies involving free-ranging deer in conifer plantations, however, found less dung during the summer than in the winter (e.g. Staines & Welch 1984; Ratcliffe in prep.; Mitchell & McCowan in press). These lower amounts are unlikely to be attributable solely to seasonal changes in decomposition and findability, and possibly differences in behaviour between free-ranging and enclosed deer are partly responsible. Wild animals frequently behave differently from those in captivity. Free-ranging deer may have fewer feeding bouts than those living in small enclosures, and possibly defaecate less frequently. The differences between these two approaches need to be resolved in order to assess the validity of the FAR technique for deer under various conditions in Britain.

Average defaecation rates of roe deer throughout the year varied from 23 pellet groups per day under favourable conditions to 17 per day in poorer habitats (Mitchell, Rowe, Ratcliffe & Hinge 1985). These are similar to the 16 per day found by Padaiga & Marma (1979) for roe deer in winter in Lithuania, but greater than the seven per day estimated by Dzieciolowski (1976b) in Poland. It is considered that the large plots (2500 m^2) used by Dzieciolowski were harder to search adequately than the smaller ones (55–100m^2) used by Mitchell, Rowe *et al.* (1985).

The FAR method has considerable merit for estimating habitat use and for assessing animal densities, especially in winter when losses are minimal and less variable. Mitchell & McCowan (in press) estimated that FAR could detect winter population densities of red deer to within 30–40%. With stratification of the habitat based on a sounder knowledge of habitat use by the deer (Staines & Welch 1984; Hinge 1986; Catt & Staines 1987) we believe that the precision can be improved.

Comparison of techniques

Ideally, techniques of assessment should be tested against populations of known size. This is seldom feasible for free-ranging populations. One way is to introduce a known number of animals into a deer-free enclosure and test various methods against this. This assumes that any animals that die or escape are either found or allowed for. Mitchell & McCowan (1980) introduced six adult red deer into an 18-ha area of fenced woodland of young Scots pine (*Pinus silvestris*). They tested drive-counting and the FAR method. As expected, drive-counts with a larger counting team spaced *c.*35 m apart were more accurate than those using fewer men spaced at 75 m. However, even in such a small area with relatively easy counting conditions compared to most planted forests, some deer were 'missed'. They attempted to extend this work to larger areas, but the vagaries of the Scottish weather meant that several planned counts had to be aborted. Such problems will continually arise when large numbers of people have to be organized well in advance of any count. We feel, therefore, that the use of drive counts is limited to small areas and is not relevant to large coniferous woodlands.

An alternative method to introducing animals is the killing of all deer following enumeration. The classic example of this approach was by Andersen (1953, 1962). In a fenced 340-ha wood at Kalø in Denmark, estimates of the roe deer population were made by several methods by experienced, professional people. These estimates suggested a population of approximately 70 deer. Over the following three months, 213 deer were shot and a possible four or five escaped. Andersen (1962) gives other examples of gross underestimates of the real populations of roe and red deer, and Pielowski & Bresinski (1982), in a similar shoot-out experiment in Poland, showed that estimates made by 'sight and intuition' also underestimated a true roe deer population by some two and half times.

Complete shoot-outs are seldom possible, an unknown number frequently escaping detection and the costs of killing the last few animals being high. Mitchell & McCowan (in press) attempted to shoot all red deer in a 520-ha fenced conifer plantation at Rannoch Forest, Perthshire. They measured the rate of faecal accumulation on fixed plots before and after the shoot-out and used the proportional decrease in dung found following the cull as an index of the decrease in the true population. So, with 30 animals shot and a decrease in dung accumulation averaging 67%, they estimated the original population to be around 45 animals. It took approximately six months to kill the 30 deer, which emphasizes the difficulties of shooting deer in dense plantations. Their vantage-point counts covered roughly 40% of the area, and the results were highly variable. The authors did not use the systematic approach of Ratcliffe (1984b), and the largest number of deer seen during any counting session was 23, around half the predicted population; most estimates were well below this figure.

A technique which mimics the shoot-out is population reconstruction from mortality data, described earlier. The method is currently being applied to red deer populations living in those forestry plantations where it is felt most mortality is from culling (Ratcliffe, this volume). Ratcliffe (in prep.) compared vantage-point counts with both the FAR and FSC methods as well as testing against a known cull, described above. He found a better fit with his adapted FSC method than with FAR, and felt that the former technique, though inherently less reliable (Mitchell, McCowan & Campbell 1983), achieved greater precision because sampling efficiency was increased and larger numbers of pellet groups were found (see also Rogers, Julander & Robinette 1958). Such comparisons still need to be made against known populations, and population reconstruction in the future will be of great value in assessing our current studies of deer in open and woodland habitats.

Status and trends of deer populations

Red deer

The range of red deer in Great Britain has been greatly reduced since prehistoric times. Deforestation of the native woodlands and persecution led to the probable extinction of all native stocks in the lowlands by the 17th century (Ritchie 1920). Only on the remote open hill lands of the Scottish Highlands and Islands were red deer able to survive. Numbers were probably

Table 4. Estimates of the Scottish red deer population.

Year	Numbers (thousands)	Source
1900 +	150	Cameron (1923)
1930	250	Parnell (1932)
< 1945	200	Anon. (1951)
1950	100	Anon. (1951)
1959	150–160	Lowe (1961)
1965	180	Stewart (1979)
1970	185	Stewart (1985)
1975	270	Stewart (1985)
1979	255	Red Deer Commission (1980)
1986	290	Stewart (in press)

Population figures after 1970 include an estimate for the numbers of red deer thought to be living in forestry plantations; before this date, estimates related to open hill ground only.

lowest around the end of the 18th century (Anon. 1922). Various intentional releases of red deer from continental and Scottish stocks occurred and there were the inevitable escapes from deer-parks. Charles II and James I (of England), for example, both introduced red deer from France into the New Forest during the 17th century (Whitehead 1964). Deer from native stocks are only confirmed in parts of Scotland, north-west England and south-west Eire (Lowe & Gardiner 1974).

During the 19th century the fortunes of red deer changed, and a rising interest in deer stalking led to deliberate attempts to increase the deer stocks in Scotland. The usual management strategy was to protect females, the reasoning being that a large stock of hinds was necessary to attract males during the shooting season. Numbers consequently rose, and the rise was accompanied by increasing pressure on agriculture, for red deer were thought to compete for grazing with domestic livestock. In recent years there has been more conflict between all deer species and commercial forestry. From a peak around 1914 (Anon. 1922) numbers of red deer apparently decreased up to the 1950s; since that time populations appear to have risen (Table 4). The current range of red deer in Scotland (Fig. 1) has increased to approximately that estimated for the Middle Ages by Ritchie (1920).

Between 1872 and 1954, no fewer than eight government enquiries were held into the so-called 'red deer problem' (Mitchell, Staines *et al.* 1977), and it is probably because of these enquiries that there has been a continued general interest in the total numbers of red deer in Scotland. For the management of a prolific but potentially damaging species, knowing total numbers has limited value. The density in different areas is much more important, for each will have its own particular management problems and environmental factors that affect deer density. Nonetheless, many attempts have been made to estimate the total stock, although until recently most relied on intelligent guesses (Table 4). The most consistent evidence comes

Table 5. Increase in red deer numbers in different areas of Scotland.

Area	Red deer numbers (date of count)			% Increase
North Ross	7 550 (1968)	10 850 (1984)	—	44
Monadhliaths	12 000 (1968)	20 350 (1977)	—	70
West Grampians	23 850 (1967)	33 100 (1983)	—	39
East Grampians	10 050 (1966)	19 300 (1975)	25 520 (1986)	154
Corrour/Ben Alder	7 350 (1968)	9 750 (1978)	—	33

Data from Stewart (1985).

Table 6. Range of red deer densities found on open hill land in Scotland.

Count area	Area (ha)	Density (deer/ km^2)
Rovie/Skibo	20 500	0.5
Skye	60 750	1.6
North Sutherland	196 350	4.3
North Ross	160 750	6.8
South Ross	177 750	13.1
West Grampians	237 650	13.9
East Grampians	118 600	14.1
Glenartney	14 150	26.9
Scalpay	1 600	31.3

Data from Stewart (1985).

from the systematic counts made by the Red Deer Commission since 1959 (Stewart 1985). These indicate a rise in numbers of red deer on open hill ground since that time (Tables 4, 5). Increases are apparently greater in the east and central Highlands than on the west (Stewart 1985), and densities vary greatly between areas (Table 6). Current analyses of the Red Deer Commission's data using a more sophisticated modelling approach suggest that the total hill-land population is now around 297K ± 40K (1 *S.E.*) animals occupying some 3 Mha of land (T. H. Clutton-Brock & S. D. Albon unpublished). In addition, estimates of red deer in Scottish commercial woodlands vary from 27K, based on culls (Stewart in press), to 50K, based on estimated densities in various habitat types (Ratcliffe, this volume).

The increase in numbers of hill-land populations in Scotland since the 1960s is thought to be due to two main factors. First, there has been an unintentional under-culling of females of around only 16% of the population (T. H. Clutton-Brock & S. D. Albon unpublished) whilst that of males is around 19%. Lowe (1969) gave evidence that a cull of 17% of the adults was possible without reducing the total population size of red deer on hill ground. Secondly, a reduction, or at least a re-distribution, of the stocks of hill sheep has led to decreased competition with subsequent beneficial effects on the red deer populations. There is some data (Osborne 1984) and much anecdotal evidence to suggest that the numbers and distribution of red deer decline in the presence of moderate to large sheep stocks, but more objective study is necessary to establish this assumption and the factors involved.

Until recently, densities of red deer were thought to be low in woodland habitats in Britain (e.g. 2–4 deer per km^2: Southern 1964), but evidence now shows that they are comparable to those found on open moorland, frequently being 5–15 per km^2, with up to 40 deer per km^2 in particularly favourable areas (Ratcliffe 1984a). Culling levels are generally low in forestry plantations when compared with the potential of the population (Ratcliffe, this volume). Rates of reproduction and survival are frequently higher in woodland red deer than in those living on open hill ground (Ratcliffe 1984b), and there seems little doubt that red deer numbers will continue to increase in forests and that they will also extend their range as more afforestation takes place.

Roe deer

Roe deer were probably extinct throughout the country by the beginning of the 18th century except in remnant woodlands in the central and northwestern Highlands of Scotland (Ritchie 1920; Whitehead 1964; Lowe 1979). The increase in woodland plantings during the 18th century led to some natural colonization within Scotland, and roe deer colonized parts of the Scottish Lowlands by 1840 (Ritchie 1920). Deliberate introductions were made in some parts and Lowe (1979) states that 'Roe deer of unknown origin were introduced into Milton Abbas, Dorset, in 1800 and the descendants of these animals, together with stock from Petworth Park in Sussex, have now recolonized much of the south and south-west of England from Cornwall to Surrey'. Roe deer from German stock were introduced into East Anglia in 1884 (Chapman, Claydon, Claydon & Harris 1985) and those found in the Lake District are thought to be of Austrian origin (Lowe 1979). The increase in commercial forestry since 1920 has led to a dramatic extension of their range and they are now common throughout southern and northern England and in Scotland (Fig. 2). Extension of range is continuing, and they now occupy many agricultural and even suburban habitats.

Densities vary according to habitat type. In Sitka spruce forests in the Scottish Borders, densities varied from around eight deer per km^2 to 25 deer per km^2 according to the age of the plantation (Loudon 1980, 1982). Numbers were highest in young forests around 10 years old, reached a nadir at 30 years, and increased again to 16 deer per km^2 in older forests (50 years). In part of Chedington Wood, Dorset, a forest predominantly composed of 10- to 26-year-old Norway spruce (*Picea abies*), estimated spring densities ranged from 33–101 per km^2 during the period 1965–1981 (A. L. Johnson unpublished). In a second rotation spruce forest of mixed-aged stands, the winter density of roe deer estimated from faecal accumulation rates suggests around 10 deer per km^2 (Staines & Welch 1984, and unpublished).

Numbers will probably continue to increase in most areas, for the annual cull of adults, coupled with natural mortality, seldom equals the known reproductive potential of the species (Loudon 1980). A low rate of mortality, coupled with the natural dispersal of young of both sexes (Bramley 1972), should ensure the extension of their current range.

Conclusions

Recent research has suggested several techniques which show promise for estimating the abundance of deer in Britain.

The current method for red deer on open ground using total counts seems most appropriate, but further tests of the method against populations of known size or using cohort analyses are necessary.

For counting deer in concealing habitats, we conclude:

1. Vantage-point counts are valuable for estimating animal densities, particularly those of red deer where terrain and forest structure allow.
2. Methods using indices of faecal abundance give useful guides to habitat usage and to densities, but both methods need more refinement, especially in statistical design, to reduce variability and to increase accuracy.
3. The relationship between roe deer density and territory size needs further study to confirm that the relationship found so far exists over a number of habitat types.
4. The method proposed by Holgate (1973) (p. 139) using known-age mortality data should also be considered and refined.
5. The method of changes in population estimates following a known cull, a known increase, or changes in the numbers of faecal depositions should be pursued as a check on other techniques.
6. Tests of accuracy and bias are needed for all methods especially using populations of known size. Population reconstruction using known-age mortality data is a useful alternative in this respect.

The range and numbers of red and roe deer in Britain are increasing. Low rates of culling and further afforestation should result in these increases continuing until they are limited by density-dependent effects.

Acknowledgements

We thank H. Kruuk and D. Burdekin for their constructive comments on the paper and S. D. Albon, T. Clutton-Brock, A. L. Johnson, B. Mitchell and D. Welch for use of their unpublished data.

References

Andersen, J. (1953). Analysis of a Danish roe-deer population (*Capreolus capreolus* (L.)) based upon the extermination of the total stock. *Dan. Rev. Game Biol.* **2**: 127–55.

Andersen, J. (1962). Roe deer census and population analysis by means of a modified marking-release technique. In *The exploitation of natural animal populations*: 72–82. (Eds Le Cren, E.D. & Holdgate, M.W.). Blackwell Scientific Publications, Oxford.

Anon. (1922). *Report of the Departmental Committee on Deer Forests, 1919.* H.M.S.O., Edinburgh.

Anon. (1951). *Report of the Committee on Cruelty to Wild Animals, 1949.* Cmd. 8266. H.M.S.O., London.

Arnold, H.R. (Ed.) (1984). *Distribution maps of the mammals of the British Isles.* Biological Records Centre, Inst. Terr. Ecol., Abbots Ripton.

Batcheler, C.L. (1973). Estimating density and dispersion from truncated or unrestricted joint point-distance nearest-neighbour distances. *Proc. N.Z. ecol. Soc.* **20**: 131–47.

Batcheler, C.L. (1975). Development of a distance method for deer census from pellet groups. *J. Wildl. Mgmt* **39**: 641–52.

Bramley, P.S. (1972). Management of roe deer (*Capreolus capreolus*) in small deciduous woods in England. *Forestry* **45**: 211–21.

Cameron, A.G. (1923). *The wild red deer of Scotland.* Blackwood and Sons, Edinburgh & London.

Catt, D.C. & Staines, B.W. (1987). Home range and habitat selection by Red deer (*Cervus elaphus*) in a Sitka spruce plantation as determined by radio-tracking. *J. Zool. Lond.* **211**: 681–93.

Caughley, G. (1977). *Analysis of vertebrate populations.* John Wiley & Sons, Chichester.

Chapman, N.G., Claydon, K., Claydon, M. & Harris, S. (1985). Distribution and habitat selection by muntjac and other species of deer in a coniferous forest. *Acta theriol.* **30**: 287–303.

Darling, F. F. (1955). *West Highland survey.* Oxford Univ. Press, London.

Dasmann, R.F. & Taber, R.D. (1955). A comparison of four deer census methods. *Calif. Fish Game* **41**: 225–8.

Dzieciolowski, R. (1976a). Estimating ungulate numbers in a forest by track counts. *Acta theriol.* **21**: 217–22.

Dzieciolowski, R. (1976b). Roe deer census by pellet-group counts, *Acta theriol.* **21**: 351–8.

Evans, H. (1890). *Some account of Jura red deer.* Private publication, Carter, Derby.

Hinge, M.D.C. (1986). *Ecology of red and roe deer in a mixed-aged conifer plantation*. Ph.D. Thesis: Aberdeen University.

Holgate, P. (1973). An estimator for the size of an animal population. *Biometrika* **60**: 135–40.

Holloway, C.W. (1967). *The effect of red deer and other animals on naturally regenerated Scots pine*. Ph.D. Thesis: Aberdeen University.

Julander, O., Ferguson, R.B. & Dealy, J.E. (1963). Measure of animal range use by signs. *U.S. Dep. Agric. For. Serv. Publ.* No. 940: 102–8.

Kelker, G.H. (1940). Estimating deer populations by a differential hunting loss in the sexes. *Proc. Utah Acad. Sci.* **17**: 65–9.

Kelker, G.H. (1944). Sex-ratio equations and formulas for determining wildlife populations. *Proc. Utah Acad. Sci.* **19–20**: 189–98.

Kruuk, H., Gorman, M. & Parish, T. (1980). The use of ^{65}-Zn for estimating populations of carnivores. *Oikos* **34**: 206–8.

Leopold, A. (1933). *Game management*. Charles Schribner's Sons, New York.

Loudon, A.S.I. (1980). *The biology and management of roe deer in commercial forests*. Forestry Commission, Edinburgh University, Edinburgh.

Loudon, A. (1982). Too many deer for the trees? *New Scient.* **93**: 708–11.

Lowe, V.P.W. (1961). A discussion on the history, present status and future conservation of red deer (*Cervus elaphus* L.) in Scotland. *Terre Vie* **108**: 9–40.

Lowe, V.P.W. (1969). Population dynamics of the red deer (*Cervus elaphus* L.) on Rhum. *J. Anim. Ecol.* **38**: 425–57.

Lowe, V.P.W. (1979). Wild and feral deer in Great Britain. *NCC/NERC Contract Report HF3/05/43*. Inst. Terr. Ecol., Grange-over-Sands.

Lowe, V.P.W. & Gardiner, A.S. (1974). A re-examination of the subspecies of Red deer (*Cervus elaphus*) with particular reference to the stocks in Britain. *J. Zool., Lond.* **174**: 185–201.

Mitchell, B. (1966). The 'unit-herd' concept and its applicability to the management of Scottish red deer. *Rep. Red Deer Commn* **1965**: 20–2.

Mitchell, B. & McCowan, D. (1980). Estimating and comparing population densities of red deer *Cervus elaphus* L. in concealing habitats. *A. Rep. Inst. terr. Ecol.* **1979**: 7–13.

Mitchell, B. & McCowan, D. (1984). The defecation frequencies of red deer in different habitats. *A. Rep. Inst. terr. Ecol.* **1983**: 15–17.

Mitchell, B. & McCowan, D. (In press). Estimating the numbers of red deer in an enclosed area of commercial forest at Rannoch, Perthshire. *A. Rep. Inst. terr. Ecol.* **1986**.

Mitchell, B., McCowan, D. & Campbell, D. (1983). Faecal depositions as indicators of site use by red deer. *A. Rep. Inst. terr. Ecol.* **1982**: 85–7.

Mitchell, B., Rowe, J.J., Ratcliffe, P. & Hinge, M. (1985). Defecation frequency in Roe deer (*Capreolus capreolus*) in relation to the accumulation rates of faecal deposits. *J. Zool., Lond. (A)* **207**: 1–7.

Mitchell, B., Staines, B.W. & Welch, D. (1977). *Ecology of red deer: a research review relevant to their management in Scotland*. Inst. Terr. Ecol., Cambridge.

Neff, D.J. (1968). The pellet-group count technique for big game trend, census, and distribution: a review. *J. Wildl. Mgmt* **32**: 597–614.

Neff, D.J., Wallmo, O.C. & Morrison, D.C. (1965). A determination of defecation rate for elk. *J. Wildl. Mgmt* **29**: 406–7.

Norton-Griffiths, M. (1978). *Counting animals.* (2nd edn). African Wildlife Leadership Foundation, Nairobi.

Osborne, B.C. (1984). Habitat use by red deer (*Cervus elaphus* L.) and hill sheep in the West Highlands. *J. appl. Ecol.* **21**: 497–506.

Overton, W.S. (1971). Estimating the numbers of animals in wildlife populations. In *Wildlife management techniques*: 403–55. (Ed. Giles, R.H. Jr). The Wildlife Society, Washington, D.C.

Padaiga, V.I. & Marma, B.B. (1979). Census of roe deer by a pellet-group count. *Ekologiya, Sverdlovsk* No. 4: 101–3. (Translated in *Soviet J. Ecol.* **10**: 355–7).

Parnell, I.W. (1932). Some notes on the natural history of the Scottish red deer. *Proc. R. Phys. Soc. Edinb.* **22**: 75–80.

Pielowski, Z. & Bresinski, W. (1982). Population characteristics of roe deer inhabiting a small forest. *Acta theriol.* **27**: 409–25.

Pucek, Z., Bobek, B., Labudzki, L., Milkowski, L., Morow, K. & Tomek, A. (1975). Estimates of density and number of ungulates. *Pol. ecol. Stud.* **1**: 121–35.

Putman, R.J. (1984). Facts from faeces. *Mammal Rev.* **14**: 79–97.

Ratcliffe, P.R. (1984a). Population dynamics of red deer (*Cervus elaphus* L.) in Scottish commercial forests. *Proc. R. Soc. Edinb.* **82B**: 291–302.

Ratcliffe, P.R. (1984b). Population density and reproduction of red deer in Scottish commercial forests. *Acta zool. fenn.* **172**: 191–3.

Ratcliffe, P.R. (In prep.). *The management of red deer in the commercial forests of Scotland in relation to population dynamics and habitat changes.* Ph.D. Thesis: University of London.

Ratcliffe, P.R. & Rowe, J.J. (1985). A biological basis for managing red and roe deer in British commercial forests. *Trans. Congr. int. Union Game Biol.* **13**: 17–21.

Red Deer Commission (1980). *Annual report for 1979.* H.M.S.O., Edinburgh.

Riney, T. (1957). The use of faeces counts in studies of several free-ranging mammals in New Zealand. *N.Z. Jl Sci. Technol.* **38B**: 507–32.

Riney, T. (1982). *Study and management of large mammals.* John Wiley & Sons, Chichester, New York etc.

Ritchie, J. (1920). *The influence of man on animal life in Scotland: a study in faunal evolution.* University Press, Cambridge.

Rogers, G., Julander, O. & Robinette, W.L. (1958). Pellet-group counts for deer census and range use index. *J. Wildl. Mgmt* **22**: 193–99.

Rowe, J.J. (1982). Roe research in relation to management in British woodlands. In *Roe and red deer in British forestry*: 25–41. British Deer Society, Warminster.

Seber, G.A.F. (1973). *The estimation of animal abundance and related parameters.* Griffin & Co. Ltd., London.

Southern, H.N. (Ed.) (1964). *The handbook of British mammals.* Blackwell Scientific Publications, Oxford.

Staines, B.W., Catt, D.C., Welch, D. & Scott, D. (1982). Red deer and forestry: current work by the Institute of Terrestrial Ecology. *Deer* **5**: 400–2.

Staines, B.W. & Welch, D. (1984). Habitat selection and impact of red (*Cervus elaphus* L.) and roe (*Capreolus capreolus* L.) deer in Sitka spruce plantation. *Proc. R. Soc. Edinb.* **82B**: 303–19.

Staines, B.W., Welch, D., Catt, D., Scott, D. & Hinge, M.D.C. (1985). Habitat use and feeding by deer in sitka spruce plantations. *A. Rep. Inst. terr. Ecol.* **1984**: 12–6.

Stewart, L.K. (1976). The Scottish red deer census. *Deer* 3: 529–32.

Stewart, L.K. (1979). The present position. The red deer population. In *The next twenty years*: 4–5. The Red Deer Commission, Inverness.

Stewart, L.K. (1985). Red deer. In *Vegetation management in northern Britain*: 45–50. (Ed. Murray, R.B.). BCPC Publications, Croydon.

Stewart, L.K. (In press). Deer in Scotland. In *The changing scene*. The Red Deer Commission, Inverness.

Strandgaard, H. (1972). The roe deer (*Capreolus capreolus*) population at Kalø and the factors regulating its size. *Dan. Rev. Game Biol.* 7: 1–205.

Van Etten, R.C. & Bennett, C.L. Jr (1965). Some sources of error in using pellet-group counts for censusing deer. *J. Wildl. Mgmt* **29**: 723–9.

Welch, D., Staines, B.W., Scott, D. & Catt, D.C. (In press). Bark-stripping damage by red deer in a Sitka spruce forest in western Scotland. I. Incidence. *Forestry* **60**: 249–62.

Whitehead, G.K. (1959). Red deer in Scotland. *The Field* **213**: 1087–8.

Whitehead, G.K.W. (1964). *The deer of Great Britain and Ireland. An account of their history, status and distribution.* Routledge & Kegan Paul, London.

Symp. zool. Soc. Lond. (1987) No. 58: 153–165

Red deer population changes and the independent assessment of population size

Philip R. RATCLIFFE

Wildlife and Conservation Research Branch
Forestry Commission
Alice Holt Lodge
Farnham
Surrey GU10 4LH

Synopsis

Estimates of fertility, survival and density were used to simulate population changes in a population of red deer in Galloway, south-west Scotland, using a Leslie-Matrix model. The growth of the population from 1965 to 1986, during a period of intensive afforestation of the range, is described. Census techniques commonly used on open range became less useful as afforestation proceeded, and were abandoned in 1977. Prior to 1980 there were thought to be less than 1000 deer present but population simulations suggested that there were over 3000. Independent assessments of population density using vantage-point counts supported the population estimates. Attempts to control the population since 1980 have apparently failed and the population continues to increase. The retrospective determination of minimum population size using cohort analysis is discussed and is considered to be an essential part of deer management. Further simulations are used to consider current management options and these indicate that a new management strategy is urgently required if managers wish to prevent further increase. Under-exploitation of forest populations of red deer appears to be common; the implications of this are discussed.

Introduction

In Scotland some 300,000 red deer (*Cervus elaphus* L.) occupy open range. Studies to determine current densities in Scottish forests suggest that there are probably more than 50,000 occupying recently established forests of exotic conifers (Staines & Ratcliffe, this volume). Research into the ecology of red deer on open range has been substantial (Mitchell, Staines & Welch

ZOOLOGICAL SYMPOSIUM No. 58
ISBN 0-19-854006-X

1977), but far less is known of their ecology in forests. A research programme was therefore initiated in 1977 to provide a biological basis for managing red deer in forests, where their impact on commercial tree crops was causing concern.

Since 1977 population dynamics and densities (Ratcliffe 1984a, b), dispersion (Catt & Staines 1987) and the impact of deer on forests (Staines & Welch 1984) have been studied. This paper uses original count data and estimates of recruitment and survival to simulate, by means of a Leslie-Matrix model (Leslie 1945, 1948), changes in a forest population of red deer in Galloway, south-west Scotland. Independent checks on population density and the retrospective calculation of minimum population size by cohort analysis were used to check the reconstructions. Current culling levels are failing to prevent population increase. Under-exploitation of forest red deer populations is considered to be common and the implications of this are discussed.

The Galloway red deer population 1965–1980

It is uncertain whether native red deer have maintained a continued presence in Galloway in historical time but a population of apparently unique quality (Jeffery 1974), which is isolated from the main range of red deer in Scotland, is present today. When afforestation in Galloway increased in the 1960s, these deer began to cause unacceptable damage to commercial tree crops, and conflicts between the objectives of control and conservation led to the formation of the Galloway red deer control scheme in 1965 (Jeffery 1974). The scheme aimed to reconcile these objectives, first by conserving the herd at about 500 individuals—a number that would ensure survival of the herd, yet would permit the development of antler trophies of high quality, and would not inflict unacceptable levels of damage on trees—and secondly by preventing the herd from extending its existing range. The present population is mainly confined to about 47,000 ha of coniferous forest in central Galloway. The history of the population is summarized by Ratcliffe (1984a).

Direct counts of the deer were made before and during 1965 when only a relatively small part of the range was covered by trees. Between 1965 and 1976 direct censuses were continued and the numbers of deer counted annually are shown in Fig. 1 (line C). These counts suggested that the population declined to less than 500 deer in 1976, by which time the numbers being culled were much larger than such a population could possibly have supported. This census was therefore discontinued.

The numbers counted in 1965, with no allowance for counting error, formed a basis for population simulations using estimates of reproductive

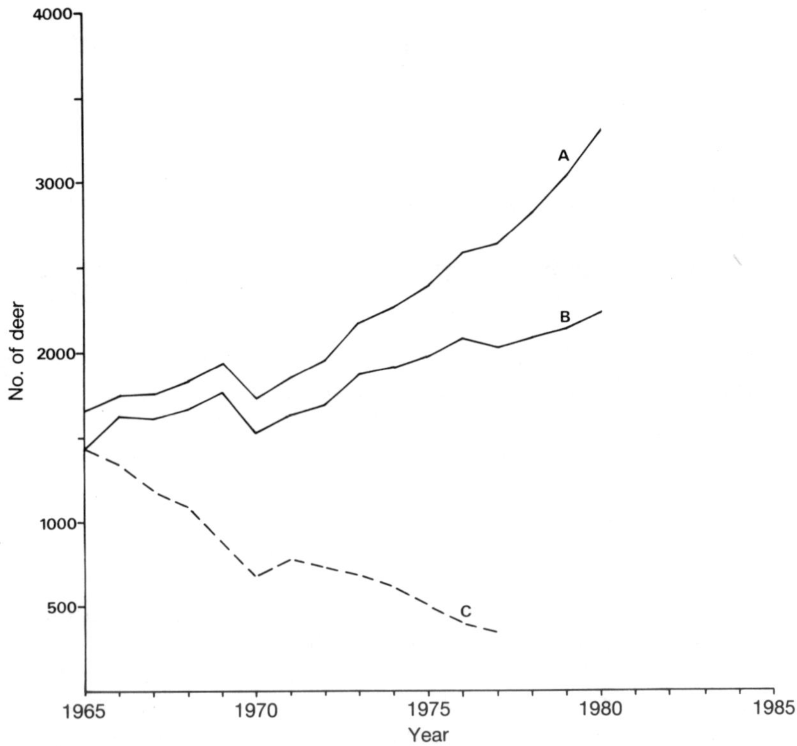

Fig. 1. Galloway red deer population projections to 1980. Conditions: birth rate, 70 calves per 100 hinds; recruitment after winter, 60 calves per 100 hinds; adult mortality, 2% per year; actual culls. A, starting values corrected for 5% misclassification of stags and calves and 10% undercount; B, actual values; C, census. After Ratcliffe (1984a).

rates and mortality and the actual numbers of juveniles and adults of each sex culled up to 1980 (Fig. 1, line B).

The Leslie-Matrix model predicts the numbers and age-structure of a population of animals after a time period, on the basis of the current age-structure and a matrix of elements that represent age-specific fecundity and mortality. In matrix notation,

$$Aa_t = a_{t+1}$$

$$a_t = (a_{t,0}, a_{t,1}, a_{t,2}, \ldots a_{t,n})$$

is a column vector representing the population structure at time t.

a_{t+1} is a column vector similar to a_t but representing the age-structure at time $t+1$.

$$A = \begin{bmatrix} F_0 & F_1 & F_2 & \dots & F_{n-1} & F_n \\ P_0 & 0 & 0 & \dots & 0 & 0 \\ 0 & P_1 & 0 & \dots & 0 & 0 \\ 0 & 0 & 0 & & P_{n-1} & 0 \end{bmatrix}$$

is a matrix describing the transition from one age-structure to another over one time interval.

F_i ($i = 0, 1, 2, \ldots, n$; $f_i \geqslant, 0$) represents number of new animals at time$_{t+1}$ born in the interval t,t_{+1} to animals in age-group i at time t.

p_i ($i = 0, 1, 2, \ldots n-1$; $1 \geqslant p_i > 0$) represents the probability that an animal age i at time t will be alive at time $t+1$ in age-group $i+1$.

The 1965 count data were subsequently adjusted to account for probable underestimates and for misclassification by identifying young males as females. This adjustment is based on the level of error discovered by Lowe (1969) when he compared this method of counting with numbers obtained by reconstructing populations from carcass and cull material. In addition, there were almost certainly some deer present in thicket plantations which were not counted in the 1965 census (Ratcliffe 1984a).

A further simulation, based upon this revised initial number, but keeping all other parameters the same, was conducted (Fig. 1, line A). This suggested a rise in numbers from around 1600 in 1965 to over 3000 in 1980. This was considered to be the most realistic description of population changes.

The Galloway red deer population 1981–1986

Forest managers were concerned by the rapid increase in deer numbers which this further simulation suggested, and its possible impact on the timber crop. They decided to increase the cull of females of breeding age from about 155 per year during 1965–80 to 380 per year, on the basis of a projected increase in the total population of around 20% per year (Ratcliffe 1984a). Even this cull would achieve only a modest decrease in numbers. This objective was not achieved, because it proved extremely hard to kill deer in dense coniferous forests, hind culls being particularly difficult during winter conditions. The actual culls since 1980 are presented in Table 1.

The number of deer killed increased from about 300 per year during 1965–80 to around 480 per year in 1980–84 but did not prevent the population increasing. Strong emphasis had been placed on the cull of females of breeding age but this only rose from 155 per year during 1965–80 to around 210 per year in 1980–84. Simulations based on a starting popula-

Table 1. Red deer cull 1980–84: Galloway central region.

Year	Calves	Stags	Hinds	Total
1980–81	80	168	210	458
1981–82	93	167	251	511
1982–83	71	184	198	453
1983–84	105	243	174	522

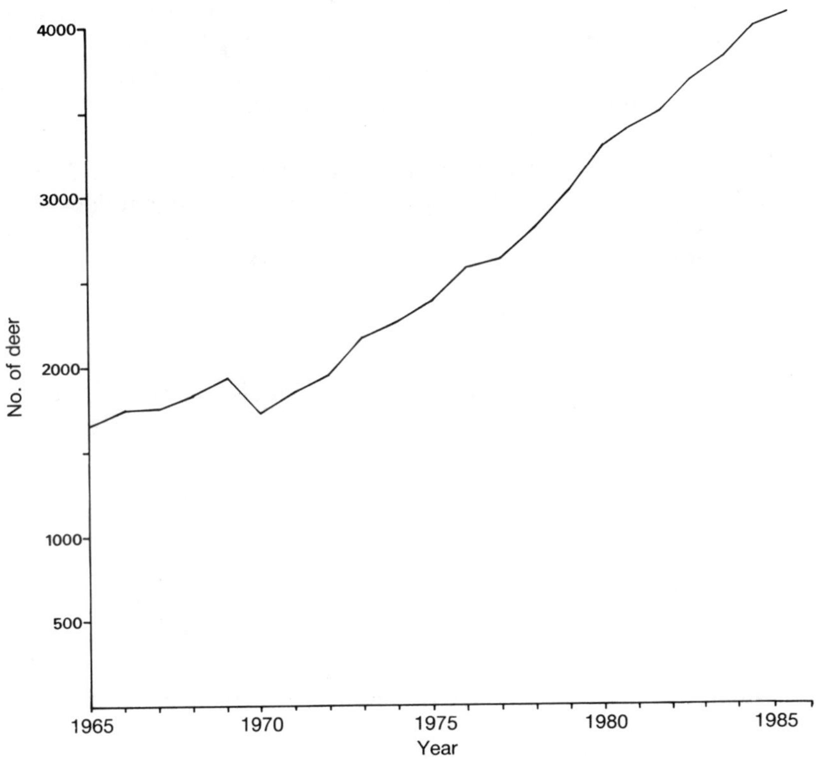

Fig. 2. Galloway red deer population projections to 1986.

tion of 3250 deer in 1980 (Fig. 1, line A), and imposing actual cull data (Table 1), suggested that the population had increased to about 4200 by the spring of 1986 (Fig. 2). This is equivalent to a density of 9.0 deer per km^2.

Independent assessment of deer density

Vantage-point counts

The population size of 4200 deer is, of course, based upon mathematical projections, which themselves are based upon some unconfirmed assumptions

Table 2. Red deer densities 1982–86: Galloway central region.

	1982	1983	1984	1985	1986
Deer per km^2	6.5	7.0	6.0	7.0	9.0

such as acceptable accuracy of initial population size and age/sex-class distributions and temporal consistency of fecundity and survival values.

Field assessments of density using vantage-point counts (Ratcliffe 1984b), were made between 1982 and 1986 at 19 locations in Galloway to provide an independent check on the projected population size. This method consists of counting and classifying stags, hinds, and calves observed in delimited areas of forest from vantage-points over periods of 2–3 h. Results of counts are presented in Table 2.

The assessment of densities from vantage-point counts (Table 2) is considered too imprecise to indicate population changes, though it does suggest that the population has increased in 1986. An average density of 6.5 deer per km^2 in 1982 suggests a total population of 3035 deer, and 9.0 deer per km^2 in 1986 is equivalent to a total population of 4230 deer.

This independent means of estimating population density supports the data provided by population simulations. An average cull of about 480 deer per year between 1980 and 1984 from a population of 4230 deer represents only 11.3%, compared with the 20% required to prevent further increase. It is clear that far more deer must be shot if control is to be achieved.

Cohort analysis

Since 1982 cull data have been collected for the reconstruction of populations and for the retrospective calculation of minimum population size. These data are not yet sufficiently complete for the whole area but reliable data from 1975 to 1984 are available from Carrick Forest in the central area (Table 3).

In Carrick Forest, by 1984, 58 deer had been shot from the 1979 cohort, 54 from the 1980 cohort and 60 from the 1981 cohort. Annual post-natal mortality is likely to be about 10–20% of calves born (P. R. Ratcliffe unpublished data) and it is clear that a large proportion of these cohorts, especially the 1981 cohort, will still be alive. This suggests that between 1979 and 1981 at least 100 calves were born per year. At least 100 hinds of breeding age (two years and above) produced these, and about 80 yearlings must have been present. There is little information available on the ratio of stags to hinds found in commercial forests, but it is unlikely that there are fewer than half as many stags as hinds present. This suggests a minimum population of about 330 deer, equivalent to 3.8 deer per km^2, and about half the density estimated for the Galloway central area as a whole. How-

Table 3. Cohort analysis, Carrick Forest.

COHORT	CULL YEAR													AGE	NO. SHOT FROM COHORT
	1971/72	1972/73	1973/74	1974/75	1975/76	1976/77	1977/78	1978/79	1979/80	1980/81	1981/82	1982/83	1983/84		
1971	9	5	5	7	2	1	0	3	0	0	0	0	0		32
1972		4	13	7	6	3	0	1	0	1	1	0	0	12	36
1973			11	12	8	7	1	4	1	1	0	0	0	11	45
1974				10	10	14	3	0	1	2	1	0	0	10	41
1975					13	9	2	3	2	0	1	0	1	9	31
1976						9	10	7	5	4	3	0	0	8	38
1977							7	13	4	9	6	3	3	7	45
1978								11	23	13	4	2	2	6	55
1979									20	24	4	6	4	5	58
1980										17	27	8	2	4	54
1981											11	30	19	3	60
1982												15	16	2	31
1983													14	1	14
1984														0	
TOTAL CULL				36	39	43	23	42	56	71	58	64	61		

ever, Carrick Forest has considerably less of the more suitable thicket-stage habitat types available than other parts of the area. The increase in the number of deer shot from each cohort between 1971 and 1981 suggests an increasing population, assuming that culling effort has been constant during this period.

The continuing accumulation of age-related data provides a factual basis for checking population estimates and is considered to be an essential component of deer management. All jaw-bones of culled deer should be systematically collected and recorded.

Management options in Galloway

Population simulations based upon the 1986 estimate of 4200 deer were conducted to determine the culling required to prevent further increase (Fig. 3). It is clear that this high-performance population requires a cull of about 20% per year to achieve even a modest decrease in numbers over the next 10 years. Culls of 16 and 18% result in a continued increase in numbers. A 20% cull of the 1986 population represents a total cull of 837 deer (Table 4) of which around 350 must be females of breeding age.

Had this type of information been available in 1970 when, following a high cull of breeding females, the population declined to about 1600 deer (Fig. 2), population control would have been relatively easy to achieve (Fig. 4). A 20% cull could have been achieved by killing only 130 females of breeding age and a total of about 315 deer (Table 4; Fig. 4).

It is perhaps unlikely, as the unsuccessful attempts since 1980 to reach the increased culling targets have shown, that deer numbers can be reduced to acceptable levels without major changes in management strategies and tactics. A cull in excess of 350 females of breeding age, during the difficult conditions imposed by Scottish winters and in a dense habitat, seems to be unattainable. It is tempting to suggest that a temporary increase in manpower for a period of one to five years during the hind-culling season might reduce deer numbers to a level at which the permanent staff could cope. Within the humane and legal constraints imposed, extended seasons for shooting hinds and the use of spotlights at night may also help. However, improvements in forest design (Ratcliffe 1985) are most likely to provide solutions.

The underestimation of deer numbers

The potential for underestimating forest deer populations is clearly very high. Studies comparing local estimates of deer numbers by forest staff with intensive vantage-point counts at other Scottish forests also demonstrate considerable underestimations (Table 5).

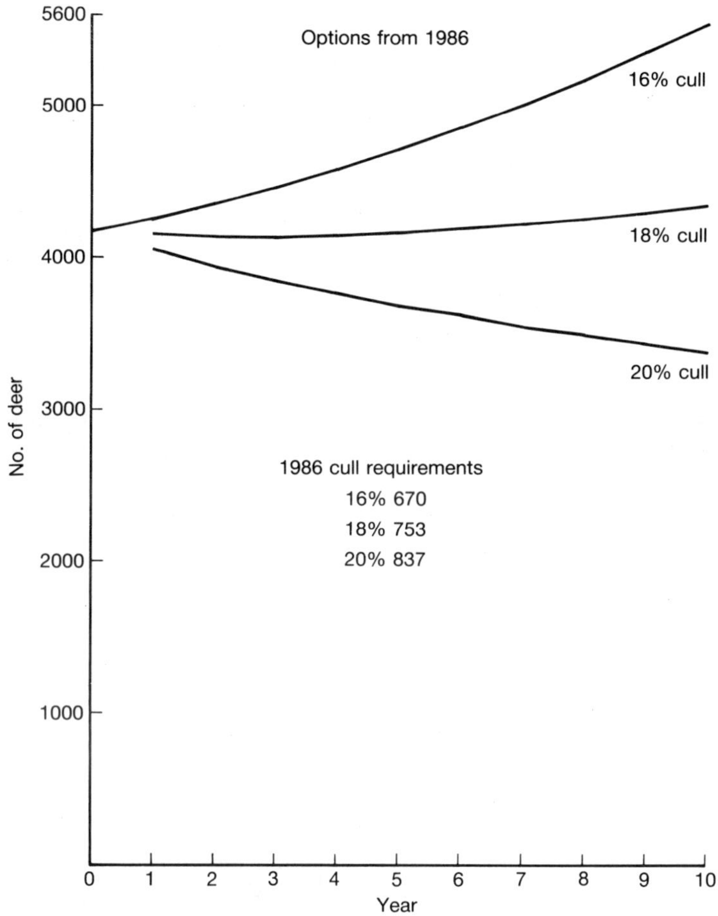

Fig. 3. Population simulations based upon the 1986 estimate of 4200 deer; projected effects of various culling options.

Table 4. Red deer culling options in 1970 and 1986: Galloway central region.

	Culling options Deer numbers	
%	1970	1986
16	253	670
18	284	753
20	315	837
22	347	924

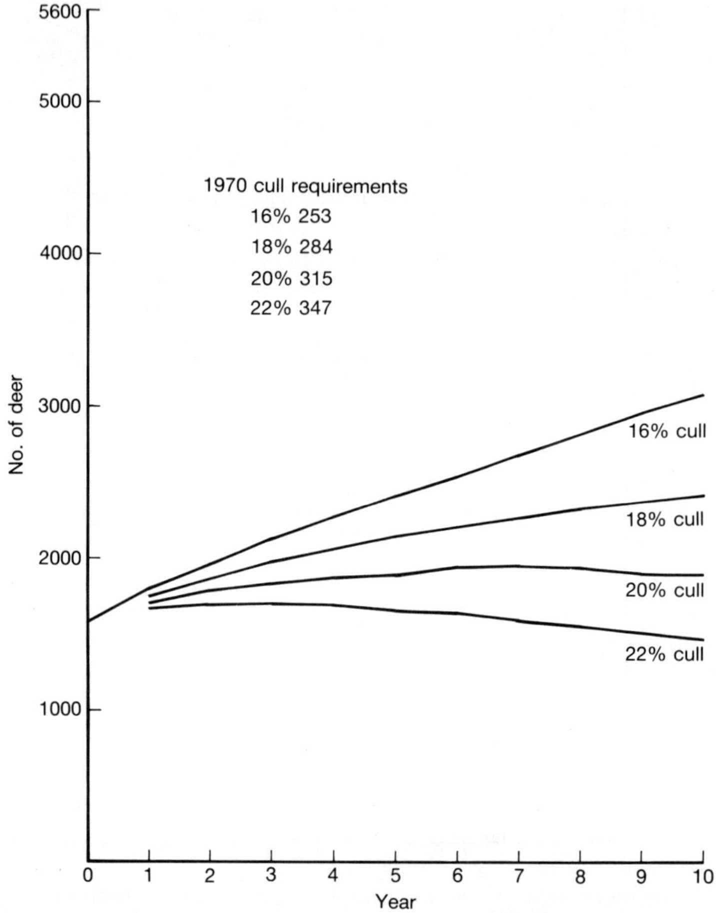

Fig. 4. Population simulations based upon the 1970 estimate of 1600 deer; projected effects of various culling options.

Table 5. Reported and determined densities of red deer in four Scottish forests.

Forest	Reported density (N/km^2)	Determined density (N/km^2)	Magnitude of underestimate
Achnashellach (Coire a Bhainne), W. Ross-shire	1.8	30	× 16.7
Glencripesdale, N. Argyll	8.8	40	× 4.5
Galloway, S. W. Scotland	1.8	8	× 4.4
Carrick, Galloway, S. W. Scotland	1.1	3.8	× 3.5

It seems likely that red deer populations occupying upland forests in Scotland are frequently underestimated, sometimes by large amounts.

Discussion

Highlighting the scale of population increase and total population size in 1980 resulted in a successful attempt to shoot more deer in Galloway. However, even the increased numbers shot represented only about 11% of the total population each year. Population simulations and density estimates suggest that a 20% cull each year is necessary to prevent further increase. Had the size of the problem been realized in 1970 when the population was only about half its present size, control would have been considerably easier to achieve.

The underestimation of deer numbers appears to be common in many Scottish forests, where checks have shown densities to vary from around three to 17 times previous subjective estimates. It seems likely that undershooting is also common, especially as densities in Galloway are lower than in most other forests studied.

Assuming that there are about 50,000 red deer occupying Scottish forests and assuming that an 18% sustained annual level of exploitation is possible, the magnitude of under-culling amounts to about 7% of 3500 deer each year. This can be interpreted as a direct loss of revenue or as an increase in the capital (total population size) which can be exploited at some later date. However, our inability to exercise control over current population levels suggests that this latter approach will only add to future problems, including the very important risk of increased damage to timber crops and sensitive plant communities.

If management objectives include harvesting deer, then management must consider how best to obtain the highest possible quality and quantity of venison or antlers of trophy quality. To achieve the maximum sustained yield of venison will demand careful consideration of the interactions between density and body weight. If antler quality is important then the optimum density is considerably lower than if managing for venison production (Caughley 1977).

The Galloway herd is considered to be of high quality, providing antler trophies of a very high standard (Whitehead 1985), and body weights are greater than in most Scottish populations. One of the objectives of management is to maintain the character and quality of the herd (Jeffery 1974). A continuing increase in numbers will result in increasing damage to trees and plant communities, and will almost certainly evoke density-dependent factors causing lower body weights and smaller antlers.

Animal populations are clearly more vulnerable to manipulation of their habitat than they are to direct attempts to remove individual animals. It seems logical, therefore, that further attempts to reduce numbers of red deer in forests should be aimed at modifying their habitat. Though improved forest design is often suggested, perhaps the prescriptions suggested by Ratcliffe (1985) do not go nearly far enough. As Caughley (1977) stated, 'The treatment of a population by changing to its detriment the key components of its habitat is the most powerful and elegant technique of population control'.

In commercial forests where the primary objective is to grow timber trees, rotation lengths, tree-spacings and other practices have been determined by the needs of silviculture. In view of this perhaps the only way in which habitat manipulation can help is by having substantial areas unplanted.

Acknowledgements

I thank Rory Putman for providing the computer program of the Leslie-Matrix model. Andrew Peace increased the flexibility of the program and contributed to discussions on population simulations. Robin Heaney provided material and assisted in the compilation of cohort data. Peter Madden (Glencripesdale), John Livingstone and many Forestry Commission rangers in Galloway helped with vantage-point counts. Tony Hinde provided data on population estimates from a number of Scottish forests. Rory Putman and Brian Staines provided helpful criticism.

References

Catt, D.C. & Staines, B.W. (1987). Home range use and habitat selection by Red deer (*Cervus elaphus*) in a Sitka spruce plantation as determined by radio-tracking. *J. Zool., Lond.* **211**: 681–93.

Caughley, G. (1977). *Analysis of vertebrate populations.* Wiley, Chichester.

Jeffery, W.G. (1974). The Galloway (red) deer control scheme. *Deer* **3**: 226–30.

Leslie, P.H. (1945). On the use of matrices in certain population mathematics. *Biometrika* **33**: 183–212.

Leslie, P.H. (1948). Some further notes on the use of matrices in population mathematics. *Biometrika* **35**: 213–45.

Lowe, V.P.W. (1969). Population dynamics of the red deer (*Cervus elaphus* L.) on Rhum. *J. Anim. Ecol.* **38**: 425–57.

Mitchell, B., Staines, B.W. & Welch, D. (1977). *Ecology of red deer: A research review relevant to their management in Scotland.* Institute of Terrestrial Ecology, Cambridge.

Ratcliffe, P.R. (1984a). Population dynamics of red deer (*Cervus elaphus* L.) in Scottish commercial forests. *Proc. R. Soc. Edinb.* **82B**: 291–302.

Ratcliffe, P.R. (1984b). Population density and reproduction of red deer in Scottish commercial forests. *Acta zool. fenn.* **172**: 191–2.

Ratcliffe, P.R. (1985). Glades for deer control in upland forests. *For. Commn Leafl.* No. 86: 1–10.

Staines, B.W. & Welch, D. (1984). Habitat selection and impact of red (*Cervus elaphus* L.) and roe (*Capreolus capreolus* L.) deer in a Sitka spruce plantation. *Proc. R. Soc. Edinb.* **82B**: 303–19.

Whitehead, G.K. (1985). Two more Galloway beauties. *Shooting Times* **4321**: 17.

Symp. zool. Soc. Lond. (1987) No. 58: 167–179

Long-term changes in New Forest deer populations and correlated environmental change

R. J. PUTMAN and
S. K. SHARMA

Department of Biology
The University
Southampton

Synopsis

The paper presents a preliminary analysis of population change of the major deer species within the New Forest between 1962 and 1985. Changes in number of fallow deer (*Dama dama* L.) roe deer (*Capreolus capreolus* L.) and sika deer (*Cervus nippon* Temminck) are examined in relation to the relative abundance of other herbivore populations over the same period and also in relation to the character of the vegetation. Changes in numbers of all deer species appeared independent of each other. Population numbers also appeared unaffected by numbers of other large Forest herbivores, except in the case of roe deer, where populations showed negative correlation with numbers of domestic cattle and ponies pastured on the Forest over the preceding three years. Roe deer numbers were also negatively correlated with the area planted with young Scots pine 0–2 m in height, with conifers of most species 15–20 m in height, and with mature broadleafed woodland. Roe deer were positively associated with Scots pine 5–15 m in height and with firs, pines other than Scots pine, and other conifers (largely western hemlock and larch) 2–5 m in height. Numbers of fallow deer showed little consistent correlation with any vegetational parameter measured, although they were positively correlated with the area of woodland ride and the area of mature firs (>20 m in height), and strongly negatively related to area of spruce 2–5 m in height. Sika deer occur only within a restricted area of the Forest. Within that area numbers were positively correlated with the area of western hemlock and larch >20 m in height. Variations in deer numbers over time and the possible biological explanation of observed correlations are discussed.

Introduction

General introduction

The New Forest in southern England is an area of some 37,500 ha of mixed vegetation, first set aside as a royal hunting preserve in the 11th century. The area currently supports populations of five deer species: red (*Cervus*

ZOOLOGICAL SYMPOSIUM No. 58
ISBN 0-19-854006-X

elaphus), roe (*Capreolus capreolus*), fallow (*Dama dama*), sika (*Cervus nippon*) and muntjac (*Muntiacus reevesi*); domestic cattle and ponies are also pastured in the Forest under ancient rights of common. This paper presents a preliminary analysis of changes in population numbers in the major deer species of the Forest since 1962. Fluctuations in number of each species and long-term change are related in each case to the abundance of other large herbivores within the New Forest over the same period and, from 1975, are considered also in relation to a number of characteristics of the vegetation of the Forest's enclosed woodlands.

The New Forest, its vegetation and the larger herbivores

Set in the South Hampshire basin, the New Forest enjoys a relatively mild and oceanic climate. Data recorded at nearby Hurn, Bournemouth, and supplied by the Meteorological Office for the period 1968–85 give a mean minimum and maximum daily temperature of 1.2 and 7.9°C respectively in January, and 11.4 and 21.6°C in July. There was an annual average of 102 days with ground frost, with the period December to April inclusive having about 15 days of frost per month. Annual rainfall varied from 510 to 980 mm, with an average of 785 mm; the rainfall was distributed throughout the year with the wettest period usually being November to January.

Of the total administrative area of 37,500 ha, some 9000 ha are now occupied by towns, villages or agricultural holdings; the remainder forms a mosaic of broadleafed and coniferous woodland, natural and artificially improved grasslands and an open 'waste' of heathland and bog. There are currently 9200 ha enclosed for commercial timber production (4850 ha coniferous, 4350 ha broadleafed or mixed woodland). In the remaining 19,000 ha of so-called 'open forest', woodlands are less abundant (*c.* 4000 ha in total) and the area is predominantly a mixture of *Calluna*-dominated dry heath (7500 ha), wetter heaths characterized by *Erica tetralix* (2000 ha) and bog (1350 ha), with 4200 ha of acid grasslands, 800 ha of which have been managed to improve their productivity by fertilizing and/or reseeding. The area is described in more detail in Tubbs (1968) and Putman (1986).

This mixture of vegetational communities supports the seven species of large herbivores noted above. Populations of red, roe, fallow and sika deer have access to the entire area (although they are primarily concentrated within the forestry enclosures) while the unenclosed Forest supports at peak times of year approximately 3500 ponies and 2000 cattle. In addition, domestic pigs are also turned out upon the Forest to forage the acorn and beech mast over the winter months. Muntjac are reported from the Forestry enclosures but their status is unknown.

Of the four main deer species, red and sika deer are essentially local in distribution, with populations restricted to limited areas of the Forest, while roe and fallow deer are widespread. All populations are managed to some extent. The cattle and ponies are true domestic animals, merely grazing at free range on the Forest, under licence; licences and control of the grazing are administered by the Court of Verderers. The Forest deer are under the jurisdiction of the Forestry Commission who attempt to maintain populations within levels compatible with commercial forestry.

Population levels of all the deer species have fluctuated markedly over the centuries, as management aims have altered. In the early history of the Forest fallow deer were undoubtedly the dominant herbivores. Although difficult to interpret, such records as are available indicate that population levels of fallow deer were maintained at around 6000–7500 animals from the 17th century to the 1850s. In 1851 the Crown gave up its rights to the Forest deer and the Deer Removal Act of that year provided for the removal of all deer from the Forest within three years of its enactment. The population was drastically reduced, though not completely eliminated, and an estimate in 1900 gave a figure of 200 head (Lascelles 1915). Since that time the population has been allowed to expand to its current level, estimated by the annual census to be about 1000 animals.

Although populations were estimated at 357 in 1670, red deer were probably extinct in the New Forest by the mid 18th century. The present population is currently estimated by the census to stand at about 70 individuals and almost certainly derives from re-introductions in 1908 and 1962. Roe deer are also presumed to have become extinct in the Forest during the Middle Ages, as indeed throughout the whole of England. Roe deer were re-introduced during the 19th and 20th centuries and probably recolonized the New Forest from 1870 onwards, spreading across from Dorset (Jackson 1980). Finally, sika deer established themselves within the New Forest area in the early 1900s as escapes from the Beaulieu Estate where they were introduced in 1904.

Responsibility for management of the Forest deer passed to the Forestry Commission in 1924; since 1960, regular annual census of populations has been attempted (Table 1). Even within this 25-year period significant changes in numbers of some species have occurred; it is these more recent changes that are the subject of the present paper.

Methods

Since 1960, the Forestry Commission has undertaken a regular annual census of the number of deer of each species present within the enclosed woodlands of the New Forest. The census is the basis on which the projected cull is calculated. The whole Forest is divided into 12 beats (ranging in size from 1470 to 2710 ha). Each beat is the responsibility of one keeper, who makes

Table 1. Census figures for deer populations of the New Forest presented by Forestry Commission keepers from 1960 to 1985.

Deer species	Year												
	1960	1961	1962	1963	1964	1965	1966	1967	1968	1969	1970	1971	1972
Roe	211	302	328	383	434	465	490	527	539	572	620	444	428
Fallow	788	857	931	933	994	1147	1036	968	855	893	1022	866	1017
Red	—	—	7	23	20	23	24	29	33	18	23	26	28
Sika	35	44	34	42	64	84	97	67	38	38	37	71	73
Deer species	Year												
	1973	1974	1975	1976	1977	1978	1979	1980	1981	1982	1983	1984	1985
Roe	435	356	375	344	363	323	354	334	309	336	281	264	261
Fallow	904	909	834	973	908	927	950	1016	996	1049	1020	1033	1006
Red	29	37	49	43	49	79	34	58	73	64	72	57	66
Sika	75	78	88	94	80	79	80	74	95	76	97	115	103

a return to the central office every year of the number of deer of each species held within his beat on a fixed date in April. In practice, this figure is reached as a compromise between the keeper's accumulated knowledge of the number of deer in the area over the previous few weeks and an actual 'single-day count' on the appointed fixed date. Counts relate to Forestry enclosures only—but since most deer are concentrated in such areas, approximate to a total census for the whole Forest area. It should be noted from the outset, however, that the census figures are not, nor are assumed to be, absolute. However, as a relative *index* of population change from year to year the census figures are reliable—since they have over that period relied on the same method and the same observers within each beat.

Surveys of the Forest's timber stocks are also undertaken on a regular basis. The Forestry Commission periodically carries out detailed inventories of all enclosed woodland blocks, recording for each the tree species present, their dates of planting and their estimated growth rate (yield class). This information is used to provide stock maps and timber production forecasts, and is continually updated as trees are felled or thinned and areas are replanted. Data collected provide details only of timber species planted and height class but estimates of areas planted are accurate to 0.5 ha. From these records can be extracted information on a number of vegetational parameters which might affect the Forest deer (Table 2).

Stock surveys are undertaken only every five to seven years and thus data are not directly available for the intervening years. An estimate of the structure of each beat in any year other than those for which specific records are available may, however, be obtained by extrapolation since survey records show, for each planting, both age and yield class; from these the rate of growth of each species represented can be accurately estimated (Rollinson 1985). A computer model developed by the Forestry Commission for

Table 2. Variables included in the candidate list for correlation with deer numbers.

Temporal analyses
1. Area in each beat of Scots pine of height 0–2, 2–5, 5–15, 15–20, >20 m
2. Area in each beat of other pines of height 0–2, 2–5, 5–15, 15–20, >20 m
3. Area in each beat of firs of height 0–2, 2–5, 5–15, 15–20, >20 m
4. Area in each beat of spruces of height 0–2, 2–5, 5–15, 15–20, >20 m
5. Area in each beat of other conifers of height 0–2, 2–5, 5–15, 15–20, >20 m
6. Area in each beat of oaks of height 0–2, 2–5, 5–15, 15–20, >20 m
7. Area in each beat of other broadleafed species of height 0–2, 2–5, 5–15, 15–20, >20 m
8. Numbers in each beat of roe, fallow, red, sika deer, current year
9. Numbers in each beat of roe, fallow, red, sika deer, previous year
10. Numbers (whole forest) of cattle, ponies, pigs, current year
11. Mean numbers of cattle and ponies over previous 3 years
12. Human disturbance

Spatial analysis: of the above variables, 1–9 *plus*

13. Total area in each beat of rides
14. Total area in each beat of open ground
15. Total area in each beat of enclosed woodland
16. Length of woodland edge
17. Area in each beat of all broadleafed species of height 0–2, 2–5, 5–15, >15 m
18. Area in each beat of all conifers of height 0–2, 2–5, 5–15, >15 m
19. Area in each beat of mixed species woodlands

forward projection of forest structure (Ratcliffe, Hall & Allen 1986) was used to 're-create' the structure of each beat in years for which no direct survey data were available; data were extracted for the same variables as those listed above (Table 2). Such generated data are only considered reliable for some five or six years either side of the base survey year, owing to the inevitable accumulation of discrepancies between the Forest profile generated by the model and the true Forest (the model incorporates a number of assumptions, which are discussed in a later section). As a result, data before 1975 are excluded as potentially unreliable.

Other data included in the analysis as potential influences upon deer populations were: total numbers of domestic stock (cattle and ponies) for which licence fees had been paid in each year, mean numbers of cattle and ponies over the preceding three-year period (figures extracted from records held by the New Forest Verderers) and the degree of human disturbance in the Forest, considered on the basis of Forestry Commission records of car park attendances and day-visitor censuses.

A series of correlation analyses were performed upon these data sets to investigate the changes in deer populations with respect to

1. changes in other herbivore populations levels over the period 1962–85;
2. the change in character of the vegetation over the period 1975–85, using the 12 different beats as separate replicates.

Such 'temporal' analyses were repeated using roe, fallow and sika deer in turn as the dependent variable. (Correlation analyses were *not* undertaken in relation to changing red deer numbers, since although whole Forest census figures are available, figures were incomplete for red deer numbers in individual beats.)

A number of additional vegetational parameters were identified which, while they may vary markedly between beats, do not vary over time (e.g. length of rides, length of woodland margin, area of open ground). A complementary analysis was thus undertaken within one year (1981) where direct survey data were available. In this case roe and fallow deer numbers were compared between different beats in relation to the variable recorded ('spatial' analysis). Neither sika nor red deer could be included in this latter analysis since they are essentially restricted in their distribution within the forest to only one or two beats.

Throughout, data for deer numbers and the independent variables included (with the exception of estimates of numbers of domestic stock, and human disturbance) are derived within the statutory enclosures of the Forest—that area enclosed for silviculture—and thus strictly relate only to some 9200 ha of the whole Forest area; nonetheless, as has been noted, this is in practice the area where the deer are primarily concentrated.

Results

No consistent trends in population number of the different deer species of the New Forest are immediately apparent from the census figures of Table 1, except in the case of sika deer, for which the census estimates—and thus perhaps actual population numbers—have steadily increased since 1960. Red deer populations were clearly extremely sparse in the early 1960s if not totally extinct. Re-introductions were made in 1962 and populations appear now to have stabilized at around the 60–70 individuals estimated by the census. Roe and fallow deer numbers in 1985 did not differ greatly from numbers in 1960, though both populations show evidence of fluctuation and, in the case of the roe deer, these fluctuations are fairly wide, with a peak in 1969–70 that was more than double the levels of 1960–62 and 1983–85.

Deer numbers in relation to populations of other herbivores

Changes observed in the numbers of the different deer species populations between 1960 and 1985 appear for the most part to be independent of each other, suggesting little interaction between the species. It is notable, however, that numbers of roe deer show significant negative correlation with grazing pressure imposed on the Forest by domestic stock. Roe deer numbers correlate with the mean numbers of cattle and ponies pastured on the Forest in the preceding three years, with $r = -0.47$, $p = 0.02$.

Deer numbers and vegetational change

When numbers of fallow deer in each Forest beat from 1975 to 1985 are considered in temporal analysis in relation to the various vegetational characteristics of the enclosures, no consistent relationships emerge. In each beat, strong positive correlations are observed between fallow deer numbers and various vegetational parameters considered, but there is no regularity of pattern, and while a number of the vegetational variables emerge with strong positive correlation in two or more of the different beats, these same variables at least as often show significant *negative* correlation with fallow deer numbers in other beats. Only one variable recorded (firs over 20 m in height) consistently shows significant positive correlation with fallow deer numbers in more than one beat. Where fallow deer numbers in 1981 were compared between beats in order to assess the influence of those variables which might differ between areas, but could not be expected to change within any one beat over time, two variables do emerge as significantly positively correlated with fallow deer population number: the total area of rides and grassy tracks ($r = 0.631$, $p < 0.05$) and the total area of enclosed woodland ($r = 0.606$, $p < 0.05$). Fallow deer showed consistent *negative* correlation with only a single variable (Table 3): area within a beat of spruce of height 2–5 m.

By contrast, a number of vegetational descriptors were consistently associated with roe deer population numbers. Ten variables showed significant positive correlation with roe deer numbers in at least one run ($p < 0.05$) and never produced negative correlations (Table 3): of these, five were significantly positively correlated with roe deer populations in two or more of the beats. These were firs, pines other than Scots pine (primarily Corsican pine, *Pinus nigra*) and 'other conifers' (largely western hemlock, *Tsuga heterophylla*, and larch, *Larix* sp.), all of height class 2–5 m, broadleafed species of height 2–5 m and Scots pine (*Pinus sylvestris*) of height 5–15 m. Roe deer showed strong and consistent negative correlation with both 'other broadleafed species' and with Scots pine of height 0–2 m, other pines of height 5–15 m and 'other conifers' of height 5–15 m. They were also negatively correlated with both spruces and broadleafed species of >20 m. In our subsequent analysis of these results we have discounted the unique correlations in view of the necessarily artificial nature of the forest profile data, and have considered only the five characters showing positive correlation in two or more replicates.

Sika deer occur only in one single beat within the Forest. Analysis within that beat suggests that the changing area of 'other conifers' of height >20 m

Table 3. List of variables emerging from temporal analyses as significantly correlated with roe and fallow deer numbers in more than one replicate.

Deer species	Variable		No. of beats correlated $p<0.01$	$0.01<p<0.05$
Roe positively correlated	'Other conifers'	2–5 m	2	
	Pines other than Scots pine	2–5 m	1	1
	Firs	2–5 m	1	1
	Broadleafed species other than oak	2–5 m		2
	Scots pine	5–15 m		2
Roe negatively correlated	'Other conifers'	5–15 m	2	
	Pines other than Scots pine	5–15 m	1	1
	Spruces	>20 m	1	1
	Broadleafed species other than oak	>20 m	1	1
	Broadleafed species other than oak	0–2 m		2
	Scots pine	0–2 m		2
Fallow positively correlated	Firs	>20 m	2	
Fallow negatively correlated	Spruce	2–5 m	3	

is the predominant factor correlating with changes in sika numbers over time in the 1975–85 period considered ($r = 0.92$; $p < 0.01$).

Discussion

Changes in number of New Forest deer populations are examined in this paper in relation to changes in number of other herbivore species or in various vegetational characteristics of their environment. Absence of detailed field records and the use of projected estimates of vegetational structure have enforced acceptance of certain assumptions and limitations in the data. The limitations of the Forestry Commission's annual census as an accurate reflection of deer population numbers have already been mentioned. In this analysis, however, since methods used have not changed over the years, we can accept the census figures at least as an index of true population change. Vegetational descriptors employed in the temporal analysis of deer population changes from 1975 to 1985 are dictated in part by the structural categories recognized by the model. It can provide data on forest structure only by tree species planted and height class, and many important vegetational features cannot be accurately assessed. Thus factors such as the presence of saplings by natural regeneration, or the quality of the ground vegetation, including abundance of such species as rose, bramble or various grasses known to be important food for deer, could not be taken into account. Similarly, it has not been possible to include in the analyses the effects of any changes in forestry practices; for instance changes in the management of rides, or variation in replanting schemes and stocking densities, or the local effects of removing brush in some areas and leaving it in others.

This lack of detailed vegetational data is unfortunate, for it is very probably just such characteristics of the ground vegetation which comprise the most important factors affecting distribution and numbers of deer (e.g. Chapman, Claydon, Claydon & Harris 1985). Ground vegetation does, however, correlate to some extent with canopy cover, and the height classes into which plantations were grouped in description of forest structure were selected in an attempt to determine categories that would broadly correspond to plantation, prethicket, thicket, polestage and mature communities—at least for conifers—and could thus be related at least subjectively to shelter or quality of forage (as, e.g., Ratcliffe & Petty 1986).

All vegetational data rely heavily on extrapolation from a single survey in 1981. The Forestry Commission's forecasting model works on the basis of supplied standard felling and thinning cycles. When an area is 'felled' by the model during a forward projection, the model then assumes the area will be replanted with the same species, unless told otherwise: we replanted Scots pine and western hemlock with Corsican pine, in line with the current practice. When working retrospectively, the model leaves areas prior to their known planting date as unplanted. By examination of a randomly selected set of 30 sites within the Forest planting (based upon comparison with manual records retained by the Commission) it was estimated that these planting assumptions result in an error of approximately 1.4% of the area per extrapolated year, an error judged acceptable given a maximum extrapolation of six years from the base year.

Finally, the entire analysis undertaken here, reliant on observed correlation, assesses only linear relationships between our chosen variables.

One additional problem as yet unaddressed is the extent to which population changes described are determined more by human intervention than by any purely environmental characteristic. None of the populations studied is in practice unrestricted; all are subjected to a regular cull. To what extent therefore is it a valid system in which to explore population changes in relation to environmental characteristics?

For the last 25 years the Forestry Commission, who are responsible for deer management within the Forest, have calculated their cull target for each species according to standard formulae. On the assumption that two thirds of the censused population of adult females (one year and over) will produce offspring in any year, in a 1:1 sex ratio, the number of individuals to be recruited to the population in that year may be calculated. Estimated losses due to poaching and road traffic accidents are then deducted, and the resultant culling level required to retain the target population established. Target populations are set at 600 individuals for fallow deer, 70 for sika deer and 300 for roe deer.

Total imposed mortality over the period 1975–85 (from accidental death, poaching and culling operations) has been of the order of 40–50% of the censused population for fallow and sika deer, which might be assumed to have a major impact on population levels. Such a calculation, however, assumes that the censused population each April approximates to the true population total and that the mortalities to be expected from poaching and road traffic accidents can be accurately assessed.

In his studies of population change in New Forest fallow deer, Strange (1976) considered that estimates for road traffic and poaching losses may well be reasonable. However, the census figure on which the cull is first calculated may, we suspect, be a considerable underestimation of true population size. Strange's figures for 1970–75 suggest that while the Forestry Commission estimated a total population of 914 fallow deer within the New Forest (mean over the five years) the true population was probably nearer 1800 (Strange 1976). Mann's (1983) figures for sika deer populations between 1979 and 1982 suggest numbers of 175–200—once again, twice the numbers given by the Forestry Commission census for the same period, of 74–95. With the assumption that the roe deer population is similarly underestimated, the true culling levels for the three species over the period 1975–85 are more likely to have been in the region of 20–25% for fallow and sika deer, with roe deer losses over the same period only some 6–10%.

An imposed mortality of even 20–25% is certainly sufficient to have a disruptive effect on fallow and sika deer populations. Clutton-Brock, Guinness & Albon (1982) found that numbers of red deer on Rhum increased substantially following the relaxation of a 17% cull in 1973, and Houston (1982) considered that human predation on *Cervus elaphus* populations in the Yellowstone National Park at an average level of 21.5% was the major factor responsible for changes in population numbers. It may thus be possible that culling in the New Forest *is* a major influence on the numbers of fallow and sika deer—and may be one reason why we were able to demonstrate rather few consistent correlations with any of the other variables we have considered (but see below for an alternative explanation). Roe deer in the New Forest are shown to suffer by comparison a far lower imposed mortality; we consider that this is likely to be no more than a minor contribution to the observed changes in numbers. Furthermore, the Commission's records reveal a similar level of mortality for the roe deer population over the entire period 1960–85. The population itself was increasing noticeably in number over the first ten years of that period, but has shown a steady decline since, suggesting that culling levels *cannot* be the only factor influencing population number.

The results of our analyses suggest that the numbers of sika deer in their localized range within the Forest (Mann 1983; Putman 1986) are positively correlated with the abundance of coniferous woodland of height >20 m. Because of their restricted distribution the analysis could not be replicated and it is possible that the correlation is purely coincidental. Sika deer numbers have consistently increased over the years since 1975, for whatever reason; over the same period one might expect the abundance of older coniferous plantings to increase purely as a function of ageing. In his study of the ecology of New Forest sika deer, Mann (1983) showed a pattern of habitat use with strong emphasis of prethicket and thicket stage coniferous woodland.

Numbers of roe deer consistently show significant positive correlation with a variety of coniferous plantings (firs, 'other pines' 'other conifers') of 2–5 m in height, plantings of Scots pine of 5–15 m in height and broadleafed species other than oak (primarily beech and sweet chestnut), once again of 2–5 m in height. Such areas will in general be relatively dense but usually with a canopy sufficiently open that they will retain a ground flora of grasses and forbs and some shrub layer of, particularly, rose and bramble. Thus they offer both cover and an abundant supply of preferred foodstuffs (see Jackson 1980 for an analysis of the diet of New Forest roe deer). Roe deer numbers were negatively correlated with Corsican pine and 'other conifers' of height 5–15 m, young Scots pine and broadleafed plantations of up to 2 m in height, and both spruce and broadleafed woodland >20 m in height. The negative correlation with Corsican pine and 'other conifers' of height 5–15 m suggests that, within a preferred height class, species of planting is important: roe deer were *positively* associated with Scots pine plantings of 5–15 m in height. Conversely, height class within a species is also important and, clearly, only when Scots pine plantations have grown beyond 2 m high, are they positively correlated with roe deer. Roe deer were also found to be negatively correlated with mean numbers of cattle and ponies grazed upon the Forest over the preceding three-year period. As classic concentrate-selectors (Hofmann 1983), roe deer are highly selective in their foraging. Particularly heavy grazing by cattle and ponies may well reduce the availability of favoured species, adversely affecting roe deer populations.

All such conclusions must for the present remain speculative, for they are based purely on observed correlation with no proof of any causal relationship. The striking overall contrast between results here for roe and fallow deer is in itself, however, reassuring and gives some measure of confidence in the relationships established. Roe deer numbers correlate regularly with a number of quite specific vegetational descriptors: particular species plantings of a particular height class (primarily coniferous plantings of height 2–5 m). No such simple correlations were established for fallow deer populations beyond a negative relationship with spruce of height 2–5 m. Fallow deer, however, show a clear and significant positive correlation with area of grazing (rides and clearings) and total area of enclosed woodland, when these variables are considered for the first time in spatial analysis; roe deer numbers show no relationship with either of these more 'general' measures. Such contrast is entirely consistent with what we know of the animals' differing ecologies, reflecting a clear distinction between a habitat specialist (roe deer) seeking particular types of woodland of particular species composition and growth stage, with total area of woodland in itself not particularly important, and a species such as the fallow deer with less specific preference for woodland type, but a greater interest in total woodland area and the associated area for grazing provided by rides and clearings.

Acknowledgements

We would like to thank the Forestry Commission (New Forest) for so freely giving us access to deer census returns and for allowing us to use their vegetational database—and we would thank the many individual keepers who have helped with additional information. We would also acknowledge the kindness of the Forestry Commission Wildlife & Conservation Research Branch (Alice Holt) in permitting use of their computing programs and facilities; in particular we would thank Phil Ratcliffe, and Laurie Tee who ran the vegetational 'forecasts' (p. 171) for us. Finally we would like to thank Phil Ratcliffe and Steve Harris for detailed and most helpful comments on the manuscript.

References

Chapman, N., Claydon, K., Claydon, M. & Harris, S. (1985). Distribution and habitat selection by muntjac and other species of deer in a coniferous forest. *Acta theriol.* **30**: 287–303.

Clutton-Brock, T.H., Guinness, F.E. & Albon, S.D. (1982). *Red deer: behavior and ecology of two sexes.* Edinburgh University Press, Edinburgh.

Hofmann, R.R. (1983). Adaptive changes of gastric and intestinal morphology in response to different fibre content in ruminant diets. *Bull. R. Soc. N.Z.* **20**: 51–8.

Houston, D.B. (1982). *The northern Yellowstone elk: ecology and management.* Macmillan, New York.

Jackson, J.E. (1980). The annual diet of the Roe deer (*Capreolus capreolus*) in the New Forest, Hampshire, as determined by rumen content analysis. *J. Zool., Lond.* **192**: 71–83.

Lascelles, G.W. (1915). *Thirty-five years in the New Forest*. Arnold, London.

Mann, J.C.E. (1983). *The social organisation and ecology of the Japanese Sika deer* (Cervus nippon) *in Southern England*. PhD thesis: University of Southampton.

Putman, R.J. (1986). *Grazing in temperate ecosystems: large herbivores and the ecology of the New Forest*. Croom Helm, Beckenham.

Ratcliffe, P.R. & Petty, S.J. (1986). The management of commercial forests for wildlife. In *Trees and wildlife in the Scottish uplands*: 177–87. (Ed. Jenkins, D.) Institute of Terrestrial Ecology, Huntingdon. (*ITE Symp*. No. 17).

Ratcliffe, P.R., Hall, J. & Allen, J. (1986). Computer predictions of sequential growth changes in commercial forests as an aid to wildlife management, with reference to red deer. *Scot. For.* **40**: 79–83.

Rollinson, T.J.D. (1985). Thinning control. *Bookl. For. Commn* No. 54.

Strange, M.L. (1976). *A computer study of the fallow deer of the New Forest*. BSc Honours thesis (Environmental Sciences): University of Southampton.

Tubbs, C.R. (1968). *The New Forest: an ecological history*. David and Charles, Newton Abbot.

Symp. zool. Soc. Lond. (1987) No. 58: 181–191

Census techniques for grey seal populations

A. J. WARD, D. THOMPSON and A. R. HIBY

Sea Mammal Research Unit
c/o British Antarctic Survey
Natural Environment Research Council
High Cross, Madingley Road
Cambridge CB3 0ET

Synopsis

Accurate estimates of grey seal (*Halichoerus grypus*) population size are required for assessment of both fisheries interactions and management policies. For most of the year grey seals are widely dispersed at sea where they are effectively uncountable. However, females come ashore to give birth at known colonies during the autumn and the pups remain ashore for approximately three to four weeks, during which time they can be counted. The total pup production combined with fecundity estimates gives a basis for estimating the adult female population size.

The numbers of pups at all the major colonies are monitored by a series of aerial photographic surveys throughout the breeding season. Because at each colony the spread of pupping dates exceeds the length of stay of individual pups, no single survey will include all the pups born. Total pup production can be estimated using one of a number of models. These vary in complexity from an empirical multiplication factor applied to a single count to a mathematical model which fits both a pupping and a mortality function to a series of pup counts. These models are discussed in relation to recent aerial survey data.

Introduction

The British Isles hold substantial populations of two species of seal, the common or harbour seal (*Phoca vitulina*) and the larger and more abundant Atlantic grey seal (*Halichoerus grypus*). Current published UK population estimates are 92,000 for grey seals and 25,000 for common seals (Anon. in press), representing about 50% and 2% of the respective world populations.

Interest in the status and dynamics of British seal populations developed

ZOOLOGICAL SYMPOSIUM No. 58
ISBN 0-19-854006-X

during the 1950s when complaints of seal predation at Scottish coastal salmon fisheries began to increase (HMSO 1963; Lockie 1959; Rae 1960; Rae & Shearer 1965). At about the same time it was noted that the Farne Islands grey seal population was increasing (Coulson & Hickling 1964). Obviously any attempt to produce management policies or predict their likely effects requires knowledge of population sizes and trends. A programme of annual censuses was established in 1967 with the aim of monitoring the size and status of all British grey seal breeding populations (Vaughan 1971).

There are two possible strategies for estimating seal numbers. Where a systematic sampling programme can be established, mark–recapture techniques may be applicable; for example, numbers of common seals in the Wash (Summers & Mountford 1975), Weddell seals in the Antarctic (Siniff, DeMaster, Hofman & Eberhardt 1977) and harp seals in Canada (Bowen & Sergeant 1983), have been estimated using these techniques. In the case of grey seals the lack of organized hunting and the practical difficulties inherent in both tagging and sampling have until now precluded the use of mark–recapture techniques. Alternatively, the whole or some known component of the population can be counted directly. Seals spend the majority of their time at sea where they are widely dispersed and effectively uncountable; however, all species spend some time ashore each year. At no time is the entire grey seal population or even a random sample of it available for counting. We are therefore restricted to estimating the size of a component of the population which must be combined with behavioural and/or biological information in order to produce a total population estimate.

The component of the grey seal population of which a census can most readily be taken is the annual pup production. Estimates of the total number of pups born, combined with measurements of fecundity rate, provide a basis for calculating adult female population size (Hiby & Harwood 1984).

This paper describes the methods used in estimating the annual pup production of the British grey seal population.

Census methodology

Ground counting technique

On each breeding beach the spread of pupping dates exceeds the length of stay of individual pups, so no single census will include all the pups born. The total pup production of any colony can be obtained directly by 'through-counting' (Coulson & Hickling 1964). From the beginning to the end of the breeding season regular pup counts are made by observers walking through the colony. On the first visit all pups are counted and dye-marked. On each subsequent visit all undyed pups are counted and then dyed, so that on each visit the number of undyed pups counted is equivalent to the number of pups born since the previous count. These newborn counts are summed to give a cumulative total of births. Because of the number and wide distribution of grey seal breeding colonies (Fig. 1), obtaining such estimates of pup production at all colonies would require an immense amount of manpower and time. This is clearly not a practical or economic option.

Our approach has been to obtain regular counts of the number of pups on each colony using vertical aerial photography.

Aerial survey technique

A twin-engined fixed-wing aircraft fitted with a 5-inch format aerial survey camera is used to carry out photographic surveys of known breeding colonies at weekly intervals. Because pup counts are based entirely on these aerial photographs it is essential that they be of extremely high quality, both to obtain reliable total counts and to distinguish, if possible, between different developmental stages. Hiby, Thompson & Ward (in press) describe development of a camera system which has enabled us to obtain very high-resolution colour photographs of the total area of each breeding colony. The main problem was to overcome the effect of ground movement, which was achieved by development of an image-motion compensation system for the camera.

Subsampling versus complete enumeration

In our survey we aim to enumerate the total number of pups present on each breeding beach. The question arises whether in general it would be more appropriate to estimate the total number of animals by subsampling the area occupied and similarly to estimate the proportions falling into various categories by subsampling the animals. In our case subsampling is not a viable strategy because the reduction in counting effort is not sufficient to justify the resulting variance in the final count. For example, even if we could arrange for the number of pups included in the subsample to be binomially distributed, then, on an island with a total population of less than 100 pups, the subsample would need to include over half the pups to obtain a final count with a coefficient of variation of less than 10%. Subsampling would be more worthwhile for larger populations, but for a subsequent analysis each seal breeding group must be dealt with as a separate entity.

Estimating pup production from aerial survey data

With the improvements in aerial survey techniques referred to above, we can produce accurate counts of the numbers of pups present on breeding

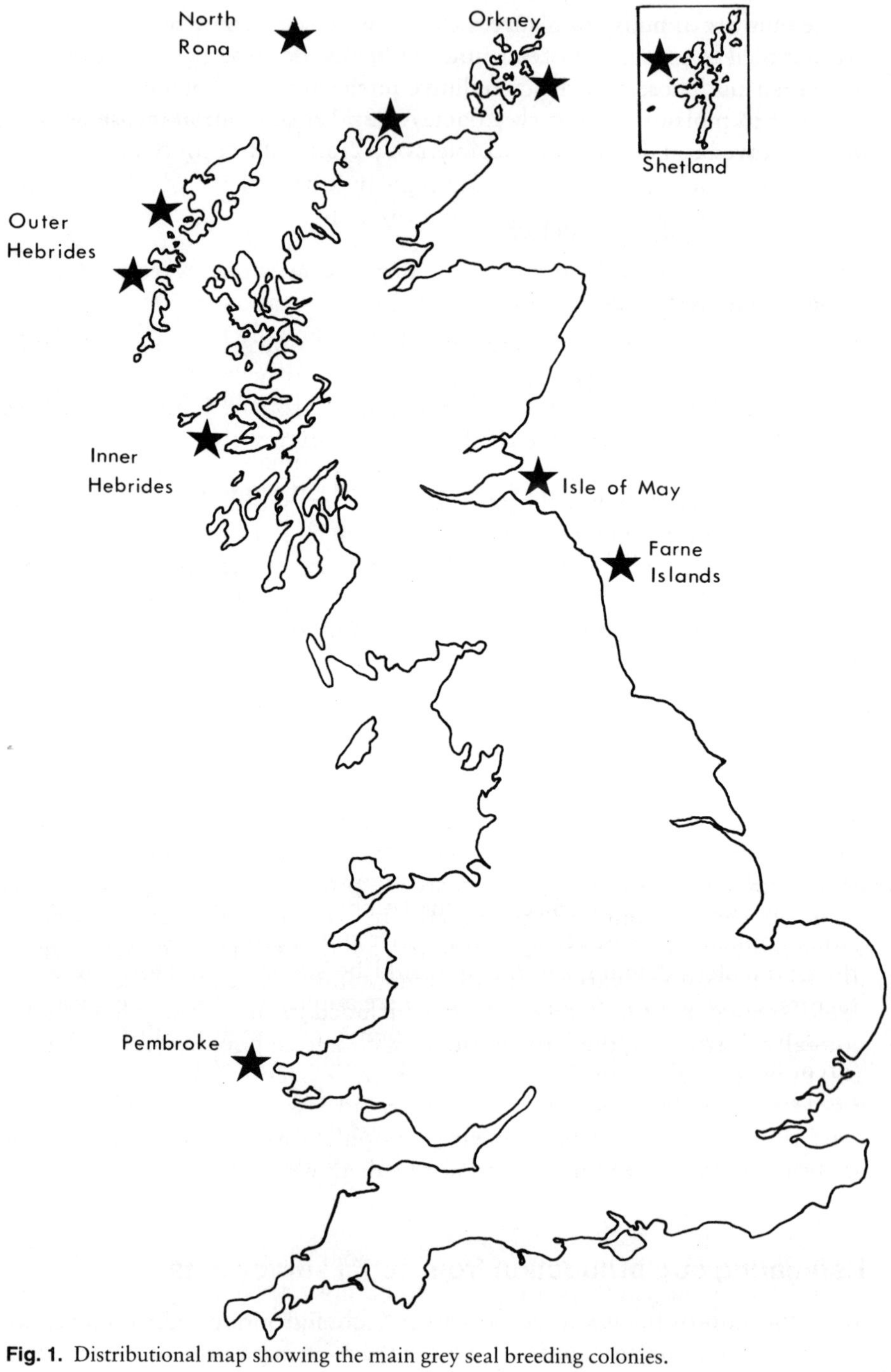

Fig. 1. Distributional map showing the main grey seal breeding colonies.

beaches at the time of each survey. It is not possible, however, to age pups with sufficient accuracy to determine whether they were included in the previous census. Therefore, the cumulative births method ('through-counting') is not applicable to aerial survey data. A number of models of varying complexity have been developed to convert pup counts into pup production.

Empirical multiplication factors

The simplest method of obtaining pup production estimates from air survey counts is to apply an empirically derived multiplication factor to the maximum count. This technique has been applied annually to aerial survey data since the late 1970s. Because aerial surveys are conducted at weekly intervals throughout most of the breeding season, the maximum air survey count must occur within three days of the time when the maximum number of pups are ashore and should be a reasonable estimate of the peak number ashore. Through-count data have been used at a number of colonies to derive factors which relate maximum number of pups ashore to the cumulative total of births (Summers 1978; Summers, Burton & Anderson 1975; Anderson & Curry 1976). Such factors can be applied to those islands and years for which only aerial survey data are available, if we can assume that the parameters defining the factors are constant within an island group and with respect to time.

The ratio of peak number of pups ashore to total number born depends on both the spread of pupping dates and the length of time individual pups remain in the colony. Clearly these parameters vary between the main island groups, producing a range of factors, e.g., 1.45 for Auskerry in Orkney (SMRU unpublished data), 1.6 for Stockay in the Outer Hebrides (Anderson & Curry 1976) and 1.8 for North Rona (Summers *et al.* 1975).

In the past it was assumed that the Orkney, Outer Hebrides and North Rona populations were reasonably distinct and that within one island group the parameters defining the factor would be similar. Therefore the above factors, derived for individual islands, were applied to whole island groups.

Examination of recent aerial survey counts for a number of islands in different years and of through-count data (Summers *et al.* 1975) shows that the ratio of maximum number of pups ashore to total pup production can vary significantly even between adjacent islands (Fig. 2) and that for individual islands the shape of the 'pups ashore curve' can vary between years (Fig. 3).

Application of inappropriate factors could lead to significant errors in estimating total pup production for individual islands. Deriving factors for each island would be expensive and would not resolve the problem of temporal variation. It is therefore necessary to develop alternative estimation techniques which can be fitted to annual air survey counts.

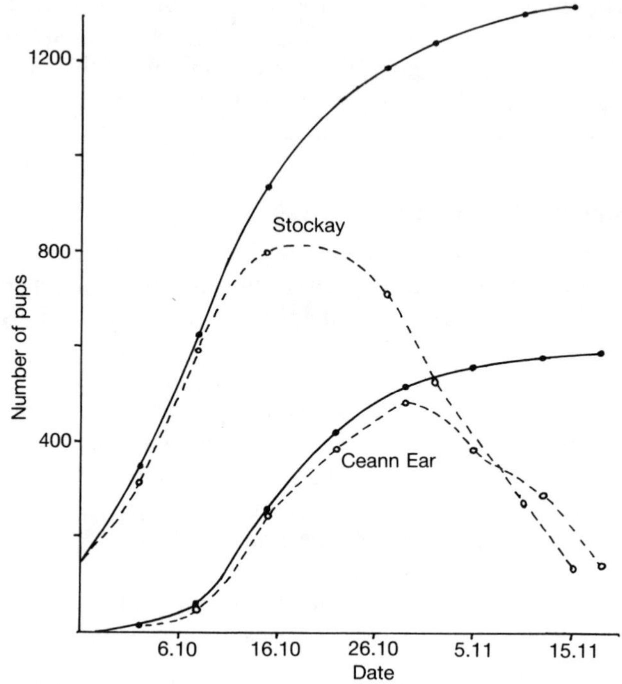

Fig. 2. Relationship between the number of pups born and the number present at any time for the Outer Hebridean islands of Stockay and Ceann Ear. The broken lines indicate the number of live pups on the beach and the continuous curves describe the distribution of cumulative births. Redrawn from Anderson & Curry (1976).

Maximum likelihood estimation

The data available from aerial surveys are the number of live pups on the beach at a sequence of times t_j, $j = 1. \ldots .M$. We may in addition be able to classify the pup counts into a number of developmental stages, but we defer consideration of this aspect until later. Let the total number of live pups at time t_j be l_j; we wish to derive the likelihood for $\{l_j\}$; however, it is useful to extend the model to include the type of data generated by through-counts where the total number of pups that die before t_j, say d_j, and the number of pups born between t_{j-1} and t_j, say n_j, are also known.

We begin by deriving the joint probability for the set of observed newborn frequencies, $\{n_j\}$. This is obtained by choosing an appropriate probability density function (PDF) for time of birth, say $f(t_b)$, and integrating this function over the intervals $t_s \rightarrow t_1, t_1 \rightarrow t_2, \ldots, t_{m-1} \rightarrow t_m$ to give the probabilities $p_1, p_2, \ldots, p_m$. Assuming time of birth is independent for different pups the joint probability for $\{n_j\}$ is equal to:

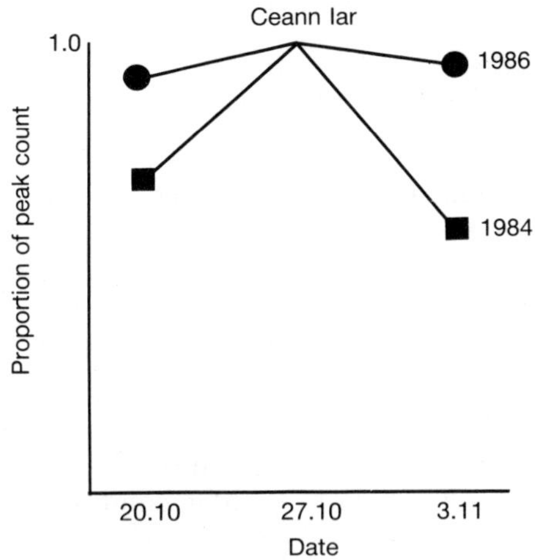

Fig. 3. Number of live pups, expressed as a proportion of the maximum number counted at one time, for the island of Ceann Iar in 1984 (■) and 1986 (●).

$$\frac{N!}{n_1!\,.\,n_2!\,......\,n_m!}\prod_{j=1}^{m} p_j^{n_j}$$

The through-count data suggest that the PDF for $f(t_b)$ is not symmetrical with respect to t_b. Originally we used a log-normal density for the random variable t_b–t_s, where the parameter t_s represents the earliest possible birth time. However, a density based on two rather than three parameters, which performs adequately, can be derived by assuming that *per capita* birth rate increases linearly with slope b from zero at time t_s. This gives:

$$f(t_b) = b(t_b - t_s)\exp(-b(t_b - t_s)^2/2)$$

We next define the joint probability of $\{l_j\}$ and $\{d_j\}$ conditional on the observed n_j values. The general approach we have used is to define for the ith pup the following four probabilities.

p_{ij} = probability that the ith pup is alive and on the beach at time of jth visit, t_j.

q_{ij} = probability that the ith pup is dead and on the beach at t_j.

pp_{ijl} = probability that the ith pup is alive and on the beach at times t_j and t_l, $l > j$.

pq_{ijl} = probability that the ith pup is alive at time t_j and dead and on the beach at time t_l.

Initially these are defined conditional on the time of birth t_b of the ith pup by assuming pups suffer a constant probability rate of dying, λ, and a probability rate of leaving which is zero up to a certain age, a^*, and constant at θ thereafter. This is consistent with results from the Isle of May (O.G. Wyile pers. comm.) shown in Fig. 4, which plots, for a sample of known-age pups, the log of the number remaining on the beach at age a, against a. This figure can be approximated by two straight-line segments meeting at age a^*—the value of a^* was observed to vary with location. The unconditional probabilities are then obtained by integrating these conditional probabilities times $f(t_b)$ with respect to t_b.

Having defined the four probabilities, which depend on the parameters t_s, b, λ, θ and a^*, it is straightforward to derive the vector of expectations and the matrix of variances and covariances for the random variables l_j and d_j, if we assume the probabilities of dying and leaving are independent for different pups.

We assume the joint probability distribution $\{l_j\}$ and $\{d_j\}$ can be adequately approximated by the multivariate normal density, which is fully defined by the vector of expectations and the variance-covariance matrix. This density for $\{l_j\}$ and $\{d_j\}$ is conditional on the observed n_j values; multiplying by the joint density for the n_j, defined above, thus gives the likelihood for the n_j, l_j and d_j combined.

In the case of through-count data, then, we can use this likelihood to obtain maximum likelihood estimates for the parameters defining birth, death and leaving. The procedure allows us to compare these values for different locations and different times and to identify those parameters which are most consistent, and those whose values cannot be predicted for a given location and year.

Moving to consideration of the aerial survey data, the likelihood for $\{l_j\}$ is a much simplified version of that derived for the through-count data but still depends upon the full complement of birth, death and leaving parameters. In addition we have the extra parameter N, the unknown total pup production at that colony. Our approach is to maximize the likelihood for $\{l_j\}$ with respect to N and those parameters which cannot be predicted in advance.

In attempting to derive the maximum information from the data the procedure described above is inevitably lengthy and expensive in computer time; however, this is not unreasonable in relation to the effort involved in obtaining the data. Two further problems may arise. First, the number of

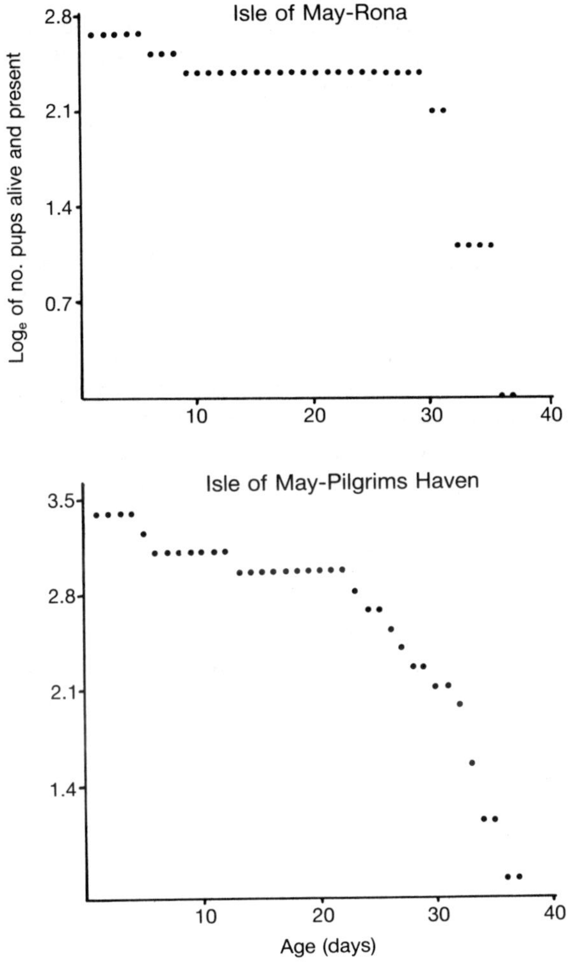

Fig. 4. Relationship between the log$_e$ of the number of pups alive and present on the breeding beach and the age of the pups. Data from known-age pups are shown for two locations on the Isle of May. (Data from O. G. Wyile, unpublished.)

parameters which cannot be set in advance may be too large to allow estimation of pup production and those parameters from the available air count data. Secondly, even for those parameters which appear, from the through-count data, to be relatively consistent, the values obtained by through-count may not be appropriate to air survey data, e.g. pup mortality or the rate of leaving may be affected by the through-count process. One solution is to increase the number of data points provided by the air counts, either by increasing the number of surveys or by classifying counts into two or more developmental stages. The latter option is now available owing to improvements in the aerial survey technique described above. There are no problems in principle in extending the likelihood to include staged counts, e.g. p_{ij} is replaced by $p_{ij}(r)$, the probability that the ith pup is on the beach, alive and in stage r at time t_j.

Conclusions

It is likely that the need for accurate estimates of the UK grey seal population will continue for the foreseeable future. Interactions of seals with commercial fisheries continue to be of concern to the fishing industry. We are not in a position to predict the likely trajectory of the population. This depends primarily on the availability of breeding sites and food—which itself depends on future developments within the fishing industry. As a consequence, any decisons concerning management of grey seals must be based on accurate monitoring, which is also the only way to assess the effect of any action taken. For example, population estimates for the Outer Hebrides and the Farne Islands following adult culls in 1972 and 1975 showed a decline in pup production in those areas much greater than would have been predicted on the basis of the number of fecund females removed. At present our monitoring of the grey seal population is based solely on annual pup production estimates. As described above, there are problems with this procedure in relation to possible variation of model parameters with time, location and the sampling technique used. It is therefore important to devise some form of independent procedure to compare with the results of the present monitoring technique. For example, we are currently investigating the possibility of conducting a large scale mark–recapture exercise based on recognition of natural pelage markings of female grey seals.

References

Anon. (In press). Seal stocks in Great Britain. Surveys conducted in 1985. *NERC News J.*

Anderson, S.S. & Curry, M.G. (1976). *Grey seals at the Monach Isles, Outer Hebrides, 1975*. ICES CM 1976/N:9 (Mimeo).

Bowen, W.D. & Sergeant, D.E. (1983). Mark-recapture estimates of harp seal pup (*Phoca groenlandica*) production in the northwest Atlantic. *Can. J. Fish. aquat. Sci.* **40**: 728–42.

Coulson, J.C. & Hickling, G. (1964). The breeding biology of the grey seal, *Halichoerus grypus* (Fab.), on the Farne Islands, Northumberland. *J. Anim. Ecol.* **33**: 485–512.

Hiby, A.R. & Harwood, J. (1984). N.E.R.C. Interactions between grey seals and UK fisheries. In *Contract report to Department of Agriculture and Fisheries for Scot-*

land: 219–34. (Ed. Harwood, J.). Department of Agriculture and Fisheries for Scotland, Edinburgh.

Hiby, A.R., Thompson, D. & Ward, A.J. (In press). Census by aerial photography—an inexpensive system based on non-specialist equipment. *Wildl. Soc. Bull.*

HMSO (1963). *Report of the consultative committee on grey seals and fisheries.* Her Majesty: Stationery Office, London.

Lockie, J.D. (1959). Grey seals and salmon fisheries on the Northumberland and Berwickshire coast. *Salm. Trout Mag.* **157**: 165–75.

Rae, B.B. (1960). Seals and Scottish fisheries. *Mar. Res.* **1960** No. 2: 1–39.

Rae, B.B. & Shearer, W.M. (1965). Seal damage to salmon fisheries. *Mar. Res.* **1965** No. 2: 1–39.

Siniff, D.B., DeMaster, D.P., Hofman, R.J. & Eberhardt, L.L. (1977). An analysis of the dynamics of a Weddell seal population. *Ecol. Monogr.* **47**: 319–35.

Summers, C.F. (1978). Trends in the size of British grey seal populations. *J. appl. Ecol.* **15**: 395–400.

Summers, C.F. & Mountford, M.D. (1975). Counting the common seal. *Nature, Lond.* **253**: 670–1.

Summers, C.F., Burton, R.W. & Anderson, S.S. (1975). Grey seal (*Halichoerus grypus*) pup production at North Rona: a study of birth and survival statistics collected in 1972. *J. Zool., Lond.* **175**: 439–51.

Vaughan, R.W. (1971). Aerial photography in seals research. In *the application of aerial photography to the work of the Nature Conservancy Council*: 88–98. (Proceedings of the Nature Conservancy Staff Seminar held in Edinburgh 14–15 October 1970). (Ed. Goodier, R.). Her Majesty's Stationery Office, Edinburgh.

Symp. zool. Soc. Lond. (1987) No. 58: 193–210

Census techniques for small populations, with special reference to the Mediterranean monk seal

A. R. HIBY and
J. S. JEFFERY

Sea Mammal Research Unit
c/o British Antarctic Survey
High Cross, Madingley Road
Cambridge CB3 0ET

Synopsis

This paper describes a pilot study designed to determine the status of the Mediterranean monk seal (*Monachus monachus*) around the Greek island of Kefallinía. Monk seals, once common around Kefallinía, are now extremely rare and this has dictated the choice of census technique used. The paper outlines development of photographic equipment for recording the presence of seals on cave haul-out sites. It presents results obtained from this equipment so far and suggests methods of analysis appropriate to data potentially available. The equipment and methods of analysis may be useful for studying and monitoring other populations of rare animals, particularly those with natural markings sufficiently distinctive to provide the potential for individual recognition.

Introduction

In recent years there has been an increase, particularly in developing countries, in the number of areas designated as wildlife reserves or sanctuaries. Though this phase of establishment of reserves may be drawing to a close (Frankel & Soulé 1981) the need for more research directed to wildlife conservation is not; indeed, it could be argued that the response by the scientific community has, as yet, failed to match the commitment of those involved in establishing the reserves.

The most important information is on the ecology of the most vulnerable (often the rarest) species; however, census of those species may also play an important role. First, a rough idea of the size of a population and its age and sex composition may indicate its viability and suggest what immediate action is required. Secondly, repeated censuses can provide a measure of the effect of any conservation steps taken or of any external influences such as

ZOOLOGICAL SYMPOSIUM No. 58
ISBN 0-19-854006-X

pollution or disturbance by tourism. Such monitoring of remnant populations may provide extremely valuable information if the census techniques used can generate indices of abundance of sufficiently low variance.

Although census techniques for exploited populations have had considerable attention, two factors limit the applicability of much of that research to remnant populations. First, many of the techniques rely, to some extent, on data derived from the exploitation itself, and secondly, exploited populations are likely to be 'large', whereas those of concern to conservationists will be 'small'. That smallness, if it results in low density, presents special problems for census: estimates based on small sample sizes have higher variance, and asymptotic properties of the chosen estimators are not available. One advantage of dealing with a small population is that the potential for using individual recognition to provide 'mark–recapture' data is increased.

Since June 1985 the Sea Mammal Research Unit has been involved in an international project of research concerning the status and biology of the Mediterranean monk seal (*Monachus monachus*) on the Greek island of Kefallinía. Mediterranean monk seals are a highly endangered species; the Greek islands of the central Aegean and Ionian Sea contain a large part of the remaining world population. For information on monk seal biology and general status see Sergeant, Ronald, Boulva & Berkes (1978), Goedicke (1981) and Marchessaux & Muller (1985). Increased levels of human activity around the resting and breeding sites may be one reason for the continuing decline of the population; for this reason a large part of our effort during the study in Kefallinía was directed to developing equipment which would allow the seals to be studied in a non-disruptive way. The data derived using this equipment do not provide a basis for a census of the seal population of Kefallinía; however, the techniques developed do have the potential for use in census of these very rare animals. This paper describes the equipment, field techniques and potential methods of analysis in the context of the monk seal project; we believe these methods may have relevance to the problems of census of rare animals in general.

Objectives and choice of field techniques

The objectives of the population studies in Kefallinía were:

1. To confirm the presence of a resident population.
2. To determine its size and geographical distribution.
3. To determine its age and sex composition.
4. To determine the degree of interchange between different parts of the island and with adjacent islands.
5. To establish a monitoring programme.

The range of 'benign' techniques available can be grouped into three classes: telemetry, active survey, and mark–recapture.

There are many examples of the value of telemetry in studies of wildlife, particularly in relation to objective 4, which itself is very important for deciding on appropriate conservation steps. However, there are considerable problems associated with attachment of transmitters to adult monk seals, and to avoid the risk of death or injury to the seals the use of telemetry for this study was rejected.

By 'active survey' we mean any procedure intended to detect a sample of animals in such a way that the probability of detection for each animal is deduced primarily from the design of the procedure. Thus quadrat, strip transect and line transect sampling are examples. In mark–recapture techniques the probability of detection is derived solely from the data, not the design of the survey itself. For the monk seal study we rejected any form of active survey. Given our ignorance of monk seal behaviour (frequency and duration of haul-out, frequency and duration of dives) it was difficult to conceive of a survey design leading to a known probability of detection. Furthermore, for animals at such low densities, the number of animals detected often turns out to be too low to provide any useful inferences. We were therefore left with some sort of mark–recapture scheme, based on recognition of naturally occurring patterns or scars to avoid the risk of death or injury associated with the application of artificial marks.

To obtain data suitable for statistical analysis it is necessary for natural markings to be recorded systematically on photographs for subsequent matching. We considered two possible ways of obtaining such photographs. The one we adopted and describe here was to design a device (a 'camera trap'), to be left at potential haul-out sites, capable of detecting and photographing any seal as it hauls out onto the beach. The other approach, previously used in a study of a resident population of monk seals in the Northern Sporades island group (Kouroutos, Papapanayotou & Matsakis 1985) is to swim silently into potential haul-out caves and take photographs using a hand-held camera of any seals hauled out. Which of these methods would be most effective was not clear. Whereas the passive 'camera trap' is favoured by a high frequency of hauling out, in the hand-held camera method the probability of detection depends on the proportion of time a seal spends at one or other of the haul-out sites. A simple model can help to evaluate the relative merits of the two methods with respect to haul-out behaviour. Assume a seal can use any of m potential sites and chooses them at random with probabilities p_1 to p_m. Suppose the mean time between successive haul-outs is t days, and the mean haul-out duration is d days. Suppose the options open are either to visit all m sites with a hand-held camera, requiring a total of n days, or to leave a camera trap at a randomly selected site. If the camera trap is left at the same site throughout the period, the probability of obtaining at least one photograph of the seal is:

$$1 - \frac{1}{m} \sum_{j=1}^{m} (1 - p_j)^{n/t}$$

The detection probability thus depends on the variability in the probabilities p_j, and is reduced if the p_j are very discrepant. It can be increased by regularly relocating the camera to a maximum of $1 - (1 - \frac{1}{m})^{n/t}$ if the camera is moved every t days. The probability of getting a photograph by visiting each cave with a hand-held camera is, to a first approximation (by ignoring the probability of encountering the seal more than once) simply d/t, the proportion of time a seal spends hauled out. Thus on this basis it is not worth using the camera trap approach unless the proportion of time hauled out is less than $1 - (1 - \frac{1}{m})^{n/t}$. For example, if we have 100 potential sites ($m = 100$) which can be surveyed over a period of 10 days ($n = 10$) and the seal hauls out daily ($t = 1$), placing a single camera trap would not be a sensible alternative to visiting all the sites if the seal spends more than a proportion $1 - 0{\cdot}99^{10}$ or 9.5% of its time hauled out. If the number of camera traps can be increased to c, the maximum detection probability becomes $1 - (1 - (1-c/m)^{n/t}$. Thus, if, in the above example, we can afford six camera traps instead of one, that approach may be worthwhile if the seal spends less than a proportion $1 - 0.94^{10}$, or 46% of its time, hauled out.

Another important consideration is which technique is less likely to cause disturbance to seals at the haul-out sites. On the few occasions when we encountered a seal hauled out in a cave in Kefallinía, the seal became aware of our approach and bolted from the cave in panic. Kouroutos *et al.* (1985) state that sleeping seals encountered during the research in the Northern Sporades were rarely woken by the approach of researchers.

Six camera traps have been constructed and deployed in Kefallinía since January 1986. The following section gives details of the development of camera housings and detectors. In outline, the technique was to mount a device on one wall of a chosen cave haul-out site to 'look' across the beach at approximately the level of the waterline so that any seal emerging from the water would be detected. Detection caused the firing of a camera also mounted on the cave wall; a delay of about 30 s was imposed between successive firings of the camera. The cameras have been deployed continuously at a small number of pre-selected sites, rather than being relocated at regular intervals as would be required to obtain data for census (see below). This was in order to test the efficiency and reliability of the equipment at sites likely to experience the highest frequencies of visits by seals, before embarking on a wider-scale application.

Development of camera/detector equipment

The objective of the equipment was to obtain high-resolution photographs of seals hauling out at chosen sites. One early approach we considered was to eliminate the need for a detector by firing the camera at regular intervals. However, we decided to use a detector since this detects the seal at the moment it reaches a certain location, thus defining the required orientation of the camera.

Detector

Inexpensive infra-red detectors, which work well for terrestrial mammals, were tested at Piperi, one of the Northern Sporades group. The unit tested was the IR 986, marketed in England by Racal. However, there was insufficient temperature difference between the seal and its surroundings to trigger the device. Therefore, the detector presently used at most of the potential haul-out sites is a simple 'trip wire' arrangement. A length of fishing line is stretched across the cave at a height of about 30 cm, parallel to and above the water's edge. When a seal encounters the line, the resulting increase in tension in the line closes a switch at one end which releases the camera shutter. In order to follow roughly the contours of the beach the line often needs to be deflected at one or more points—this is achieved by running short spur lines to convenient attachment points to pull the main line into a suitable configuration. One problem with the trip wire detector is that it may cause disturbance. However, in tests with captive grey seals, similar lines did not elicit any reaction when they were encountered.

A more sophisticated optical detector was developed for this study by Spirkl and Ries Technology in Munich, West Germany, and this is also in use on some sites. A very faint light beam is directed from the detector, mounted in a housing on one wall of the cave, onto a 'cat's-eye' reflector on the opposite wall and is thus reflected back to the detector. Here it is focused onto a light-sensitive element. A break in reception of the reflected beams initiates closure of an electronic switch which releases the camera shutter. The principle of phase-lock amplification is used to allow the device to operate with a light beam of far lower intensity than used in commercially available units. This allows the device to operate on battery power and avoids any risk of the beam being visible to the seals. It also means that any variation in the total amount of incident light received by the detector, other than that caused by a break in the beam, does not initiate shutter release, so the detector is equally effective in daylight and in darkness. The maximum one-way range is 40 m, which is adequate for all the potential cave haul-out sites in Kefallinía.

The optical detector carries no risk of causing disturbance and, effectively installed, detects seals with a lower probability of failure than is inherent in the trip-wire arrangements, which may be fouled by pieces of flotsam or be broken. However, it is far more time-consuming to install and is therefore more suitable for semi-permanent installation than for frequent relocation. To avoid either type of detector firing the camera when triggered merely by waves surging up the beach, they are wired to the camera system via a float level switch, located on the cave wall slightly above the water surface. This breaks the circuit linking the detector and camera when an incoming wave causes the water level to rise and open the switch before it can trigger the detector.

Camera

A Canon T70 camera with 50mm lens and 'databack' was chosen. The camera has an integral power winder and electrical shutter release.

All photographs are taken using flash illumination. Some of the caves have some daylight penetrating back as far as the beach but most are too dark for daylight photography and, of course, artificial illumination is required to record seals hauling out during the night. The use of flash illumination carries a risk of disturbing seals at the haul-out sites. The photographs obtained to date using the camera traps provide no evidence of reaction by seals to the camera flash; however, photographs showing a fright reaction would not be expected because the shutter release mechanism has been modified to maintain a minimum delay between successive photographs. It is clear from the camera trap photographs that some seals are unaffected at least to the extent that they are not dissuaded from hauling out.

Infra-red sensitive film can be used in conjunction with an IR filter covering the flash tube but photographs of sufficient resolution to permit individual recognition are then more difficult to obtain except from the very small cave sites. The film type used in the camera traps is Ilford XPI which is a high-speed and high-resolution monochrome film with a very wide range of acceptable illumination levels. This is important because it is not possible to predict how far from the camera and flash the seal will haul out.

The flashgun used (a Vivitar 283) is extensively modified to use a high-voltage external battery pack to supply the capacitor instead of the usual charging circuit. This is because the capacitor must remain charged at all times, and the usual continuous charging involves an unsupportable battery drain.

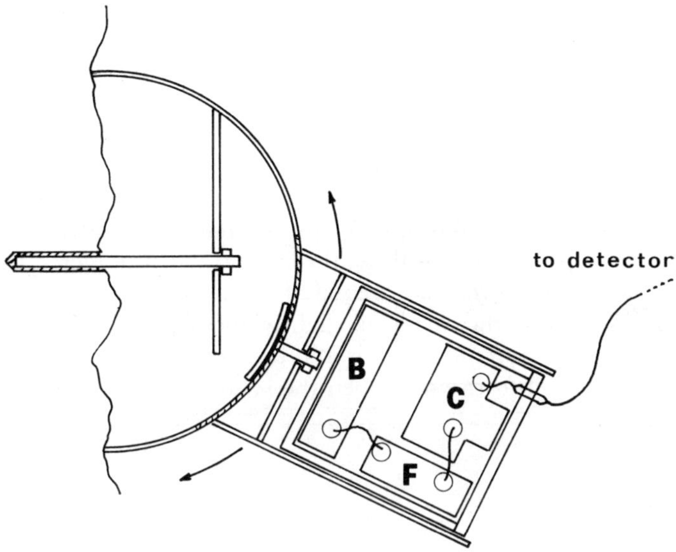

Fig. 1. Vertical cross-section of the camera housing and attachment system. C, camera; F, flashgun; B, high-voltage flash unit battery. The hatched section of the hemisphere illustrates the extent of the slot cut to allow lateral movement of the steel tube.

Housing and attachment systems

In designing the housing/mounting system for cameras and detectors we attempted to meet the following objectives.

1. Protection of the cameras and detectors from moisture and mechanical damage during storms, when some of the caves are completely flooded.
2. A system of attachment to the cave wall strong enough to avoid cameras and detectors being swept away by waves, but simple enough for quick installation and adjustment to minimize the time spent in haul-out sites.
3. A system of attachment allowing rapid camera relocation.
4. A camera housing design allowing rapid access to cameras for film changing and servicing.
5. Use of surplus materials to reduce costs.

The housing (Fig. 1) for the camera, flashgun and high-voltage battery is constructed of ABS plastic drainage pipe with a removable clear 'perspex' front port. The attachment system is constructed in 3mm gauge galvanized mild steel. It consists of a hemispherical section held onto the cave wall by a single bolt, and a cylinder held to the surface of the hemisphere by a single bolt. When slackened, the second bolt can move along a slot cut into the hemisphere so that the cylinder can move across the surface of the

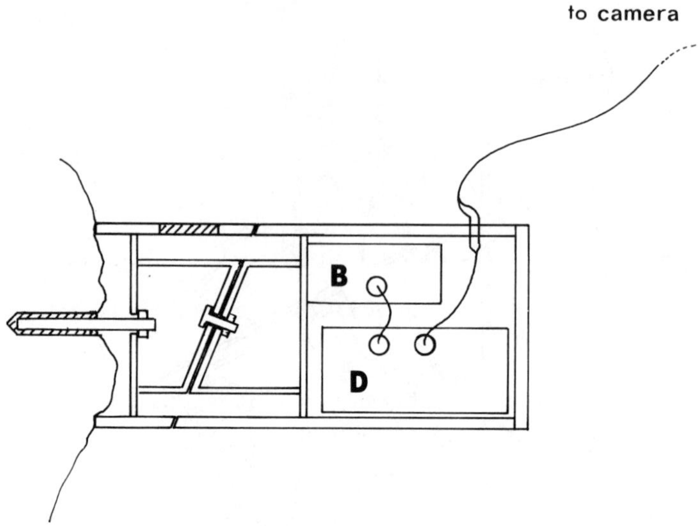

Fig. 2. Vertical cross-section of the optical detector housing and attachment system. D, detector; B, battery. The hatched section of the part bolted to the rock wall illustrates the port cut to allow access to the bolt joining the two parts of the housing/attachment system.

hemisphere. The hemisphere itself can be rotated about its fixing bolt, thus the cylinder can be adjusted to point in any direction. The plastic camera housing slides into the metal cylinder and is retained by a hardened steel security clamp. To permit servicing of the camera and film changing the whole camera housing is exchanged for another housing containing a serviced and reloaded camera. That operation requires only 1 or 2 min to complete. Installation of the attachment system requires about 30 min. The cave wall has to be drilled to accept the hemisphere fixing bolt which is retained in the wall using a resin bonding system which requires 10 min to set. Expanding bolts cannot be used as they fracture the rock wall. Relocation of the attachment system from one wall bolt to another and adjustment require only a few minutes.

The housing attachment system for the optical detector is shown in Fig. 2. The housing is constructed from the same plastic pipe as the camera housing with a similar transparent front port. At its rear, the pipe is cut off at an angle. The off-cut is retained to the cave wall by a single bolt, and the two pipe sections are then rejoined using a single bolt running normal to the cut faces. When the second bolt is slackened, via an access port on the side of the detector housing, the two pipe sections are free to counter-rotate, causing them to join at an angle varying with that rotation. The off-cut section is free to rotate on the cave wall, and so the detector housing can be adjusted

Fig. 3. Locations of the potential seal haul-out sites selected for study using the camera equipment.

to point in any direction. For the detector, that adjustment is critical, because the light beam has to be aimed to hit the cat's eye reflector attached to the opposite wall of the cave.

Results and qualitative inferences available

From January to August 1986 up to six cameras were located in caves selected as potential haul-out sites on the basis of the proximity of previous sightings, or the presence of tracks or faeces. The aim was to test the effectiveness of the equipment rather than generate data for statistical analysis. The position of these caves is shown in Fig. 3, and the details of camera coverage obtained are shown in Table 1. Total coverage obtained was 432 camera days, i.e., about one third of the potential 1,260 days from the six camera systems available. This reduction in coverage was due to a combination of logistic problems involving the boats and engines used to reach the caves, and the problem of theft of equipment from some caves during the tourist season in June and July. Coverage after the end of the tourist season in August was increased to near the potential available.

Table 1. Relative usage of selected caves, January to August 1986.

Cave No.	Seal visits/Days of camera coverage
39	4/65
45	8/160
43	0/54
36	0/54
35	1/10
I1	1/11
I1a	4/78
Totals	18/432

Table 1 also provides an indication of the relative importance of the chosen sites as seal haul-outs. The data illustrate the difficulty in prejudging the suitability of potential sites from their location or topography. For example, Cave 39, yielding four visits in 65 days, is inside the harbour at Assos, and therefore not remote from human presence. Cave 43 had no visits during 54 days of coverage despite its apparent suitability as a refuge during the peak of the tourist season, given its remote location and hidden entrance.

Table 2 demonstrates the presence of at least four seals in the northern Kefallinía/Ithaca region during the first half of 1986. Most photographs are of seal D, a 'large' male. Records from Caves 39, 45 and Ithaca (I) 1A define a minimum extent of the range of this animal. Its regular appearance at Caves 45 and Ithaca 1A, and the absence of other seals at those locations, suggest that it is the only seal 'resident' in the northern part of the channel separating Kefallinía from Ithaca. The visits listed in Table 2 occurred during both day and night with a peak of activity between 8 pm and midnight. Some durations of stay were surprisingly short, less than 30 min, which suggests that one reason for visiting caves may be to check for presence of other seals, as well as to haul out for rest. Seal D has a heavily scarred neck and head, perhaps resulting from aggressive interactions with other seals over many years. The average duration of stay for seals encountered during visits by researchers to caves is likely to be longer than the mean for the population as a result of size-biased sampling: a visit by a researcher is more likely to occur during a long visit by a seal than a short one.

Methods of quantitative analysis

Mark–recapture technique for small populations

No useful statistical analysis of the results presented in the previous section is possible, for even if more data had been obtained during this period, the

Table 2. Seals photographed in chronological order.

Seal	Cave Number	Date
A	39	6.2.86
B	39	17.2.86
C	39	23.2.86
U	45	4.3.86
D	45	5.3.86
U (NA, NC)	45	5.3.86
D	45	13.3.86
D	39	16.3.86
D	45	24.3.86
U (ND)	35	25.3.86
D	45	25.3.86
D	45	26.3.86
D	45	27.3.86
U	I1	14.4.86
D	I1a	15.4.86
D	I1a	19.4.86
D	I1a	28.4.86
D	I1a	13.5.86

Letters A to D represent distinct seals.
U represents an unidentified seal.
NA, NC and ND represent unidentified but 'not A', 'not C' and 'not D' respectively.

method of deployment of the cameras was designed to test the equipment rather than to fulfil the assumptions required for statistical analysis. In this section we consider how camera traps would need to be deployed to generate data suitable for analysis, and consider what methods of analysis would be appropriate. We also consider briefly some aspects of mark–recapture data of particular relevance to rare animals in general. The approach in this section is based on the classification of models given by Otis, Burnham, White & Anderson (1978), for closed populations.

Otis *et al.* (1978) recognize three main types of unequal capture probabilities: 'heterogeneity', meaning inherent differences between the capture probabilities of different animals; trap response; and temporal variability in capture probabilities affecting all animals equally. We suggest that for monk seals, and perhaps many other rare animals, the effect of the first two types of variation in capture probability cannot be dealt with effectively at the stage of data analysis. A robust procedure for estimating the size of a population subject to heterogeneity has been proposed (Burnham & Overton 1978) but it is also emphasized, in Otis *et al.* (1978), that for 'adequate estimation of population size' we require an experiment in which 'many animals are caught a relatively large number of times'. It is also intuitively clear that the need to base a model of heterogeneity solely on capture histories will make heavy demands on the data. An obvious exception is when the variation in capture probability is associated with, say, sex or developmental stage, which can be determined from the photographs. As to trap response, although the situation in which all animals are equally affected by capture can be analysed, this is not likely to apply to monk seals—if, for example, seals are frightened by the camera flash there is no reason to assume that they will all be affected to the same degree. In general we would have to deal with model M_{bh} or M_{tbh} rather than M_b or M_{tb} (in the notation of Otis *et al.* 1978).

If these types of variation in capture probabilities cannot be dealt with at the analysis stage then the only option is to try to eliminate them at the sampling stage. In the case of trap response, that means reducing the potential disturbance of 'capture'—the implication of this for the design of the camera traps has been considered in the previous section. We can try to reduce a major potential source of heterogeneity—that due to differences in accessibility of traps as influenced by individual home ranges—by relocating the traps on a random basis. For example, for each new sampling effort we could relocate our c available camera traps at c potential sites randomly selected from the total number of potential sites. That does not guarantee equal capture probabilities, because animals hauling out more frequently and/or using a greater variety of sites will have a higher probability of being photogaphed, but it does reduce the extreme variation resulting from individual home ranges if traps are not relocated. In this context the technique of swimming into selected caves, to obtain photographs using a hand-held camera, has a distinct advantage over the camera trap approach, for then equal capture probabilities are assured by selecting sites at random if all seals spend the same proportion of time hauled out.

To select sites at random it is not essential to enumerate all potential cave haul-out sites in advance. For example, the coastline could be divided into a number of sections, and those sections regarded as potential sites, provided each section is sufficiently small to cover all caves in that section when it is chosen. Sites should not be chosen by taking points at random along the coastline and choosing the nearest cave, because this causes capture probabilities for animals in areas containing a high density of potential sites to be reduced in comparison to those in low-density areas.

It may appear extreme or even eccentric to take such steps in sampling design before determining the existence of trap response or heterogeneity from the data; however, given small sample size, it may not be possible to identify such problems even if they exist.

If the population can be regarded as closed for the duration of the experiment, and we can assume that heterogeneity and trap response have been eliminated as far as possible, then we are left, in general, with model M_t of Otis *et al.* (1978), and a *ML* estimate for population size N which maximizes the log-likelihood function:

$$ln(N!/(N - M_{t+1})!) + \sum_{j=1}^{t} n_j \, ln\,(p_j) + \sum_{j=1}^{t} (N - n_j)\, ln\,(1 - p_j)$$

M_{t+1} represents the number of distinct seals photographed during the t sampling occasions, n_j is the number of distinct seals photographed on the jth occasion and p_j is the capture probability on the jth occasion. The *ML* estimate for p_j is n_j/N. A formula for estimating the variance of $\hat{N}$ is available but would not be reliable in the case where random relocation of traps is used to eliminate heterogeneity, for even if this procedure assigns an equal capture probability p_j to each animal, it does not ensure independence.

A simplifying strategy is available in the case of rare animals if the probability of obtaining more than one capture per sampling occasion is negligible, i.e., $n_j = 0$ or 1. In that case each animal has, independently, an equal probability $1/N$ of being the next photographed. If K successive photographs show S distinct animals then the *ML* estimate for N maximizes the likelihood:

$$\frac{K}{\prod_i f_i! \prod_i (i!)^{f_i}} \; N!/N^K(N - S)!$$

This likelihood is the joint probability for the set of frequencies f_i, where f_i is the number of animals appearing on exactly i photographs. The sufficient statistics are $K = \Sigma i f_i$ and $S = \Sigma f_i$.

Maximizing the likelihood with respect to N leads to the equation

$$\sum_{i=0}^{s-1} (N - i)^{-1} = \frac{K}{N}$$

which can be solved iteratively for N. The same model has been considered by Lewontin & Prout (1956) for estimation of the number of classes in a population. For this very simple model exact 95% confidence limits are available: a lower limit N_L is the value of N for which the probability of observing S or more distinct animals among the K photographs is α_1, and an upper limit N_u is the value of N for which the probability of observing S or fewer distinct animals among the K photographs is α_2; α_1 plus α_2 equals 0.05. (An obvious choice is $\alpha_1 = \alpha_2 = 0.025$, but if N_L is equal to the number of distinct animals photographed then $\alpha_2 = 0.05$ is the appropriate value for determining N_u). The required probabilities can be easily obtained because this mark–recapture scheme is an example of the classical occupancy problem where it is required to calculate the probability that exactly S cells will be occupied (S distinct animals photographed) when K balls (photographs) are distributed at random among N cells (population of animals). Using the theory developed for the occupancy problem (Feller 1950) we obtain N_L and N_u as the population levels satisfying:

$$\sum_{T=s}^{\min(K,N_L)} \binom{N}{T} \sum_{r=0}^{T} (-1)^r \binom{T}{V} \left(1 - \frac{N-T+V}{N}\right)^K = \alpha_1$$

and

$$\sum_{T=1}^{S} \binom{N}{T} \sum_{r=0}^{T} (-1)^r \binom{T}{V} \left(1 - \frac{N-T+V}{N}\right)^K = \alpha_2$$

J. E. Cooke (pers. comm.) has derived an explicit approximation to the ML estimate and has also considered the case of unequal capture probabilities. If S is greater than $K/2$ then

$$N = \frac{S(K-1)}{2(K-S)} + \frac{K+S+1}{12}$$

is within 1 of the ML estimate. He considered effect of unequal capture probabilities by using a gamma distribution to model the variation in the probability rate of capture. This leads to the following generalization for the equation defining the ML estimate of N:

$$\sum_{i=0}^{S-1} (N-i)^{-1} = ln(1 + KV/N)/V$$

where V is the square of the coefficient of variation of the catchability. This provides, for given K and S, one measure of the sensitivity of the estimate for N to failure of the assumption that the probability of capture is constant at $^1/_N$ (i.e. that V equals 0).

Problems unique to mark-recognition data

In the treatment above it has been assumed that each animal photographed is uniquely identifiable, as would be the case if seals were tagged. There are two ways in which this assumption is violated in practice. First, not all seals are equally well-marked, so that we may wish to classify photographs into two groups: those showing seals which, if photographed again from a similar position, are certain to be identified and a second group which, although distinguishable from all members of the first group, are not sufficiently well marked to be certain of identification on later photographs. The analyses must then be extended to incorporate this classification. Secondly, each photograph shows only one view of the animal—in the case of the monk seal camera traps, which are mounted low on the cave walls to one side of the beach, the view is of either the left or right side. Occasionally we will be able to identify marks on both sides but in general it will not be possible to match photographs of different sides of the same seal. One solution to the second problem is to deal with left-hand and right-hand views as entirely distinct data sets leading to separate analyses and estimates. If possible, sampling should be arranged so as to try to take photographs from a standard viewpoint. To deal with the first problem, we generalize the likelihood functions to include, in the case of the M_t model, the number r_j of animals first photographed on occasion j with markings sufficiently distinctive to be certain of identification in any later photographs (so that there are $(n_j - r_j)$ animals either with markings previously 'noted' or without distinctive markings) and the number C_{ij} of animals photographed on occasion j whose markings were first noted on occasion i. We obtain:

$$ln\left(\frac{N!(N-r_1)!(N-r_1-r_2)!\ldots\ldots(N-r_1-r_2-\ldots-r_{t-1})}{(N-n_1)!(N-r_1-n_2+c_{12})!(N-r_1-r_2-n_3+c_{13}+c_{23})!\ldots(N-r_1-r_2-\ldots-r_{t-1}-n_t+c_{1t}+c_{2t}+\ldots+c_{t-1,t})!}\right)$$

$$+\sum_{j=1}^{t} n_j ln(p_j) + \sum_{j=1}^{t} (N-n_j)ln(1-p_j)$$

This reduces to the previous likelihood if $r_j = n_j$.

One problem with this approach may be in determining the value of n_j if some seals are indistinguishable and may appear more than once. In the case of the occupancy problem approach the relevant part of likelihood is generalized to:

$$\frac{N!}{N^{\Sigma if_i}(N-S)!} \left(\frac{N-S}{N}\right)^{K-\Sigma if_i}$$

where $K - \Sigma if_i$ is the number of photographs of seals not considered to be sufficiently well marked for future identification.

It should be noted that, if population census is the main objective, a bias in favour of photographing well-marked animals must be avoided, for otherwise capture probabilities for 'marked' and 'unmarked' animals will differ. In this context the use of camera traps has an obvious advantage over hand-held cameras in avoiding any possible bias.

Open population analysis for monk seals

With long-lived animals such as monk seals, population closure during an experimental period of a few months should not be violated to a significant degree by birth or natural mortality, but migration to adjacent areas may be a problem. It is probably most reliable to check this possibility directly—taking the island of Kefallinía as an example, movement to the adjacent islands of Lefkas and Zakinthos could be identified by locating camera traps on those islands in caves nearest to Kefallinía. If migration does occur the approach by Darroch (1961) could be used to quantify the migration rates and estimate population size, provided sufficient data were available.

Note that if the main objective is to census the central island only, then it is not necessary to assume that marked and unmarked animals have identical capture probabilities except on that central island. Thus it is only necessary to modify sampling to avoid heterogeneity on the central island, and only a subset of potential sites need be monitored on the adjacent islands. It is necessary, however, that the probability of capture on the adjacent islands be identical for animals marked on the central and adjacent islands. Consequently 'marking' on the adjacent island should not be restricted to the mainland sites only, to avoid the danger that all animals marked on the adjacent islands are resident at the monitored sites whereas those arriving from the central island are transients.

Concluding remarks

Our experience with the monk seal study suggests that systematic collection of mark-recognition data provides the best potential for census of these very rare animals. We suspect that the same conclusion applies to many other situations. One method of collecting such data is by use of camera traps, and again we suspect that technique has application to other situations where animals visit a finite number of potential sites—monitoring water holes in arid regions is an obvious example. More active methods of collecting this type of data may also be available; an excellent example is afforded by the studies based on photographing fluke patterns on humpback whales (Hammond 1986). There may, in certain situations, be a potential for combining the mark–recognition approach with that of active survey. That is, if the probability of capture can be deduced from the design of the survey then that known probability can be incorporated into the mark–recapture analysis. It is certain that if we are to monitor the progress of remnant populations effectively the potential for any possible source of information will need to be fully investigated.

Acknowledgements

We gratefully acknowledge the assistance of all the researchers involved in the fieldwork in Kefallinía and particularly Aliki Panou, Harold Ries and Hilde Lemmer (from the University of Munich), Winifred Denk from Cornell University, Ithaca, New York State, and Jonathan Gordon from the University of Cambridge. The effectiveness of the optical detector equipment is the result of painstaking development work by Wolfgang Spirkl and Hertz Ries of Spirkl & Ries Technology, Reumanstr. 100, 8000 Munchen 40, West Germany. We thank Justin Cooke of the University of York for providing advice on some aspects of the mark–recapture analyses. The project was funded by the Natural Environment Research Council and the Commission of the European Communities under contract ENV 853 UK. We are also most grateful to the International Fund for Animal Welfare for a grant allowing us to extend the duration of the project and evaluate the field techniques used.

References

Burnham, K.P. & Overton, W.S. (1978). Estimation of the size of a closed population when capture probabilities vary among animals. *Biometrika* **65**: 625–33.

Darroch, J.N. (1961). The two sample capture-recapture census when tagging and sampling are stratified. *Biometrika* **48**: 241–60.

Feller, W. (1950). *An introduction to probability theory and its applications*. John Wiley & Sons, New York.

Frankel, O.H. & Soulé, M.E. (1981). *Conservation and evolution*. Cambridge University Press, Cambridge.

Goedicke, T.R. (1981). Life expectancy of monk seal colonies in Greece. *Biol. Conserv.* **20**: 173–81.

Hammond, P.S. (1986). Estimating the size of naturally marked whale populations using capture-recapture techniques. In *Behaviour of whales in relation to management*: 253–82. (Ed. Donovan, G.P.). IWC, Cambridge.

Kouroutos, V., Papapanayotou, D. & Matsakis, J. (1985). L'état de la population du phoque moine. In *Programme "Sporades du Nord"*: 45–73. (Project report to Directorate General for the Environment, Consumer Protection and Nuclear Safety, Commission of the European Communities).

Lewontin, R.C. & Prout, T. (1956). Estimation of the number of different classes in a population. *Biometrics* **12**: 211–23.

Marchessaux, D. & Muller, M. (1985). *Le phoque moine*, Monachus monachus, *distribution, statut et biologie sur la côte Saharienne*. Centre d'Etude des Resources Animals Marines, Université d'Aix-Marseille 111, Rue H. Poincaré 13397—Marseille-Cédex 13.

Otis, D.L., Burnham, K.P., White, G.C. & Anderson, D.R. (1978). Statistical inference from capture data on closed animal populations. *Wildl. Monogr.* No. 62: 1–135.

Sergeant, D., Ronald, K., Boulva, J. & Berkes, F. (1978). The recent status of *Monachus monachus* the Mediterranean monk seal. In *The Mediterranean monk seal*: 31–54. (Eds Ronald, K. & Duguy, R.) (*UNEP tech. Ser.* **1**.) Pergamon Press, Oxford.

Symp. zool. Soc. Lond. (1987) No. 58: 211–223

Estimating pup production of elephant seals at South Georgia

P. ROTHERY and
T. S. McCANN

British Antarctic Survey
Natural Environment Research Council
High Cross
Madingley Road
Cambridge CB3 0ET

Synopsis

The paper reports the results of a survey to estimate pup production of southern elephant seals (*Mirounga leonina*) at South Georgia in 1985. The cows ashore were counted at least once during the breeding season at every breeding beach on the island. A mathematical model for the haul-out pattern is developed using detailed counts from five other studies. The model enables observed counts to be 'corrected' and thereby used to estimate the size of the haul-out population. When the model was applied to the survey data, an observed maximum total of *c.* 87,600 breeding cows yielded an estimated total of 102,000 pups. Errors of estimation are evaluated and discussed. Some suggestions for further observations in the field and alternative methods of analysis are put forward.

Introduction

A recent paper (McCann 1985) on the size and status of southern elephant seal (*Mirounga leonina*) populations reviewed and re-assessed the available information on the various island stocks. However, at three of the four major breeding sites (South Georgia, Macquarie Island and Heard Islands) a census of southern elephant seals had not been taken since the 1950s (Laws 1960; Carrick & Ingham 1962). The establishment in Antarctic waters of commercial fin fisheries, which have been implicated in the decline of the elephant seal population at Kerguelen (Van Aarde 1980; Pascal 1985), and the long time-lapse since the previous surveys make new stock assessments timely.

In 1985, a survey was conducted on South Georgia. Every breeding beach on the island was visited at least once during the breeding season. Adult females were counted rather than pups, since the latter are difficult to count accurately, especially in dense breeding groups. This paper reports the results of the survey and describes a method for correcting the counts to

ZOOLOGICAL SYMPOSIUM No. 58
ISBN 0-19-854006-X

estimate pup production. The approach uses a simple mathematical model of the haul-out process and improves upon the existing methods of Van Aarde (1980) and Pascal (1981). The model is tested using detailed counts from earlier studies and then applied to the survey data.

Methods

South Georgia (54°S, 36°W) is an extremely mountainous island about 170 km long. Approximately 60% of the land area is covered in permanent ice and snow and many glaciers calve directly into the sea, thus preventing travel on land around the coast. Elephant seals breed on beaches around the entire coast and in order to visit these a yacht was chartered, from which the survey was conducted.

The yacht proved to be a very versatile and efficient workboat. Most counts were made either from the deck of the yacht or from the crow's nest, where an observer's eyes were about 15 m above sea level. The yacht was under power while working and would cruise along the edge of the shore, often only 20 m from the beach, while the counts were made. There were few beaches which could not be surveyed from the yacht; at such beaches landings were made to count from the shore.

When counting, observers normally used 8 × 30 power binoculars even when the boat was very close to shore. These had the effect of reducing the field of vision so that in dense groups of females the eyes were concentrated on a small number of females at any one time. Females were usually distributed along the shore in linear 'harem crowds' rather than spread evenly or randomly across the beach, and this made counting easier and more accurate.

Whenever possible, female groups were counted more than once, but most groups were counted once only on any particular day, as the movement of the yacht along the coast prevented recounts. Counts from the boat were compared with counts made on land. It is not possible to say whether boat counts or shore counts were the more accurate, as this depended to a large extent on the topography of the beach and the state of the sea.

Mathematical model

The number of elephant seal cows ashore during the breeding season displays a marked seasonal pattern building up to a peak in mid to late October and then declining (Pascal 1979, 1985; Van Aarde 1980; McCann 1985). From a detailed series of counts at one area, McCann (1985) estimated that at the peak approximately 95% of cows were ashore. Within 10 days of the peak this number fell to less than 75%. A crude estimate of the haul-out population size using a single count will generally underestimate by an amount depending on when the count was made, and McCann's (1985) results show that the bias could be large. The following mathematical model, which describes the haul-out process and the seasonal pattern of counts, provides a basis for calculating appropriate correction factors.

The model assumes that the times at which cows first come ashore follow a normal distribution with mean μ, measured in days from some convenient date, and standard deviation σ. Furthermore, it is assumed that all animals haul out for the same length of time, S. If N_H denotes the size of the haul-out population in a given area and $n(t)$ cows are counted ashore at time t, then on average

$$n(t) = N_H p(t,\mu,\sigma,S) \tag{1}$$

where $p\ (t,\ \mu,\ \sigma,\ S)$ is the expected proportion of cows hauled out at time t. This is made up of all the animals hauling out before time t, less those back at sea which hauled out before $t - S$. For normally distributed haul-out times

$$p(t,\mu,\sigma,S) = \Phi\left(\frac{t-\mu}{\sigma}\right) - \Phi\left(\frac{t-S-\mu}{\sigma}\right) \tag{2}$$

where Φ denotes the cumulative distribution function of the standardized normal distribution.

Equation 2 describes a curve, symmetrical about $t = \mu + S/2$ at which the maximum value of $2\Phi(\frac{S}{2\sigma}) - 1$ occurs. Thus, the proportion of the total haul-out population represented at the peak depends on the ratio of the time spent ashore to the standard deviation of the time coming ashore. In the extreme situation of all cows coming ashore together, corresponding to $\sigma = 0$, Equation 2 has a 'top-hat' shape with a flat peak lasting S days. Thus, σ is a measure of synchrony in the haul-out process.

For a particular beach in a given year the parameters μ, σ and S will be unknown and need to be estimated. In what follows, the observed counts will be used to estimate μ and σ but for S we follow McCann (1985) and assume that each female remains ashore for 30 days, comprising a seven-day pre-partum period and 23 days for lactation. Note that for a model population in which 95% of the cows are out at the peak, a 30-day period ashore implies a standard deviation of 7.5 days.

Other models for correcting elephant seal counts have been proposed by Pascal (1979, 1981, 1985) and by van Aarde (1980). Both approaches are based on empirical descriptions of the observed haul-out pattern. Pascal (1981) used a Gaussian, or normal, density curve to mimic the percentage of females ashore whereas van Aarde adopted a parabolic model. *A priori*, the parabola is a poor representation since negative values are possible and, as expected, the model is a bad fit at the beginning and end of the haul-out period. Pascal's model is a more plausible description of the haul-out curve but lacks correspondence with the haul-out process. Furthermore, the model is fitted to the observed percentage of cows ashore and therefore is applicable only when the size of the haul-out population is known. In this paper, and more generally, this is not the case. The present model, derived via the distribution of the times at which the cows come ashore, adds a different approach to existing methods and extends them. The model gives a good description of the data, contains parameters with a simple biological interpretation, and provides a flexible framework which could accommodate additional complexity—for example, a variable length of stay ashore—if the need arose.

Analysis of detailed counts

Relatively detailed series, consisting of nine or more counts, were available from the areas Dartmouth Point, 1976 and 1977, Hestesletten, 1951 and 1962, and Paul Beach, 1954. Before attempting an analysis of the survey data we turn to these more extensive observations at particular areas, first to test the applicability of the model, and secondly to assess its usefulness for estimating the size of the haul-out population.

For the parameter values μ and σ pertaining to a particular area, Equation 1 specifies a straight line relationship between $n(t)$ and $p(t, \mu, \sigma, S)$ which passes through the origin with slope equal to N_H, the size of the haul-out population. Least squares estimates of μ and σ can be obtained by choosing those values which minimize the residual variation about the regression line. The results of this exercise are summarized in Fig. 1a–e which illustrates the fit of the model to the five sets of data. In each case, the observed counts are expressed as a percentage of the estimated haul-out population size. The overall features of the haul-out distribution are reasonably well captured by the model. Estimated parameter values are given in Table 1a. The analysis suggests variation in the peak haul-out dates and possibly in the standard deviations of the times at which the animals come ashore, although the latter are more difficult to ascertain since they are tied to variations in S. The estimated maximum percentage ashore varied from 87–96%. Patterns at Dartmouth Point, 1977 (Fig. 1a) and Paul Beach, 1954 (Fig. 1e) appear to show similar, systematic departures from the model, dipping below it about one week before the estimated peak and rising above the curve afterwards. The effect, however, is not apparent at Dartmouth Point, 1976 (Fig. 1b) nor at Hestesletten (Fig. 1c and d), although there are relatively few observations at the latter.

With regard to the survey data, it is relevant to ask how well a single count, when corrected by using Equation 1, estimates the haul-out popula-

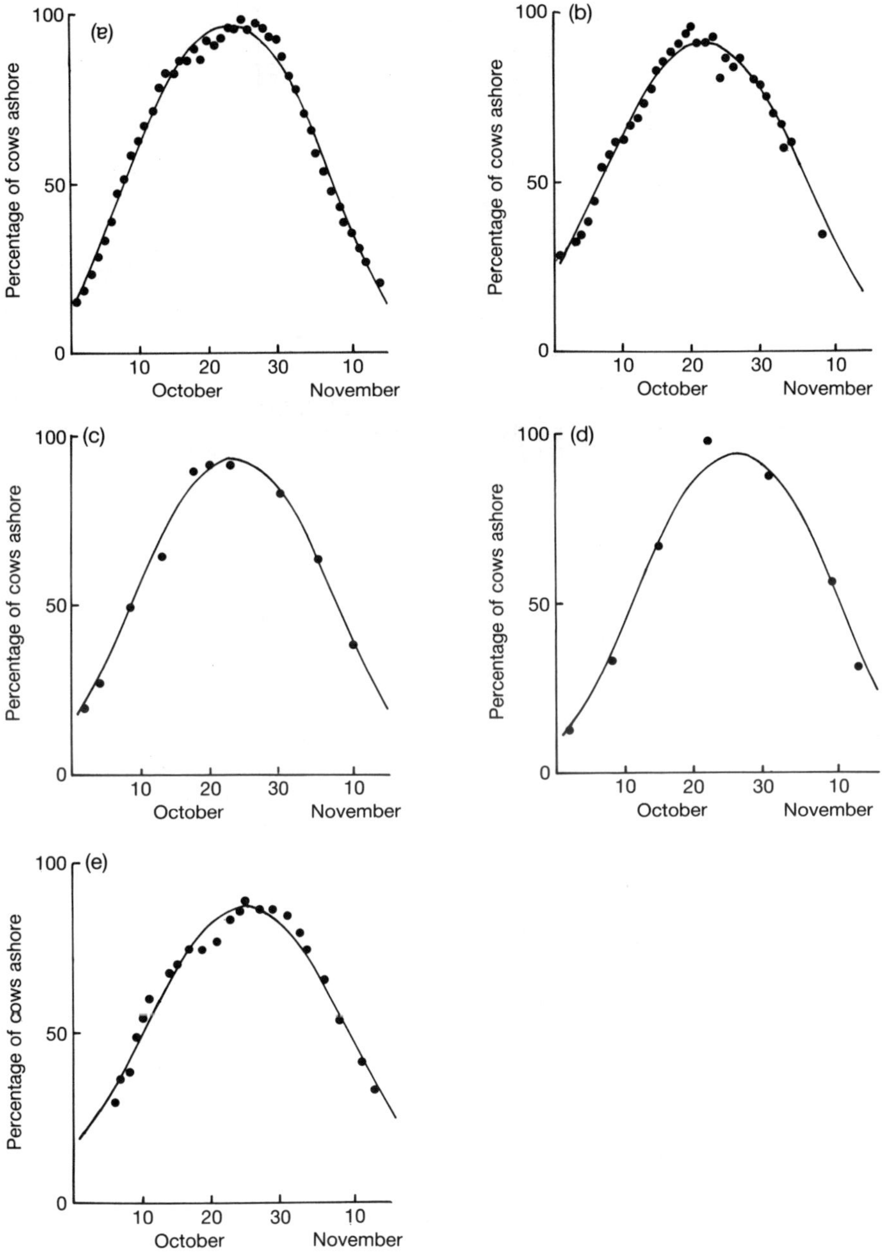

Fig. 1. Observed (●) and fitted (———) haul-out distributions: (a) Dartmouth Point, 1977; (b) Dartmouth Point, 1976; (c) Hestesletten, 1962; (d) Hestesletten, 1951; (e) Paul Beach, 1954. Estimated parameter values given in Table 1.

Table 1. (a) Estimates of haul-out population parameters for five study areas; (b) an assessment of the accuracy of using a single corrected count to estimate the size of the haul-out population.

Area	(a) Haul-out population parameter estimates					(b) Assessment of accuracy of a single corrected count				
	n	Population size (N_H)	Peak day from 1st October	Standard deviation (σ)	Percentage ashore at peak	Corrected counts n	Corrected counts Mean	Corrected counts Coefficient of variation	Maximum pup count	Date of last pup count
Dartmouth Point 1977	47	1030	23.1	7.3	96	31	1030	4%	1081	23/11
Dartmouth Point 1976	43	442	21.7	8.9	91	29	444	5%	425	8/11
Paul Beach 1954	23	409	25.2	10.0	87	17	414	5%	377	13/11
Hestesletten 1962	12	1815	23.6	8.4	93	6	1789	8%	1860	10/11
Hestesletten 1951	9	936	26.2	8.1	94	6	936	8%	1113	19/11

tion size. An assessment of this is possible for the above detailed counts by examining the variation in the corrected counts. In this exercise we restrict counts to those within 15 days of the peak date since most of the survey counts fall in a 30-day period from mid-October to mid-November. Table 1b gives means and coefficients of variation of corrected counts using the parameter estimates in Table 1a.

Dartmouth Point, 1977, provides the most reliable yardstick for validating the method because daily counts of cows and pups, both weaned and dead, are available. McCann (1985) reports a maximum count of cows, weaned pups and dead pups of 1,081, of which 34 were dead. Depending on whether the cows from dead pups stay ashore or not, this puts the haul-out population of cows at 1,047–1,081. Observed counts during the 15 days either side of the estimated peak date ranged from 484 to 1,001. When corrected, they yielded a mean of 1,030 with a coefficient of variation (CV) of 4%. Thus, in this favourable situation where relatively good estimates of haul-out parameters are available, the method appears relatively precise but underestimates by 2–5%.

The other areas are more difficult to assess because of the paucity of pup counts. For Dartmouth Point and Paul Beach no obvious underestimation is apparent. At Hestesletten in 1962 the estimated bias is at least 4% and in 1951 an even larger value of 16% emerges. However, with so few observations and a CV of 8% there is considerable uncertainty in these bias estimates.

The above analysis shows that the model developed in the previous section provides a reasonable description of the five haul-out patterns of elephant seal cows. Although no statistical tests have been applied, the findings suggest differences in haul-out distributions, but whether these differences reflect variation from beach to beach or from year to year, or both, is not clear. The results of Table 1 show the potential bias and precision of estimation using a single corrected count.

Analysis of the 1985 survey data

Beach-to-beach variation

Counts in the 1985 survey were made throughout the breeding season but, of the individual beaches, 54 were visited only once, 67 twice and 23 on three or four days. Therefore a detailed analysis along the lines of the previous section is not possible. Progress can be made by assuming that the same haul-out pattern prevails at the different areas, or at least over certain stretches of coastline. The assumption is capable of limited examination using those areas with three days' counts. For each area, the model given by Equation 1 can be fitted exactly to yield estimates of the haul-out population size, the peak haul-out date and the standard deviation of the haul-out

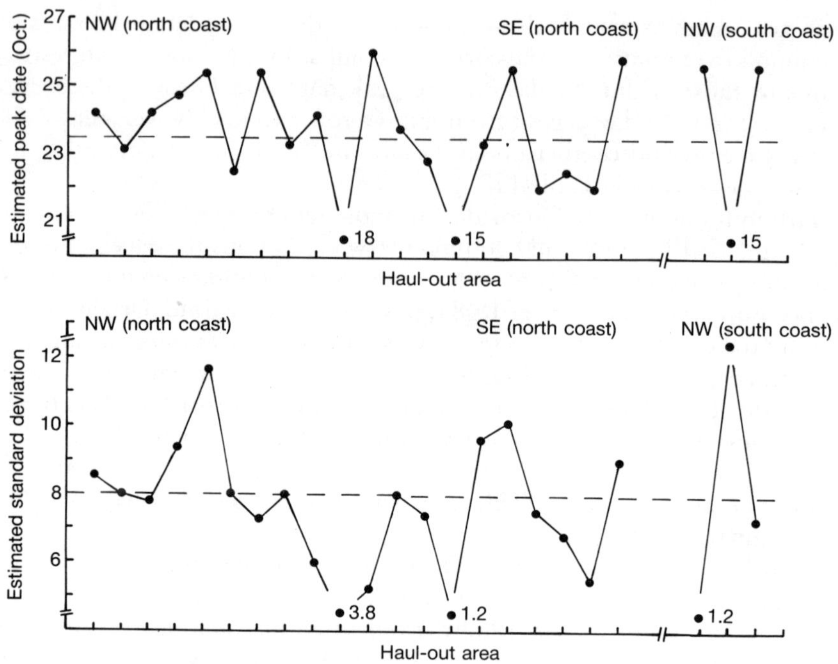

Fig. 2. Estimated peak dates and standard deviations of the haul-out distributions at selected survey areas with three counts. Information displayed schematically to look for pattern of variation around the coast. Note broken axes. Broken lines denote median values.

times. The estimates are then examined in relation to the locations of the beaches around the coast to look for pattern. The findings are represented schematically in Fig. 2. There are marked fluctuations in parameter estimates even between nearby areas, but some variation is to be expected by virtue of the scatter about the lines (see Fig. 1). Three estimated peak times occur unrealistically early, before 18 October, probably because of the noise in the data. Overall, however, there are no obvious trends on a large scale and no suggestions for separating particular sections of the island. Using medians to summarize the figures suggests a peak haul-out date of 23–24 October, with a standard deviation of the times of haul-out of eight days.

The possible dependence of timing and synchrony of the pattern on haul-out population size was also examined, by calculating correlation coefficients between the parameter estimates. Peak date showed no relationship with population size ($r = 0.06$) when all areas were considered: with an outlying area of more than 6,000 cows removed, the observed correlation is negative but not statistically significant ($r = -0.30$, $p > 0.10$). Larger haul-out populations appeared neither more nor less synchronous ($r < 0.28$), with or without the very large outlying group.

Estimating pup production

The analysis of the survey data proceeds by assuming a single underlying haul-out pattern specified by the mean (μ) and standard deviation (σ) of the times at which the animals come ashore. The approach is to estimate μ and σ and then use Equation 1 to correct either the single count or the largest of the two or three counts at each area. The use of the corrected maximum count requires some qualification, since in some situations overestimation could conceivably occur—viz. counts taken on days symmetrical about the peak date which differ only by chance variation. This presented no particular problems for the survey data, since an unambiguous maximum was apparent except for only one case of which the effect is negligible.

The proposed procedure to estimate μ and σ necessarily uses those areas with at least two counts and in particular utilizes only the two largest counts. The relatively few discarded counts occurred early in the season when numbers were very low and any loss of efficiency incurred by omitting them is considered to be small and acceptable when offset against the simplicity of the analysis based on ratios of two counts.

For the *i*th area let R_i denote the ratio of the second largest count to the largest. For the model given by Equation 1, the approximate mean of R_i does not depend on the size of the haul-out population at the area and is given by

$$E(R_i) = \frac{p(t_{1i},\mu,\sigma,S)}{p(t_{2i},\mu,\sigma,S)} = \varrho_i,\ i = 1,\ldots,k$$

where t_{1i} and t_{2i} are the times of the two counts.

For the actual values of μ and σ the regression of R_i on ϱ_i is a straight line through the origin with slope equal to one. Therefore, scatter diagrams of R_i against ϱ_i, as shown in Fig. 3, for a range of values of μ and σ, can be used to obtain initial parameter estimates. These have then been refined by solving, iteratively, the Generalized Least Squares equations (McCullagh 1983). Briefly, if the variance of R_i is A/w_i, the corresponding equations are:

$$\sum_{i=1}^{k} w_i \frac{\partial \varrho_i}{\partial \mu}(R_i - \varrho_i) = 0$$

$$\sum_{i=1}^{k} w_i \frac{\partial \varrho_i}{\partial \sigma}(R_i - \varrho_i) = 0$$

These yield approximately unbiased estimates for μ and σ with approximate variance-covariance matrix

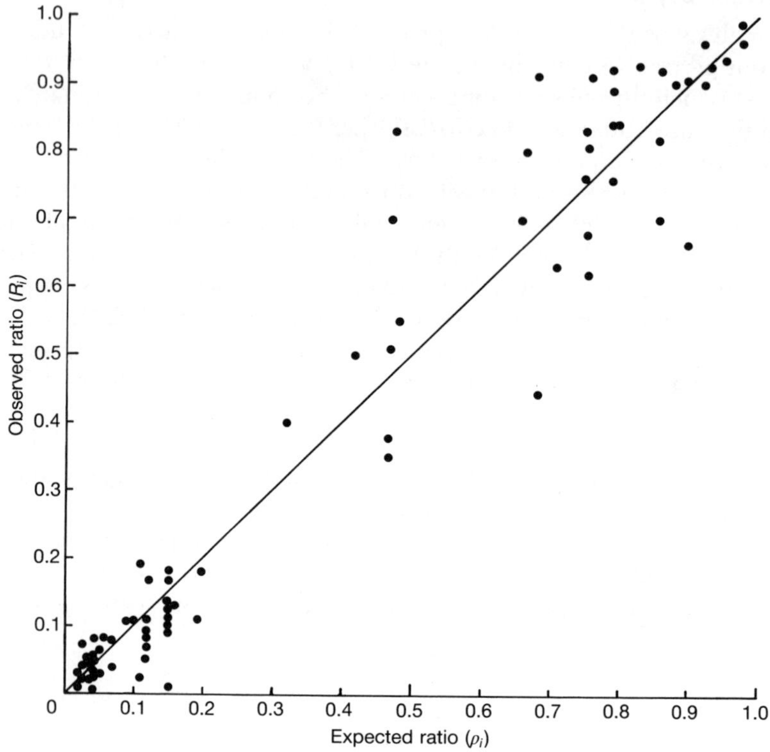

Fig. 3. Observed ratios of the second largest to the largest count (R_i), against their expected values (ϱ_i) for an estimated peak date of 23 October and standard deviation of the haul-out distribution of 7.7. Line corresponds to $R_i = \varrho_i$.

$$A\begin{pmatrix} \sum_{i=1}^{k} w_i\left(\frac{\partial \varrho_i}{\partial \mu}\right)^2 & \sum_{i=1}^{k} w_i \frac{\partial \varrho_i}{\partial \mu}\frac{\partial \varrho_i}{\partial \sigma} \\ \sum_{i=1}^{k} w_i \frac{\partial \varrho_i}{\partial \mu}\frac{\partial \varrho_i}{\partial \sigma} & \sum_{i=1}^{k} w_i\left(\frac{\partial \varrho_i}{\partial \sigma}\right)^2 \end{pmatrix}^{-1}$$

An estimate of A is $\sum_{i=1}^{k} \hat{w}_i(R_i - \hat{\varrho}_i)^2/(k-2)$, where 'hats' indicate values calculated using estimates of parameters.

To seek an approximate set of weights, w_i, the residuals about the line in Fig. 3 were examined in relation to the values of $\hat{\varrho}_i$ and the estimated haul-out population sizes of the different areas. The analysis showed more variable residuals for intermediate values of ϱ_i; and less variation for larger haul-out populations. This suggests that the observed counts vary about the underlying haul-out pattern in proportion to $N_H p(1 - p)$, where p is the expected proportion of the cows ashore when the count was made. The weights used in the analysis have therefore been derived on this assumption using the approximate variance of the ratio of two uncorrelated variables (Kendall & Stuart 1977: chapter 10).

For the survey data, the above procedure yields an estimated mean of 38.0 (*S.E.* = 0.4) and standard deviation of 7.7 (*S.E.* = 0.3). The constant of proportionality, A, is estimated as 28.2.

The final step in the analysis is to estimate the size of the total haul-out population for South Georgia and to assess its precision. If the count to be corrected at the ith area is n_i, observed at time t_i, then the estimated total population size is given by

$$\hat{N}_{HT} = \sum_{i=1}^{k} \hat{N}_{Hi} = \sum_{i=1}^{k} \frac{n_i}{p(t_i,\hat{\mu},\hat{\sigma},30)}$$

where N_{Hi} is the size of the haul-out population for the ith area. The variance of $\hat{N}_{HT}$ is given approximately by

$$\sum_{i=1}^{k} \frac{1}{p_i^2} V(n_i) + V(\hat{\mu}) \left[\sum_{i=1}^{k} \frac{N_{Hi}}{p_i} \frac{\partial p_i}{\partial \mu} \right]^2$$

$$+ V(\hat{\sigma}) \left[\sum_{i=1}^{k} \frac{N_{Hi}}{p_i} \frac{\partial p_i}{\partial \sigma} \right]^2$$

$$+ 2\, COV(\hat{\mu},\hat{\sigma}) \left[\sum_{i=1}^{k} \frac{N_{Hi}}{p_i} \frac{\partial p_i}{\partial \mu} \right] \left[\sum_{i=1}^{k} \frac{N_{Hi}}{p_i} \frac{\partial p_i}{\partial \sigma} \right]$$

The first term in the above expression measures errors of estimation arising from the observed variation in counts about the prevailing haul-out pattern. Each element in the summation is the variance of a single count corrected by a factor $1/p_i$. The previous analysis of detailed counts suggests that this contribution can be estimated by using the earlier, empirical finding that the coefficient of variation of corrected counts is approximately 5%.

The last three terms in the expression for $V(\hat{N}_{HT})$ reflect variation due to estimating the parameters μ and σ. An estimate of this component is obtained by substituting estimates of μ, σ and N_{Hi} in the formula.

In numerical terms, the above algebra and analysis amounts to an estimated total of 102,000 cows with a standard error of 2,100. Thus, the crude estimate of *c.* 87,600 based on uncorrected counts has been increased by 16%.

Discussion

The required correction for the counts emerges as relatively small. In this sense the model vindicates the design of the 1985 survey which has been successful in obtaining a count relatively close to the peak haul-out date at each area.

Superficially, the derived 95% confidence limits of 102,000 ± 4,200 seem surprisingly small when expressed as 4% of the estimated total population size. However, relative to the estimated numbers not actually counted the figure is 29%. Furthermore, the quoted errors need to be qualified. First, the analysis using detailed data of Dartmouth Point suggests that the method may underestimate by as much as 5% despite the apparent small coefficient of variation. Secondly, the assumption of a single haul-out pattern implicit in the method is capable of very limited checking from the data. Third, the derivation of the errors is based on a somewhat tentative model for the variability of the counts around the seasonal build-up. Specific sources of error, and, in particular, errors of counting, have not been identified and it would be useful to obtain some separate information on this. An interesting observation is that the above error of 4% is comparable to the figure of 2% quoted by Pascal (1979) for the Kerguelen elephant seal population, although details of the derivation of his figures are not given in the paper.

The above comments call for additional data to validate and develop the model. Some suggestions are: more detailed counts at particular areas on both cows and pups; observations on the behaviour of cows in respect of the variability in the pre-partum and lactation periods and the response to pup loss; and counts at several areas in the same season, perhaps over a range of haul-out population sizes.

The present analysis corrects maximum counts to estimate haul-out numbers and uses ratios of counts to estimate the haul-out pattern. This *ad hoc* approach is largely tailored to the present survey data which usually have only one or two counts at each area. A more general framework is currently being developed. This augments the average pattern specified by Equation 1 with a plausible error structure for the counts, which incorporates some serial dependence, and estimates the population parameters accordingly using the Generalized Least Squares equations. This leads to a weighted linear regression based on Equation 1 to estimate haul-out population sizes, as opposed to the more arbitrary use of a corrected maximum count.

Acknowledgements

We thank B. C. Osborne for his excellent work in the field, and D. Brown, J. P. Croxall, L. Hiby and M. D. Mountford for constructive comments on the manuscript. We also thank Dr R. M. Laws and W. N. Bonner for the use of their counts from Hestesletten and Paul Beach. Finally, we would especially like to thank Jerome and Sally Poncet for their enthusiasm and hard work which combined to make the survey a success.

References

Carrick, R. & Ingham, S.E. (1962). Studies on the southern elephant seal, *Mirounga leonina* (L.). V. Population dynamics and utilization. *CSIRO Wildl. Res.* **7**: 198–206.

Kendall, M.G. & Stuart, A. (1977). *The advanced theory of statistics* **2**. (4th edn) Griffin, London.

Laws, R.M. (1960). The southern elephant seal (*Mirounga leonina* Linn.) at South Georgia. *Norsk Hvalfangsttid.* **49**: 466–76; 520–42.

McCann, T.S. (1985). Size, status and demography of southern elephant seal (*Mirounga leonina*) populations. In *Studies of sea mammals in south latitudes*: 1–17. (Eds Ling, J.K. & Bryden, M.M.). South Australian Museum, Adelaide.

McCullagh, P. (1983). Quasi-likelihood functions. *Ann. Statist.* **11**: 59–67.

Pascal, M. (1979). Essai de dénombrement de la population d'éléphants de mer (*Mirounga leonina* (L.)) des îles Kerguelen (49°S, 69°E). *Mammalia* **43**: 147–59.

Pascal, M. (1981). Évolution numérique de la population d'éléphants de mer (*Mirounga leonina* L.) de l'Archipel des Kerguelen au cours des 30 dernières années. *CNFRA Publs* No. 51: 119–35.

Pascal, M. (1985). Numerical changes in the population of elephant seals (*Mirounga leonina* L.) in the Kerguelen Archipelago during the past 30 years. In *Marine mammals and fisheries*: 170–86. (Eds Beddington, J.R., Beverton, R.J.H. & Lavigne, D.M.). George Allen & Unwin, London.

Van Aarde, R.J. (1980). Fluctuations in the population of southern elephant seals *Mirounga leonina* at Kerguelen Island. *S. Afr. J. Zool.* **15**: 99–106.

Symp. zool. Soc. Lond. (1987) No. 58: 225–245

Techniques for estimating the size of whale populations

P. S. HAMMOND

Sea Mammal Research Unit
c/o British Antarctic Survey
High Cross, Madingley Road
Cambridge CB3 0ET

Synopsis

Estimates of the size of whale populations are required for rational conservation and management, both to monitor the recovery of populations which have been reduced to low levels by heavy commercial exploitation and to calculate catch quotas for those species which are still exploited. Population estimation is particularly difficult for marine species such as whales which spend most of their time underwater. This can cause a problem with population definition, present practical difficulties in experimental design and carrying out censuses or surveys, and affect the interpretation of the results of some estimation methods.

Techniques of estimation are described through the examples of four populations: line transect sampling of the Southern Hemisphere minke whale, capture–recapture analyses of photographs of natural markings of the North Atlantic humpback whale, land-based visual census of the California gray whale, and the combination of visual and acoustic ice-based censuses of the Western Arctic bowhead whale. The size of populations of those species which are characterized by predictable, generally coastal migratory behaviour can be estimated by methods which make use of this, such as censuses and capture–recapture methods. It is no coincidence that these species were also the first to suffer heavy commercial exploitation, and these methods are frequently used to monitor their recovery. Line transect sampling is the estimation technique commonly used for pelagic, widely dispersed populations which include those species exploited commercially in the modern whaling era.

Introduction

The rational conservation and management of a whale population requires, among other things, an estimate of its size. There are two closely related reasons why this is so. The first is to monitor the recovery of a population which has been heavily exploited in the past but is now not hunted for commercial purposes. If no whales are being taken then our interest is largely to observe the rate of recovery. If, however, the population is subject to aboriginal subsistence whaling then an additional and most important consider-

ZOOLOGICAL SYMPOSIUM No. 58
ISBN 0-19-854006-X

ation is whether or not the population is able to sustain any hunting and, if so, how much. The second reason is to calculate a catch quota for a population which is still exploited commercially. This has become less and less important over the years as more and more populations have become protected from commercial whaling by the International Whaling Commission (IWC).

Whales present several practical problems in population estimation and the fact that they live in the sea contributes to these. Perhaps the most fundamental problem is that of distribution and population definition. It is often difficult to know the extent of a population because any physical boundaries which may delimit one from another are less apparent in the ocean. Compounding this is the tendency for whale populations to be widely dispersed. Even if we can define a population to our satisfaction, the physical extent of that population may present considerable practical problems for surveys. Populations which cannot be censused from land require ships or aircraft to carry out a sampling survey. These research platforms are expensive to operate and maintain and the amount of effort expended on such a survey is frequently determined by financial constraints rather than by the experimental design. Another problem is that whales spend most of their time underwater and can only be seen at the surface for a small proportion of the time. Not only can this increase the amount of study time needed but it can also affect techniques of population estimation by causing violation of assumptions. This particularly applies to line transect sampling.

Techniques of population estimation can be divided into two categories: total enumeration and sampling. For whales, counting all the animals in a population is only a feasible option if the entire population is known to migrate very close to a suitable land-based census point. Even where this is the case and a census is attempted, there are several reasons why animals may not be seen and these have to be accounted for. Total enumeration by census is described below through the examples of the California gray whale and the Western Arctic bowhead whale.

Many sampling methods of population estimation rely on the physical removal of animals from the population. In the past, large numbers of whales have been taken from many of the world's populations and the placing of marks in whales which were then recovered via the whaling industry used to be widespread. However, population estimates using mark–recapture techniques on these data have frequently been criticized because of failure of the necessary assumptions and lack of precision. Data from the whaling industry have also been used extensively to calculate trends in relative abundance using catch-per-unit-effort indices. Data dependent upon the exploitation of whales can be important in some applications of determining abundance but this paper concentrates on estimation methods which do not rely on whales being killed. Now that the days of large-scale whaling have long gone, these methods are more relevant to the present and the foreseeable future.

There are two ways to estimate the size of a live whale population by sampling. The first is to search a sample of the area inhabited by the population and to extrapolate sample density to the entire area. This can be achieved using line transect sampling and is described below through the example of the Southern Hemisphere minke whale. The second way is to sample the animals themselves by identifying them individually and to employ capture–recapture techniques of analysis to calculate population size. This method is described below through the example of the North Atlantic humpback whale.

The Southern Hemisphere minke whale

Following the sequential depletion and subsequent protection of blue, fin and sei whales, the larger *Balaenoptera* species, in the Southern Ocean, whalers turned their attention to the smallest member of the genus, the minke whale, *B. acutorostrata.* About 3,000 whales were taken in the 1971–72 Antarctic season and, since then, annual catches have mostly been between 6,000 and 8,000 whales. Although the IWC has now enforced a moratorium on all commercial whaling, some countries, as is their right under the Whaling Convention, have lodged formal objections to this and whaling continues on this population.

Southern Hemisphere minke whales are widely dispersed throughout the Southern Ocean. They migrate northwards towards the tropics in the winter to breed, returning southwards in the summer to feed in high latitudes. It is at this latter time that the whales are easiest to survey because they tend to aggregate along the edge of the pack ice around the Antarctic continent.

In the 1978–79 season a series of cruises to assess Southern Hemisphere minke whales in the six Antarctic Areas defined by the IWC was initiated as part of the International Decade of Cetacean Research (IDCR). These cruises have continued annually, completing a circumnavigation of the Antarctic continent in 1983–84. Recent cruises have returned to reassess the whales in key Areas. In the first two cruises, the emphasis was on marking whales to be recaptured from animals killed subsequently by the whaling fleets. Since 1980–81, however, the emphasis has been on sightings in the form of line transect surveys and no marking has taken place since 1984–85. Hammond (1986a) gives a brief history of these cruises.

Method of estimation

Line transect sampling was developed to estimate the density of flushing birds, such as partridge, and has become a well-used terrestrial technique. It is only relatively recently that cetaceans have been surveyed using this technique. Much of the theoretical and experimental work on the applicability of line transect methodology to populations of whales has been undertaken to improve the estimates calculated from data collected on the Southern Hemisphere minke whale cruises.

The aim of a direct sampling technique is to estimate the density of animals in a sample area inhabited by the population and to extrapolate this to the entire area. In line transect sampling, observers search ahead and to the sides of the survey platform as it traverses the study area along predetermined transect lines. The number of animals (or groups of animals), n, seen in the sample area of length L and half-width W depends upon the average probability of detecting them, P, within the sample area. Formally, density can be descibed as

$$D = \frac{n}{2WLP}.$$

In a strip transect survey, it is assumed that $P = 1$. This will rarely, if ever, be a valid assumption and the only way to test its validity is to collect data on the perpendicular distance of each animal (or group of animals) from the transect line so that P can be estimated from these data. Burnham & Anderson (1984) provide convincing arguments against using strip transects in the absence of perpendicular distance data. If animals are distributed uniformly with respect to perpendicular distance and all animals in the vicinity of the transect line are seen with unit probability then Burnham & Anderson (1976) have shown that

$$f(0) = \frac{1}{WP}$$

where $f(0)$ is the probability density function of the perpendicular distances evaluated at the origin. An estimate of $f(0)$ can be made by fitting a suitable function to these data. The estimator for density is then

$$\hat{D} = \frac{n\,\hat{f}(0)}{2L}.$$

If density of groups has been estimated it must be multiplied by an estimate of mean group size, $\bar{g}$, to obtain density of animals. Multiplying by the size of the study area, A, will then give population size, N, so that

$$\hat{N} = \hat{D}\,\bar{g}\,A.$$

Burnham, Anderson & Laake (1980) provide a comprehensive review of line transect sampling.

Application of method

In order that sample density is representative of the true density in the study area, it is important to place the transect lines in a way that will ensure a representative sample. In the first several surveys this was not done; one vessel followed the ice edge where the minke whales were densest and the other covered the area away from the ice in a square grid pattern. Hammond (1986a) has discussed ways of appropriately analysing these data. In recent surveys, though, there have been three vessels traversing saw-tooth patterns of transects in strata at different distances from the ice edge, which does give a more representative sample.

Cooke (1987) has presented a new method of survey design which does not require the surveyed area to be covered uniformly. This is an important development because it is very difficult to ensure uniform coverage close to the ice edge where minke whales are densest. The placement of cruise tracks is so determined that it is possible to calculate the probability of covering any given point in the survey area as a function of its geographical position. Total population size is then estimated as the sum of the reciprocals of the coverage probabilities of all the groups of whales seen. Note that this method does not obviate the need for estimating the average probability of detecting a group; this is required to calculate the coverage probabilities.

Perpendicular distances from groups of minke whales to the transect line are calculated from measurements of radial sighting distance and the angle subtended from the transect line to the sighted group. Up until the 1983–84 cruise, angles were measured approximately with a compass ring and distances were either calculated from speed and time to reach a sighted group or estimated visually. Data were rounded up to convenient values and the rounding of angles to 0° created artificially high numbers of sightings exactly on the transect line. The problems in analysis caused by this were lessened by a process known as 'smearing', which attempts to reverse these errors by smoothing the distributions of angles and radial distances according to a set of conditions determined from experiments. Hammond (1984a) discusses the relative merits of different methods of 'smearing'. Since then, angles have been measured using a specially designed angle board and distances with the aid of binoculars containing graticules, thus improving accuracy on both counts. Buckland (1987) has recently investigated the performance of several different functions for estimating $f(0)$ from 'smeared' and 'unsmeared' perpendicular distance data collected on the minke whale cruises. A better way to deal with rounding of data to convenient values would be to model the process, but suitable data from the field are currently unavailable.

The estimation of mean group size also creates problems. First, there is the definition of a group. This is easy if there are only a few animals swimming closely together but very difficult if there are many whales dispersed over a large area. Secondly, larger groups are easier to detect at greater perpendicular distances from the transect line, so that a simple average of all sighted groups is likely to be an overestimate of mean group size. What is needed is an estimate of mean group size on the transect line itself where all groups are assumed to be detected regardless of size. This can be approximated by regressing group size against perpendicular distance or by taking the mean of the sizes of all groups sighted within a given distance of the transect line (Butterworth & Best 1982; Hammond 1986a).

The major problem in applying line transect sampling to sighting surveys of whales is that the assumption that all whales in the vicinity of the transect line are seen is certainly violated; like all whales, minkes spend a large proportion of the time underwater. This causes an underestimate of population density equal to the proportion of whales missed on the transect line. It may also cause an overestimate of mean group size, because those groups remaining undetected are likely to be smaller than those which are sighted. Attempts to estimate the proportion of whales which are missed on the transect line have concentrated on experiments using two vessels searching in parallel (e.g. Butterworth, Best & Hembree 1984) and, more recently, using two independent observers on one vessel. These experiments have been hampered by the problems of group definition and determining which groups have been seen by both observers. As a result, the Scientific Committee of the IWC has been unable to produce an acceptable estimate for the proportion of minke whales missed on the transect line and currently it is assumed that none are missed.

Line transect sampling also assumes that target animals are stationary before detection. However, whales are continually moving and this creates a problem if the speed of the survey vessel approaches that of the whales. The faster the whales move relative to the survey platform, the more chance an individual group has of coming within sighting distance of the vessel. To avoid a significant positive bias, it is important that survey speed should be considerably greater than whale speed (Hiby 1982). There is also the possibility that whales may react to the survey vessel before they have been detected, by swimming either away or towards the vessel and, therefore, the transect line. However, this is not thought to be a problem for Southern Hemisphere minke whales (Leatherwood, Awbrey & Thomas 1982).

The most recent population estimates for Southern Hemisphere minke whales of takeable size (>8.2m) are given in IWC (1986: 77). These are presented in Table 1.

An alternative approach to estimating the number of Southern Hemisphere minke whales within the framework of a sightings survey is the

Table 1. Population estimates for Southern Hemisphere minke whales > 8.2m (from IWC 1986:77) for each Antarctic Area, with coefficients of variation (CV).

Area	Estimate	CV
I	36,064	0.181
II	25,163	0.163
III	50,115	0.180
IV	37,926	0.127
V	71,617	0.268
VI	36,010	0.151
Total	256,895	0.092

recent development of cue-counting (Hiby 1985). This method does not make the assumption that all whales are seen on the transect line. Instead, it calculates the density of whale sighting cues, e.g. blows, by recording the radial distance of each cue from the vessel and assuming that all cues will be seen in the vicinity of the survey vessel. Density of cues is then converted to density of whales using independent data on rates of blowing. This method was used successfully in an aerial survey for fin whales in the North Atlantic (Hiby, Martin & Fairfield 1984). Experiments carried out on the Southern Hemisphere minke whale cruises indicate that it is also a potentially useful method for ship surveys (Ward & Hiby 1987). Further work is being conducted to assess whether it can be considered as a viable alternative to traditional line transect methods for this population.

The North Atlantic humpback whale

Humpback whales, *Megaptera novaeangliae*, have been exploited in the western North Atlantic since at least the 18th century by aboriginal peoples and during the late 19th and early 20th century by commercial whalers as well. Population size prior to the start of commercial whaling in 1865 has been conservatively assessed as at least 4,700 with at least 6,770 whales taken between 1866 and 1931 (Mitchell & Reeves 1983). In recent years about 20 whales have been removed annually from the population by aboriginal hunting and entrapment in fishing gear. North Atlantic humpback whales migrate between winter breeding grounds in the West Indies and summer feeding grounds in northern latitudes. On the breeding grounds there appears to be complete mixing of the population but whales show strong site specificity to particular feeding areas.

One of the more noticeable features of the humpback whale is the variable black and white pattern on the ventral surface of the tail flukes (Fig. 1). The uniqueness of this pattern and the shape of the trailing edge of the flukes allow individual animals to be identified (Katona, Baxter, Brazier,

Fig. 1. Photograph of the tail flukes of a humpback whale taken off the west coast of Iceland in 1982 (courtesy of A. R. Martin).

Kraus, Perkins & Whitehead 1979). Researchers have used this natural tag to study the movements and behaviour of the whales. A large number of fluke photographs taken by many contributors in the North Atlantic are stored and catalogued by S. K. Katona at the College of the Atlantic, Bar Harbor, Maine, USA. The earliest photographs date back to 1968 but the majority have been taken since 1978. In July 1985 there were 3,219 individually identified whales in the catalogue. These data have been used to estimate population size using capture–recapture techniques.

Method of estimation

Capture–recapture estimators of population size are based on the assumption that the proportion of marked animals recaptured in a sample of the population is equivalent to the proportion of marked animals in the total population, N. If n_1 animals are marked in the first sample and m_2 of these are recaptured in a second sample of size n_2 we have the relationship

$$\frac{n_1}{N} = \frac{m_2}{n_2} \quad \text{and} \quad \hat{N} = \frac{n_1 n_2}{m_2}$$

where $\hat{N}$ is the simple two-sample Petersen estimator for the size of a closed population. If we allow the population to entertain births, deaths and permanent immigration and emigration, then multiple samples are necessary to estimate population size. The most widely-used open population model is the Jolly–Seber model which allows population size to be estimated for each sampling occasion except the first and last. This model uses a general form of the Petersen equation for the ith sample

$$\hat{N}_i = \frac{\hat{M}_i n_i}{m_i}$$

where $\hat{M}_i$ is an estimate of the total number of marks present in the population at time i. The estimate $\hat{M}_i$ is obtained by assuming that two groups of animals, those marked at time i, the S_i, and those marked up to but not including time i, the $(M_i - m_i)$, will be recaptured in the same proportion subsequent to time i. Thus, if Z_i of the $(M_i - m_i)$, and r_i of the S_i are recaptured after time i, we have

$$\frac{Z_i}{M_i - m_i} = \frac{r_i}{S_i} \quad \text{so that} \quad \hat{M}_i = \frac{S_i Z_i}{r_i} + m_i.$$

Substituting into the general Petersen equation above, we obtain

$$\hat{N}_i = n_i \left[1 + \frac{S_i Z_i}{r_i m_i} \right].$$

Estimates of survival rate and number of births can also be obtained from the Jolly–Seber model.

There are several texts giving extensive descriptions of capture–recapture methods. Seber (1982) gives a statistical treatment and Begon (1979) is perhaps more suitable for the biologist.

Application of method

It is important that the sampled animals are representative of the entire population. The standard capture–recapture models are based on the general assumption that all individuals in the population have the same characteristics and behave in the same way. Thus each animal should have

Table 2. Independent Petersen estimates, with coefficients of variation (CV), for the North Atlantic humpback whale population, 1978–1983 (Balcomb, Katona *et al.* 1986).

Year	Estimate	CV
1978	3,503	0.36
1979	5,005	0.20
1980	4,783	0.30
1981	6,582	0.32
1982	10,712	0.28
1983	7,756	0.16

the same probability of being captured on any one sampling occasion and have the same probability of surviving from one sample to the next (this is implicit as unity in the Petersen model because births and deaths are assumed not to occur). The available data for the North Atlantic humpback whale span several years. Births and deaths were clearly occurring during this period, so we would automatically think of an open population model such as the Jolly–Seber as the most appropriate method of analysis. However, although we can consider the annual sample of photographs taken on the breeding grounds to be representative, those taken on all feeding grounds combined were certainly not representative. Not all feeding grounds were sampled each year and the data were not always collected during scientific studies. Hammond (in preparation) is investigating the use of open population models on these data.

An alternative is to consider a series of independent Petersen estimates using data collected at the feeding grounds as the first sample and data collected at the breeding grounds approximately six months later as the second sample. A valid population estimate still results if the assumption of a representative sample is relaxed for one of the two samples, as long as complete mixing occurs between samples. There are no births between the two samples and, if deaths only are occurring, the Petersen model gives a valid estimate of population size at the time of the first sample. The current best population estimate for North Atlantic humpback whales is a weighted (by inverse of variance) mean of independent Petersen estimates for 1978 to 1983 (Balcomb, Katona & Hammond 1986) (Table 2). The weighted mean is 5,561 whales with coefficient of variation 0.103.

Using natural markings to identify whales in capture–recapture studies has several advantages over the placing of marks. There is no physical contact with the animals, they do not have to be killed in order to be recaptured and marks cannot be lost. In addition, marking cannot alter the probability of capture or of survival, which often happens in studies of small mammals and fish. The disadvantages are that the natural markings may change with time and that identification takes longer and is more prone to errors.

For North Atlantic humpbacks it has been shown that fluke patterns can change in young animals but there is no evidence that the patterns continue to change in adults, so this should not affect population estimates. Problems associated with identification increase as the number of photographs increases if identification is made visually, as it is for North Atlantic humpbacks. It may now be time to consider a computer-based system for analysing the photographs.

A problem which probably affects all capture–recapture studies is that of heterogeneity of capture probabilities as a result of inherent differences in individual behaviour. This results in negatively biased population estimates. In the case of the photographic data for the North Atlantic humpback whale this manifests itself in three different ways. Heterogeneity can be introduced if some individuals are consistently less available to be seen in the sample area than others. This can only be minimized by careful choice of sample area and timing of the sample. Heterogeneity can also be introduced if individuals consistently display specific behaviour which results in inherent difference in their availability to be photographed. In humpbacks the tendency to show flukes can be variable, e.g. cows with young calves rarely 'fluke-up' on the breeding grounds, but this is not thought to introduce a significant bias. The third potential source of heterogeneity is if photographs are included in the sample on the basis of how easy it is to identify the animal. This is not a problem for North Atlantic humpbacks where measures of photographic quality are the only criteria for inclusion used. Hammond (1986b) provides an extensive review of the applicability of capture–recapture techniques for estimating the size of naturally marked populations of whales.

The California gray whale

The California gray whale, *Eschrichtius robustus*, has been exploited for many centuries by native peoples in the North Pacific. It was hunted extensively by Yankee whalers between 1845 and 1874 when an estimated 7–8,000 animals were killed along the coast of California, reducing population numbers from perhaps as many as 15,000 to perhaps as few as 4,000 (Henderson 1984). Modern commercial whaling accounted for about another 1,000 animals between 1914 and 1946 (Reeves 1984). Since 1969, an average of 166 whales has been taken each year by the Soviet Union for aboriginal subsistence purposes.

The California gray whale migrates between feeding grounds in the Bering and Chukchi Seas and breeding grounds off the coast of Baja California, Mexico. Gray whales travel very close to the shore, especially during their southbound migration. This allows censuses to be taken from strategic points on the coast. Such censuses are relatively cheap when compared to the expense of using boats or aircraft as research platforms and can therefore be conducted more often than other types of population survey. They are thus an ideal way of monitoring the recovery of an exploited population.

The first such census for California gray whales was conducted in 1885 but it was not until 1946 that systematic censuses began. The major effort was a series of 13 consecutive annual censuses made from Monterrey, California, from 1967 to 1980 (Reilly, Rice & Wolman 1983). Surveys were also conducted from 1978 to 1981 from Newport, Oregon (Herzing & Mate 1984) and from 1977 to 1979 from Unimak Pass, Alaska (Rugh 1984).

Method of estimation and its application

The basic method of surveying whales from the shore is to choose a suitable location past which the entire population is thought to migrate and to count all the whales between the dates which are thought to delimit the period of migration. There are several conditions under which a whale could pass a census point and not be seen and these need to be accounted for in the calculation of a population estimate.

Firstly, whales may pass before and/or after the period of the census. This was accounted for in the Monterrey censuses by fitting a predictive model to the daily counts and extrapolating it to include the 'tails'. Estimates of population size were increased by about 5% as a result of this. At Unimak Pass a similar method was employed but at Newport a correction was considered unnecessary because of the length of the census.

Secondly, whales may pass too far offshore to be seen by the census observers. Aerial surveys have shown that offshore distribution changes during migration. At Unimak Pass, practically all whales passed within 1 km of the shore, but at Monterrey an estimated 20% of the populaion migrated past out of range of shore-based observers, and at Newport this estimate was 38%. A recent census in 1985–86 at Monterrey showed, however, that the offshore distribution had changed since the 1980 census so that the 1967–80 correction factor could not be applied.

Thirdly, poor visibility may result in whales within range being missed. Data from all shore censuses of gray whales have shown that the conditions of visibility and the difference between the number of whales expected and those actually seen are highly correlated. The data were therefore corrected using either a predictive model, as at Monterrey, or by interpolating between periods of good visibility as at Unimak Pass and Newport.

Fourthly, whales may be missed because the probability of seeing them is a decreasing function of distance from the observer. The gray whale censuses are, in fact, analogous to one-sided line transect surveys, with whales moving past a stationary census point instead of a survey platform moving through the whales. The only difference is that the distribution of gray whales offshore is not uniform with respect to distance. All of the gray whale censuses have assumed that within a fixed distance all whales will be seen, i.e. have made the same assumptions as for a strip transect. As for strip transects, this is not a reasonable assumption and Cooke (1986) has suggested that independent counts made at the same time could evaluate the proportion of whales missed as a function of distance offshore. Double-counting experiments conducted during the 1985–86 Monterrey census have shown that this is indeed a problem which should be investigated further.

Finally, whales can only be counted during daylight hours and they may pass the census locations at a different rate during the night. Data collected in two experiments at Unimak Pass using night goggles at dusk showed a slight slowing to 73% of the daytime rate in one case and no differences in the other (Rugh 1984). More recently, a telemetry study of 18 individuals during the 1985–86 Monterrey census showed no difference between daytime and night-time migration rates (Swartz, Jones, Goodyear, Withrow & Miller 1987). These results suggest that daytime counts can simply be multiplied proportionally to estimate the daily numbers of gray whales passing a census point.

The census conducted at Unimak Pass in November–December of 1978 estimated that 17,000 gray whales passed the census point. The 1980–81 census at Newport in December–January gave an estimate of 17,670. The 1979 Monterrey census from mid-December to early February estimated 15,650 whales. These estimates are all for the southward migration and are all corrected for the factors described above. The closeness of these estimates is encouraging but this is probably as much a reflection of the similarity of the census techniques at the three sites as a confirmation of the accuracy of the estimation procedure.

The Western Arctic bowhead whale

The bowhead whale, *Balaena mysticetus*, has been hunted by the Inuit Eskimos of Alaska for thousands of years. In the mid 19th century, commercial whalers discovered the Western Arctic bowhead population and by 1910 there were too few whales left for whaling to remain viable. The population size at this time has been estimated to have been between 1,200 and 3,800 out of an initial population of 14–20,000 (Breiwick, Eberhardt & Braham 1984). Inuit hunters continue to take whales from this population and the current strike quota set by the IWC is an average of 26 per year (not all 'struck' whales are landed).

Western Arctic bowheads migrate from winter breeding grounds in the Bering Sea through the Bering Straits to the north-west coast of Alaska, which they follow in the nearshore leads of open water through the ice. It is along this stretch of their migration that the whales are subject to most of the Inuit hunt. The whales pass very close to Point Barrow and then enter the Beaufort Sea, their summer feeding area. In the autumn, they return to the Bering Sea passing close to the north coast of Alaska, where further whaling occurs. The proximity of the northbound migratory pathway to the shorefast ice at Point Barrow provides an opportunity for the bowheads to be counted as they move through the ice. Censuses have been conducted annually at this location since 1978.

Method of estimation and its application

The method of enumeration for the bowhead whale at Point Barrow is similar in principle to that for the California gray whale. Most of the same problems apply: the possibility of migration before and/or after the census period; migration out of the visual range of the observers; poor weather conditions affecting the counts; the probability of seeing a whale decreasing with distance from the census point. The question of night-time migration rates does not arise because at the time of the census (mid April–mid June) there is continuous daylight at Point Barrow. An additional factor is the condition of the lead system through which the whales pass. A well-defined, stable, open lead close to the shorefast ice presents no problems but the sea ice is continually moving and leads may close and open up in different places, sometimes at considerable distances from the shorefast ice. It is not uncommon for the lead system to close up altogether.

Recognizing that some whales will be missed by a single census point regardless of conditions, the bowhead census is conducted using two counting locations, known as perches. These are located close enough together on the shorefast ice that it can be assumed that each perch has the same probability of detecting each approaching whale. The proportion of whales missed by both perches is then calculated using the so-called 'removal' method (Seber 1982: 318) by treating the whales seen at each perch as independent samples from the population. The n_1 whales seen at the first perch are 'removed' from the population by notifying the second perch that they have already been seen and n_2 of the whales missed by the first perch are then counted at the second perch. An estimate of the proportion of whales missed altogether is simply n_2/n_1. So that whales seen from the first perch can be passed on to the second perch, the position of each whale is determined from its distance and bearing from the perch. The use of theodolites increases the accuracy to which positions can be fixed.

Raw counts of whales made during periods of acceptable visibility are divided by an estimate of the proportion of whales missed calculated for the same periods. Within each day, counts corrected for missed whales are then divided by the proportion of time lost to conditions of unacceptable visibility. For days with less than one hour of acceptable visibility, the daily count is estimated from an autoregressive time-series model which interpolates between those days where data are available. Daily counts for the period of the census are summed to give a minimum population estimate. It is a minimum estimate because no account is taken of whales which may pass before and/or after the census or of whales which may pass outside the visual range of the observers. These estimation methods are described more fully by Zeh, Ko, Krogman & Sonntag (1986).

The length of the censuses is determined in part by ice conditions. Once the shorefast ice begins to break up, the census must be terminated for reasons of safety even though some whales may still be passing. Aerial surveys have been conducted to investigate the offshore distribution of bowheads during the census but the data are few and variable. For example, in 1985, whales were found as far as 17 km from the census points and 54% were more than 4 km offshore, but in 1984, no whales were found beyond 4 km (Nerini & Rugh 1986). Hammond (1984b) has suggested that line transect sampling techniques could be used to estimate the number of bowheads passing Point Barrow by incorporating data from aerial surveys to estimate the true distribution of whales across the lead.

In 1983, the technique of locating whales acoustically using hydrophones under the sea was investigated. Since then, acoustic location of whales has become an integrated part of the census. Initially it was hoped to correlate whale sounds with visual sightings but Ko, Zeh, Clark, Ellison, Krogman & Sonntag (1986) showed that this was not possible and, instead, proposed a system of joining up locations into 'whale tracks' using an algorithm based on swimming direction and speed.

Individual sounds are located by calculating the difference in the time that the sound arrives at each of three hydrophones. For each of the three pairs of hydrophones there is a line of possible locations of the sound. The intersection of the three lines gives an exact position. The algorithm which has been developed to create 'whale tracks' makes the maximum number of connections between sounds possible given a range of allowed swimming speeds and swimming directions. This gives a minimum count of whales from the recorded sounds. The algorithm is fairly sensitive to the value of minimum swimming speed supplied but robust to values of maximum swimming speed and the limits of swimming direction (Sonntag, Ellison, Clark, Corbit & Krogman 1986).

The acoustic recording of bowhead whale sounds has resulted in three important breakthroughs in the censusing of bowhead whales. Firstly, it has shown that bowheads pass by the perches even when the lead system is completely frozen over. The breathing holes found in the ice show how the whales achieve this (Krogman, George, Carroll, Zeh & Sonntag 1986).

Secondly, by leaving hydrophones under the ice after the ice begins to break up and becomes unsafe, it has shown that whales do pass after the census is forced to finish. Thirdly, it has shown that whales regularly pass by out of the visual range of the perches. In the latter case, minimum numbers of whales can be added to the minimum visual count in the calculation of population size. The current best estimate is 4,417 made up of 3,242 whales estimated visually out to 3 km and a further 1,175 whales estimated to have passed beyond 3 km calculated from aerial and acoustic data (IWC 1986: 108–9).

The visual estimate of 3,242 whales was calculated from data collected in 1978 and 1982. The variability in ice conditions from year to year means that there are not always sufficient data for a population estimate to be calculated. In 1986, however, there were favourable ice conditions and the acoustic census was in full operation. The analysis of these data should result in an improved estimate. A recent approach, developed by Gentleman & Zeh (1987), is to analyse all the data simultaneously using capture–recapture techniques. In this application, whales detected visually compose the first sample and whales detected acoustically compose a second independent sample from which the visually detected whales are 'recaptured'.

Discussion

There are some general observations which can be made about methods of population estimation for whales. Two basic classes of whale can be identified. The first contains those species which follow well-defined migration pathways, usually along a coast for at least part of the route, and which congregate at specific sites to breed or feed. Thus, the entire population, or a defined part of it, shows predictable behaviour for a well-documented period of time in a limited but accessible area and estimation methods which make use of these characteristics can be used. In the case of the California gray whale and the Western Arctic bowhead whale, migration past specific points at specific times allows a method of estimation based on counting animals as they go past. In the case of the North Atlantic humpback whale, the aggregation of the population in known breeding and feeding areas allows the surveyor access to take photographs of natural markings. The same is true for humpbacks in the North Pacific (e.g. Baker, Herman, Perry, Lawton, Straley & Straley 1985).

Similar considerations apply to the right whale populations in the northwest Atlantic (Kraus, Moore, Price, Crone, Watkins, Winn & Prescott 1986), off Argentina (Whitehead, Payne & Payne 1986) and off South Africa (P. B. Best personal communication). Their occurrence close to land at predictable times and places allows the systematic taking of aerial photographs of the pattern of callosities on the head. This pattern uniquely identifies each individual, so that capture–recapture techniques can be applied. It is no coincidence that the gray, bowhead, humpback and right whales are also those which were the first to be hunted by commercial whalers. Their accessibility and predictability made them ideal targets for the early whaling industry. Thus, the methods of population estimation which make specific use of predictable behavioural characteristics are those which are being used to monitor the recovery of the earliest exploited whale populations.

The second class of whale contains those species which can be described as pelagic. This class includes all the balaenopterids, the blue, fin, sei, Bryde's and minke whales. These species also migrate between breeding and feeding grounds but they tend to disperse widely. They spend most of their time in the open ocean and it is only where there is a conveniently situated piece of land that a part of the population, usually undefined, is available to coastal whaling operations. Consequently, the whole population is not available to a census from land. In addition, those natural markings which balaenopterids do possess do not contain very much information and/or are not easily photographed. Therefore, capture–recapture techniques cannot readily be used for population estimation, unless this involves firing marks into animals and recapturing them via whaling. In the absence of whaling, the only practical way to estimate population size for these species is to employ line transect sampling techniques, as has been done for the Southern Hemisphere minke whale. Other balaenopterids surveyed using line transect sampling include North Atlantic fin whales off Iceland (Hiby, Martin & Fairfield 1984), and off Spain (Sanpera & Jover 1986), Bryde's whales in the western North Pacific (Miyashita 1986a), and off South Africa (Best, Butterworth & Rickett 1984), and humpback and fin whales off Newfoundland/Labrador (Hay 1982).

This second class of whale could also include many of the odontocetes. The sperm whale (Laake & Hammond 1984), Baird's beaked whale (Miyashita 1986b), the killer whale (Hammond 1984c) and the pilot whale (Hay 1982) have all been surveyed using line transect sampling. Dolphin populations have also been surveyed in this way, notably in the extensive studies of the spotted, spinner and common dolphins in the eastern tropical Pacific (e.g. Smith 1981; Hammond & Laake 1983). Hammond (1986c) has reviewed the subject of line transect sampling of dolphin populations. Killer whales have also been extensively studied off the coasts of British Columbia, Canada (Bigg 1982) and Washington, USA (Balcomb, Boran & Heimlich 1982) using photo-identification, which has led to estimates of the size of local populations calculated by summing all the individually recognized animals.

In summary, techniques of population estimation have been successfully applied to many populations of whales. Future research will no doubt improve the methodology and the means of practical application. In addition, further work on relatively unexplored techniques such as the use of towed arrays of hydrophones, as used by Thomas, Fisher, Ferm & Holt (1986) for small whales and dolphins and proposed by Whitehead & Gordon (1986) for sperm whales, can be expected in the future.

Acknowledgements

I am grateful to Lex Hiby and Greg Donovan for constructive critical comment.

References

Baker, C.S., Herman, L.M., Perry, A., Lawton, W.S., Straley, J.H. & Straley, J.H. (1985). Population characteristics and migration of summer and late-season humpback whales (*Megaptera novaeangliae*) in southeastern Alaska. *Mar. Mamm. Sci.* **1**: 304–23.

Balcomb, K.C., Boran, J.R. & Heimlich, S.L. (1982). Killer whales in Greater Puget Sound. *Rep. int. Whal. Commn* **32**: 681–5.

Balcomb, K.C., Katona, S.K. & Hammond, P.S. (1986). Population estimates for the North Atlantic humpback whale using capture–recapture methods on photo-identified whales. *Rep. int. Whal. Commn* **36**: 110.

Begon, M. (1979). *Investigating animal abundance—capture–recapture for biologists*. Edward Arnold, London.

Best, P.B., Butterworth, D.S. & Rickett, L.H. (1984). An assessment cruise for the South African inshore stock of Bryde's whales (*Balenoptera edeni*). *Rep. int. Whal. Commn* **34**: 403–23.

Bigg, M.A. (1982). An assessment of killer whale (*Orcinus orca*) stocks off Vancouver Island, British Columbia. *Rep. int. Whal. Commn* **32**: 655–66.

Breiwick, J.M., Eberhardt, L.L. & Braham, H.W. (1984). Population dynamics of western Arctic bowhead whales (*Balaena mysticetus*). *Can. J. Fish. aquat. Sci.* **41**: 484–96.

Buckland, S.T. (1987). An assessment of the performance of line transect models on modelling IDCR cruise data, 1978/79–1984/85. *Rep. int. Whal. Commn* **37**: 277–9.

Burnham, K.P. & Anderson, D.R. (1976). Mathematical models for non-parametric inferences from line transect data. *Biometrics* **32**: 325–36.

Burnham, K.P. & Anderson, D.R. (1984). The need for distance data in transect counts. *J. Wildl. Mgmt* **48**: 1248–54.

Burnham, K.P., Anderson, D.R. & Laake, J.L. (1980). Estimation of density from line transect sampling of biological populations. *Wildl. Monogr.* No. 72: 1–202.

Butterworth, D.S. & Best, P.B. (1982). Report of the Southern Hemisphere minke whale assessment cruise, 1980/81. *Rep. int. Whal. Commn* **32**: 835–74.

Butterworth, D.S., Best, P.B. & Hembree, D. (1984). Analysis of experiments carried out during the 1981/82 Antarctic minke whale assessment cruise in Area II. *Rep. int. Whal. Commn* **34**: 365–92.

Cooke, J.G. (1986). On the net recruitment rate of gray whales with reference to interspecific comparisons. *Rep. int. Whal. Commn* **36**: 363–6.

Cooke, J.G. (1987). Estimation of the population of minke whales in Antarctic Area IV W in 1984/85. *Rep. int. Whal. Commn* **37**: 273–6.

Gentleman, R. & Zeh, J.E. (1987). A statistical model for estimating the number of bowhead whales, *Balaena mysticetus*, passing a census point from combined visual and acoustic data. *Rep. int. Whal. Commn* **37**: 313–27.

Hammond, P.S. (1984a). An investigation into the effects of different techniques of smearing the IWC/IDCR minke whale sightings data and of the use of different models to estimate density of schools. *Rep. int. Whal. Commn* **34**: 301–7.

Hammond, P.S. (1984b). On the application of line transect sampling to the estimation of the number of bowhead whales passing the Point Barrow ice-camps. *Rep. int. Whal. Commn* **34**: 465–7.

Hammond, P.S. (1984c). Abundance of killer whales in Antarctic Areas II, III, IV and V. *Rep. int. Whal. Commn* **34**: 543–8.

Hammond, P.S. (1986a). On the post-stratification of sightings data from the 1978/79–1982/83 IWC/IDCR Southern Hemisphere minke whale assessment cruises and survey design for future cruises. *Rep. int. Whal. Commn* **36**: 225–37.

Hammond, P.S. (1986b). Estimating the size of naturally marked whale populations using capture–recapture techniques. *Rep. int. Whal. Commn* Spec. Issue **7**: 253–82.

Hammond, P.S. (1986c). Line transect sampling of dolphin populations. In *Research on dolphins*: 251–79. (Eds Bryden, M.M. & Harrison, R.M.). Oxford University Press, Oxford.

Hammond, P.S. (In prep.). *North Atlantic humpback whales: estimating population size from photo-identification data using variations of the Jolly–Seber open population model.*

Hammond, P.S. & Laake, J.L. (1983). Trends in estimates of abundance of dolphins (*Stenella* spp. and *Delphinus delphis*) involved in the purse-seine fishery for tunas in the eastern Pacific Ocean, 1977–81. *Rep. int. Whal. Commn* **33**: 565–88.

Hay, K. (1982). Aerial line-transect estimates of abundance of humpback, fin and long-finned pilot whales in the Newfoundland-Labrador area. *Rep. int. Whal. Commn* **32**: 475–86.

Henderson, D.A. (1984). Nineteenth century gray whaling: grounds, catches, kills, practices and depletion of the whale population. In *The gray whale*: 159–86. (Eds Jones, M.L., Swartz, S.L. & Leatherwood, S.). Academic Press, Orlando.

Herzing, D.L. & Mate, B.R. (1984). Gray whale migrations along the Oregon coast, 1978–1981. In *The gray whale*: 289–307. (Eds Jones, M.L., Swartz, S.L. & Leatherwood, S.) Academic Press, Orlando.

Hiby, A.R. (1982). The effect of random whale movement on density estimates obtained from whale sighting surveys. *Rep. int. Whal. Commn* **32**: 791–3.

Hiby, A.R. (1985). An approach to estimating population densities of great whales from sightings surveys. *IMA J. Math. appl. Med. Biol.* **2**: 201–20.

Hiby, A., Martin, A. & Fairfield, F. (1984). IDCR cruise/aerial survey in the north Atlantic 1982: aerial survey report. *Rep. int. Whal. Commn* **34**: 633–44.

IWC (1986). *Thirty-sixth report of the International Whaling Commission.* IWC, Cambridge.

Katona, S.K., Baxter, B., Brazier, O., Kraus, S., Perkins, J. & Whitehead, H. (1979).

Identification of humpback whales by fluke photographs. *Behav. mar. Anim.* **3**: 33–44.

Ko, D., Zeh, J.E., Clark, C.W., Ellison, W.T., Krogman, B.D. & Sonntag, R.M. (1986). Utilization of acoustic location data in determining a minimum number of spring-migrating bowhead whales, unaccounted for by the icebased visual census. *Rep. int. Whal. Commn* **36**: 325–38.

Kraus, S.D., Moore, K.E., Price, C.A., Crone, M.J., Watkins, W.A., Winn, H.E. & Prescott, J.H. (1986). The use of photographs to identify individual North Atlantic right whales, *Eubalaena glacialis*. *Rep. int. Whal. Commn* Spec. Issue **10**: 145–51.

Krogman, B., George, J.C., Carroll, G., Zeh, J. & Sonntag, R. (1986). Preliminary results of the 1985 spring ice-based census of the bowhead whale, *Balaena mysticetus*, conducted near Point Barrow, Alaska. *Rep. int. Whal. Commn* **36**: 343–52.

Laake, J.L. & Hammond, P.S. (1984). Estimates of sperm whale density in the eastern tropical Pacific, 1974–1982. *Rep. int. Whal. Commn* **34**: 255–8.

Leatherwood, S., Awbrey, F.T. & Thomas, J.A. (1982). Minke whale response to a transiting survey vessel. *Rep. int. Whal. Commn* **32**: 795–802.

Mitchell, E. & Reeves, R.R. (1983). Catch history, abundance and present status of northwest Atlantic humpback whales. *Rep. int. Whal. Commn* Spec. Issue **5**: 153–212.

Miyashita, T. (1986a). Sightings estimate for the Bryde's whale stock in the western North Pacific. *Rep. int. Whal. Commn* **36**: 249–52.

Miyashita, T. (1986b). Abundance of Baird's beaked whales off the Pacific coast of Japan. *Rep. int. Whal. Commn* **36**: 383–6.

Nerini, M.K. & Rugh, D.J. (1986). Distribution of bowhead whales in the vicinity of Point Barrow, Aslaka: aerial observations 1984 and 1985. *Rep. int. Whal. Commn* **36**: 359–61.

Reeves, R.R. (1984). Modern commercial pelagic whaling for gray whales. In *The gray whale*: 187–200. (Eds Jones, M.L., Swartz, S.L. & Leatherwood, S.). Academic Press, Orlando.

Reilly, S.B., Rice, D.W. & Wolman, A.A. (1983). Population assessment of the gray whale, *Eschrichtius robustus*, from California shore censuses, 1967–80. *Fish. Bull. U.S. natn. ocean. atmos. Admn* **81**: 267–82.

Rugh, D.J. (1984). Census of gray whales at Unimak Pass, Alaska: November–December, 1977–1979. In *The gray whale*: 225–48. (Eds Jones, M.L., Swartz, S.L. & Leatherwood, S.). Academic Press, Orlando.

Sanpera, C., & Jover, L. (1986). Results of the 'Ballena 4' fin whale sighting cruise. *Rep. int. Whal. Commn* **36**: 253–5.

Seber, G.A.F. (1982). *The estimation of animal abundance and related parameters.* (2nd edn). Charles Griffin, London.

Smith, T.D. (1981). Line transect techniques for estimating density of porpoise schools. *J. Wildl. Mgmt* **45**: 650–7.

Sonntag, R.M., Ellison, W.T., Clark, C.W., Corbit, D.R. & Krogman, B.D. (1986). A description of a tracking algorithm and its application to bowhead whale acoustic location data collected during the spring migration near Point Barrow, Alaska 1984–85. *Rep. int. Whal. Commn* **36**: 299–310.

Swartz, S.L., Jones, M.L., Goodyear, J., Withrow, D.E. & Miller, R.V. (1987).

Radio-telemetric studies of gray whale migration along the California coast: a preliminary comparison of day and night migration rates. *Rep. int. Whal. Commn* 37: 295–9.

Thomas, J.A., Fisher, S.R., Ferm, L.M. & Holt, R.S. (1986). Acoustic detection of cetaceans using a towed array of hydrophones. *Rep. int. Whal. Commn* Spec. Issue **8**: 139–48.

Ward, A.J. & Hiby, A.R. (1987). Analysis of cue-counting and blow rate estimation experiments carried out during the 1985/86 IWC/IDCR minke whale assessment cruise. *Rep. int. Whal. Commn* **37**: 259–62.

Whitehead, H. & Gordon, J. (1986). Methods of obtaining data for assessing and modelling sperm whale populations which do not depend on catches. *Rep. int. Whal. Commn* Spec. Issue **8**: 149–65.

Whitehead, H., Payne, R. & Payne, M. (1986). Population estimate for the right whales off Peninsula Valdes, Argentina, 1971–1976. *Rep. int. Whal. Commn* Spec. Issue **10**: 169–71.

Zeh, J.E., Ko, D., Krogman, B.D. & Sonntag, R. (1986). Statistical considerations in estimating the number of bowhead whales, *Balaena mysticetus*, from ice-based visual census data. *Rep. int. Whal. Commn* **36**: 317–23.

Symp. zool. Soc. Lond. (1987) No. 58: 247–262

The use of spraints to monitor populations of otters (*Lutra lutra* L.)

J. W. H. CONROY and
D. D. FRENCH

*Institute of Terrestrial Ecology,
Banchory,
Scotland*

Synopsis

This paper summarizes the results of a study of sprainting of otters *Lutra lutra* in Shetland, gives details of methods to monitor faeces (spraint) numbers and tests the assumption that numbers of otter spraints are an index of otter population size.

Sprainting on land was seasonal, most spraints being found in late winter/spring, fewest in summer. Otters sprainted more frequently on some coasts than others; exposed rocky coasts had most spraints, sandy coasts fewest. Overall, a relationship existed between spraint numbers and the numbers of otters seen but to determine this relationship required an understanding of otter sprainting behaviour; it did not exist on short lengths of coast, small islands or some types of coast.

We tested two methods for detecting changes in spraint numbers. Each used counts of spraints from more than 50 monitoring sites distributed throughout the study area.

Because of seasonality and variations in the numbers of spraints at individual sites, only changes in spraint numbers of 30% or more could be detected above the normal variation in spraint density.

Seasonal fluctuations in spraint density, and differences in the numbers of spraints counted in different areas, similar to those found in our Shetland study, have been reported in other otter studies.

In the light of our findings, we conclude that, until such time as the relationship between otter numbers and spraint numbers has been established, spraint surveys are not a good method for monitoring otter populations. Repeated surveys may give an indication of changes in spraint numbers and distribution (and also possibly otter distribution), but they do not necessarily reflect similar changes in otter numbers. Spraint surveys also require better planning than has been exercised until now: they should be done by experienced observers, repeat counts carried out at the same time of year and sources of possible errors quantified.

Introduction

The aims of this paper are to give details of methods to monitor changes in spraint numbers, and, from these results, to test the assumption that

ZOOLOGICAL SYMPOSIUM No. 58
ISBN 0-19-854006-X

numbers of otter spraints are an index of the relative abundance of otters and to discuss the use of spraints as a method to monitor changes in otter numbers. The data are from our study of otter sprainting in Shetland.

During the 1950s evidence from otter hunts in the U.K. suggested that otter numbers had declined markedly (Lloyd 1962; Anon. 1969, 1974; Chanin & Jefferies 1978; Jefferies, Chanin & Lenton 1984). In subsequent years, similar reductions were reported from much of the otter's European range (see Muffling 1977; Reuther 1980).

Base-line data on otter numbers were therefore needed for conservation and management plans. Because otters are difficult to observe directly, indirect methods have been used to monitor their populations. In several countries, e.g. Norway (Myrberget & Froiland 1972; Heggberget & Myrberget 1979, 1980) and the Netherlands (van Wijngaarden & van de Peppel 1970; Veen 1985) questionnaires giving details of the numbers of otters seen, trapped and killed were used to obtain these base-line data. In some cases, these suggested declines in otter populations. For example, Erlinge (1972, 1980) suggested that otter populations showed no sign of recovery over 10 years in Sweden.

In the British Isles, otter signs, in particular their spraints, have been used to ascertain the status of otter populations (Joint Otter Group 1977, 1979), and four national spraint surveys were made (Wales: Crawford, Jones & McNulty 1979; England: Lenton, Chanin & Jefferies 1980; Scotland: Green & Green 1980; Ireland: Chapman & Chapman 1982). Spraint surveys were subsequently widely used in Europe and North Africa, and in the study of *Lutra provocax* (Chehebar 1985) and *Aonyx capensis* (Arden-Clarke 1986).

However, these surveys rely on the assumption that otter numbers or 'status' and spraint numbers are simply correlated. Erlinge (1967, 1968) suggested a seasonal pattern might exist, and Macdonald & Mason (1983a) accepted that seasonal fluctuations in spraint numbers might affect the interpretation of spraint surveys and commented that they might account 'for the large proportion of unexplained variation in the statistical analysis.' They further acknowledged (Macdonald & Mason 1985) that because several factors could influence sprainting, it cannot be used as a direct indication of population size, but then stated 'there is a general relationship between the level of marking and the success of the population, with fragmented declining populations leaving fewer signs than healthy breeding ones.' No one has tested the validity of this statement.

In addition, for reasonable interpretation of sprainting data, one needs to know the ease with which spraints can be found in different types of terrain and to compare the abilities of different observers to find them.

In this paper we present some results from a study of otter sprainting in Shetland (Conroy & French unpublished; Conroy, French & Jenkins in prep; Conroy & French in prep; Jenkins & Conroy 1982), give details of a programme to monitor changes in spraint numbers, and examine the assumption that spraint density can be used as an index of otter numbers. We then discuss the potential value of using otter spraints to monitor otter populations and ask:–

What sample size is needed to monitor fluctuations in numbers of spraints?

Can spraint surveys of different habitats be compared?

What is the relationship between spraint density and otter numbers?

The study area

The study area (Fig. 1), centred around the oil terminal at Sullom Voe, stretched from Fethaland to Lunna Ness on Mainland, from North Taing to Ramna Geo on Yell and included the islands in Yell Sound. This was approximately 250 km of coastline and varied from sandy beaches to cliffs over 100 m high. This area was chosen because otters in Shetland are at risk from oil spills (Baker, Jones, Jones & Watson 1981), and we were asked to develop a monitoring programme to determine the effect that any future oil spill would have on the otter populations.

Methods

General

Details of some of the methods used and their assumptions have been documented (Conroy & French unpublished). To summarize:–

1. The study area was divided into 106 sections of coast, each approximately 2 km long.
2. Seven different types of coast were identified and each 2 km section was allocated to one of these types (Table 1).
3. The main study area, 22 sections chosen at random, but in proportion to the occurrence of the coastal types and with a minimum of two sections of each type, was visited 18 times, and the remaining sections at least three times, between January 1981 and July 1983.
4. All spraints were counted from the high tide mark to 15 m inland. Sites further inland, for instance on the banks of lochs and rivers, known to be used by otters, were also investigated.
5. Records were kept of all the otters that were seen.
6. We assumed that, during the study, the otter population in the area remained relatively stable.

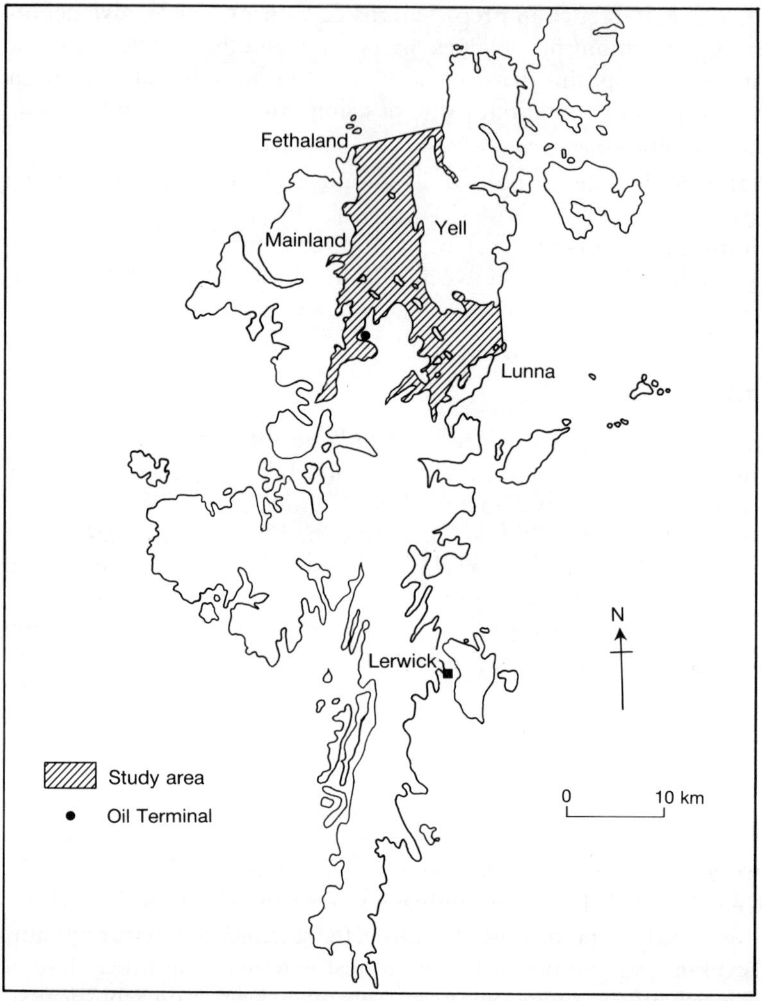

Fig. 1. Shetland, showing the position of the detailed study area.

The monitoring programme

Fifty-four 50-m lengths of coast distributed throughout the study area were chosen as monitoring sites, with some concentrations in areas where large numbers of spraints and spraint sites were found (e.g. Lunna, Fig. 2). The monitoring sites were those areas where most spraints were found between January 1981 and July 1983, but with no two monitoring sites less than 1 km apart.

Because spraint numbers varied from year to year, it was not possible to predict their spatial distribution or abundance over long periods (more than

Table 1. The main coastal types, their frequency of occurrence and numbers sampled in the detailed study area.

Coastal type	No. of sections	No. sampled
1. Cliff-backed (>5 m high), sheltered coast	8	2
2. Cliff-backed (>5 m high), exposed coast	22	5
3. Rocky/boulder coast, sheltered with a medium steep gradient	9	2
4. Rocky/boulder coast, sheltered with a gentle gradient	14	2
5. Rocky/boulder exposed coast	12	2
6. Sand/shingle/pebble coast, sheltered with a medium steep gradient	8	2
7. Other sand/shingle/pebble coasts	31	7

Exposed coast is coast with more than 2 km of fetch.
Inclines of >30° have a medium steep gradient, and <30° a gentle gradient.

one year). It was therefore essential to maintain regular counts of spraints at a number of monitoring sites in order to estimate how many spraints should be counted at any group of sites in the normal situation. These estimates depend on the total spraint numbers each year (which could vary by a factor of almost 2:1 between 1982 and 1983), and also on the variation between sites in different seasons and in different years.

Two methods were tested, each comparing the numbers of spraints counted at a group of sites thought to have changed because of an oil spill or other incident (the 'affected' sites) with numbers counted at 'unaffected' sites.

The Ratio Method

This estimated, from counts at unaffected sites, the spraint numbers one would expect to find at affected sites should no change have occurred, and compared this value with the actual count from the affected sites. To do this, we calculated the ratio r between the affected and unaffected sites for each visit prior to the time of the suspected change. R, the mean value of r from several sample times, was used to test whether or not there had been a significant change in the population (Conroy & French unpublished).

The Rates of Change Method

This compared the changes from counts before an incident to counts after it, on affected sites, with the corresponding change in unaffected sites. We used a Mann-Whitney U-test to see whether the changes in affected sites were significantly lower numerically than those in unaffected sites. A rank-order test was used to 'damp-down' the effects of erratic single sites.

The Ratio Method tests whether a particular count is different from some expected value at that time. The Rates of Change Method tests whether counts at affected sites are decreasing in relation to those at unaffected sites over two or more sample times.

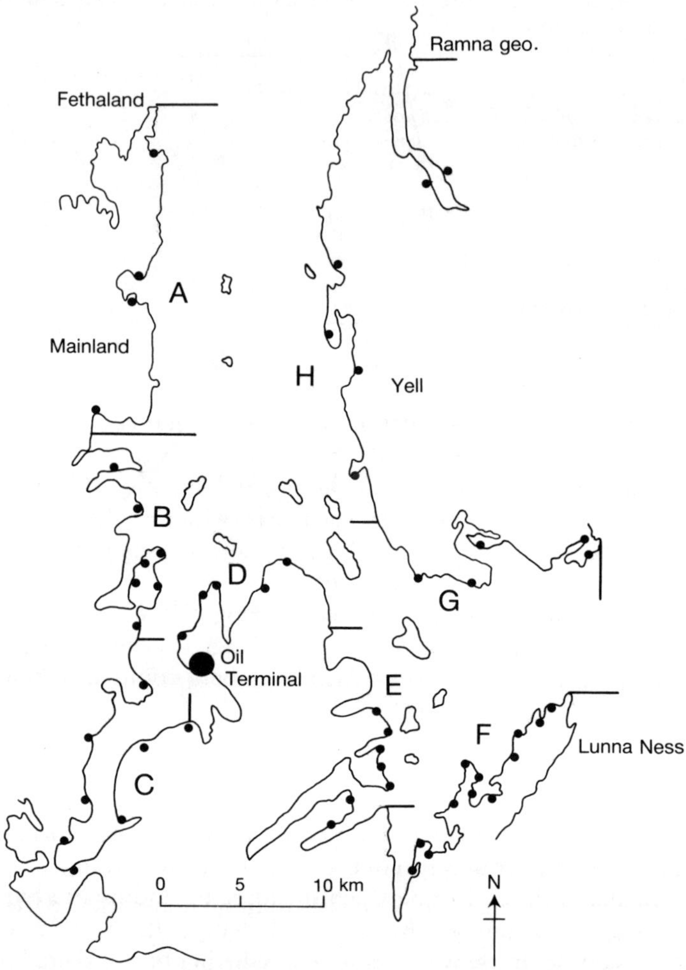

Fig. 2. The distribution of the monitoring sites (●) and the eight contiguous lengths of coast (A–H) used in the monitoring programme.

Results

Seasonal trends and preferences for sprainting in the detailed study area

Sprainting on land occurred throughout the year, with varying intensity, most spraints being counted in late winter/early spring and fewest in summer (Fig. 3). This seasonal pattern was repeated on all seven types of coast. The total numbers of spraints counted in different years varied markedly, the peak count in 1983 being 58% of the peak count in 1982. There were differences in the numbers of spraints counted on the different types of coast

Fig. 3. The total numbers of spraints counted on Shetland and on the River Severn. (Severn data from Mason & Macdonald (1986: fig. 2: 13)).

(Fig. 4). The coasts were ranked using a Jonckheere S-test (Jonckheere 1954). Non-rocky coasts had significantly fewer spraints than rocky coasts ($P < 0.001$), and in the latter case, significantly more spraints were found on exposed rocky coasts ($P < 0.001$).

More otters were seen on rocky coasts than on non-rocky coasts (Conroy & French in prep).

The relationship between numbers of spraints and numbers of otters seen

We calculated the number of otters seen/km/h, and the average number of spraints counted per 200 m distance. These activities were then scored from 0–5 (Conroy & French unpublished). The relationship between spraint numbers and otters was significant for the Mainland ($r_s = 0.783$; $t = 10.542$; $P < 0.001$) and Yell ($r_s = 0.685$; $t = 5.322$; $P < 0.001$) but not for the smaller islands ($r_s = 0.402$; $t = 1.458$; $P > 0.01$). This relationship is, however, not simple, since, because of the seasonal fluctuations in sprainting, scores for spraints must be corrected for the time of counting. For example, 50 spraints counted in February would have a lower score than 50 spraints counted in July. A knowledge of the sprainting pattern throughout

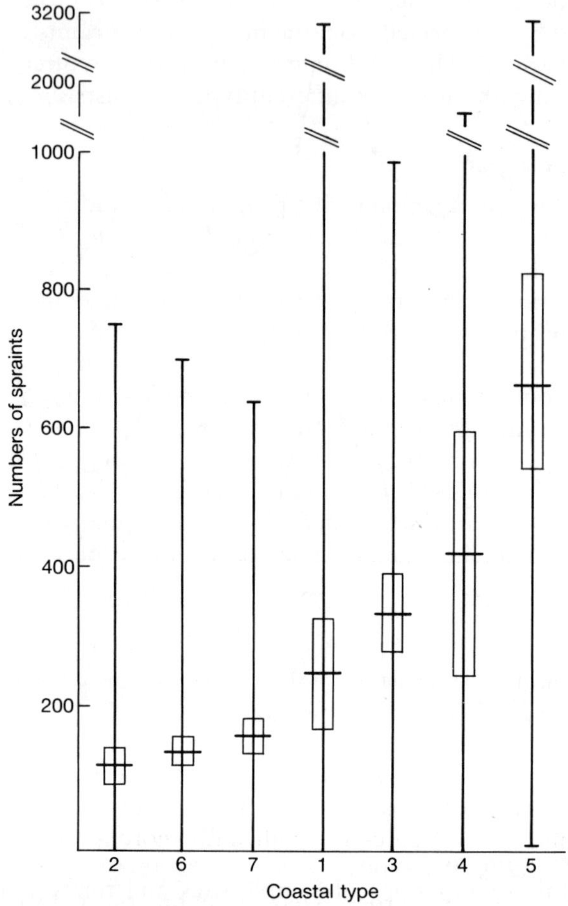

Fig. 4. Mean numbers of spraints/200 m (± 1 *s.e.*) and range for each coastal type for all visits to the detailed study area. For details of coastal types see Table 1.

the year is necessary to correct for seasonality. The scores for spraint counts used in the above calculations were corrected for the time of counting.

The correlation between the numbers of otters seen and the numbers of spraints counted holds only over relatively long lengths of coast (> 40 km); there was no significant relationship between the numbers of otters seen and the numbers of spraints found on the islands in Yell Sound, nor over relatively short stretches of coast (*c.* 7 km) (Kruuk, Conroy, Glimmerveen & Ouwerkerk 1986).

A Multivariate Analysis of various indices of otters and otter signs, including spraints, followed by a classification of component scores by coastal type, identified some types of coast where more otters were seen than one would have expected from the numbers of spraints counted (Conroy & French in prep). This was confirmed by watching otters, and following animals fitted with radios, on different types of coasts.

The monitoring programme

To see how accurately the monitoring programme could estimate counts of spraints, we developed a series of tests around the 54 monitoring sites.

The amount of change detectable in any monitoring programme is determined by the variance and sample size of the data. The numbers of spraints at individual spraint sites were extremely variable. Although, overall, there was a coherent seasonal pattern from all the monitoring sites, individual sites could be out of phase with the rest, and the numbers of sprints counted at individual spraint sites could vary by ± 200%.

Because of this variability, we initially chose a change of 30% from the 'normal' level as a reasonable minimum detectable target. To see if we could always detect a 30% drop we needed to know how many sites it would be necessary to sample; how many times it would be necessary to sample before an incident; and whether the method would work for all combinations of sites.

The Ratio Method

This method was tested by seeing whether we could predict the 'natural' variation in spraint numbers to within 30%. If so, we could be reasonably confident that any difference of more than 30% was a 'real' effect of an incident.

Number of sites required

We tried several combinations of sample sites and times and found that using more than 50 monitoring sites we could detect changes of 30% or more. With fewer sites this was not possible; also fewer sites gave an inadequate cover of Yell Sound.

Number of sample times

We found that the ratio R based on five or six bi-monthly visits prior to an incident gave an estimate to within 30%. Because of seasonality and year-to-year fluctuations in spraint numbers, calculations of R from more or less than five or six visits were less accurate.

Site combinations

There are about 1.8×10^{16} possible combinations of affected and unaffected sites. Obviously it is not practicable to test for all of these. Instead we tested a series of combinations which covered a range of situations of interest. We used two scenarios.

The first was an incident similar to the oil spill in Sullom Voe from the '*Esso Bernicia*' (Richardson unpublished). In this, there were many widely scattered, heterogeneous areas affected by oil, few areas were unaffected, and many of the effects were localized on numerous short stretches of coast. In the initial tests the number of affected sites varied from 12 to 36, and 14 of the 18 tests were within 30%. We then tested smaller groups down to five affected sites; the results were still to within 30% of the estimate.

The second scenario was of a localized incident affecting only one or two long contiguous stretches of coast. For this the coast was divided into eight zones (A–H in Fig. 2), with the number of sites in the zones varying from four to 12, and we tested each one against all the others. All tests were within 30% of the estimate (8/11 to within 20%).

The Rates of Change Method

This was tested by applying an artificial change to observed spraint counts after a hypothetical oil spill, and seeing what level of change was detectable at 95% significance.

Number of sites required

Using all 54 sites gave satisfactory results. We did not test reduced numbers because we did not wish to rely on only one method of detecting change, and we already knew that we needed this number of sites for the Ratio Method.

Number of sample times

In theory only one count before and one after an incident are needed. This gives a 10% chance of mistakenly recording a change as significant. The chance can be greatly reduced by using two counts before and two counts after an incident. (With our data this number and timing of samples eliminated the error completely.)

Site combinations

Because the Ratio Method could adequately detect changes for contiguous groups, we used only the '*Esso Bernicia*' data to test the Rates of Change Method. We tested a large number of combinations, including those used in the tests of the Ratio Method and, except between March and May, a 30% change at affected sites was always detectable at the 95% significance level. This applied whether the 30% decrease was applied uniformly to all sites or with realistic stochastic variation about a 30% mean (for fuller details see Conroy & French unpublished).

From the above tests, we concluded that:–

1. For either method to operate satisfactorily, a total of at least 50 monitoring sites was needed.

2. Using both methods, a change of 30% or more could always be

detected except during the main drop in spraint numbers between March and May.

3. A minimum of five sites needed to be affected or unaffected for the system to operate.

4. For the Ratio Method, calculating the ratio *R* from the five previous visits (e.g. over 10 months) gave the best results. (This allowed for the seasonal fluctuations in the numbers of spraints.)

5. Although only one count before and one after an incident are needed for the Rates of Change Method to operate, it is better to use two or more counts on either side of the incident. This lessens the errors due to seasonal cycles.

Discussion

From our studies in Shetland, where we assumed a relatively constant otter population, we established that:–

1. There were seasonal variations in the numbers of spraints counted on land.

2. There were annual fluctuations in the total numbers of spraints counted.

3. There were some types of coast where more spraints were found than others.

4. There was a significant correlation between the numbers of spraints counted and the numbers of otters seen over long stretches of coast, but only when the data had been corrected for seasonality. The relationship was not significant over short lengths of coast, and there were also some types of coasts where more otters were seen than would have been expected from the number of spraints counted.

5. Because of the natural variation in spraint numbers, the monitoring programme could only detect changes of 30% or more in spraint numbers.

The cause of the decline in spraint numbers in summer was thought to be changes in behaviour associated with breeding, rather than a decline in the numbers of otters using the coasts in summer. In Shetland, otters spraint less frequently on land in the summer, when most pups are born (Kruuk, Conroy & Moorhouse, this volume).

Seasonal patterns of sprainting are not unique in Shetland. For at least part of most long-term studies of otters, very similar seasonal cycles are found, e.g. on the River Severn (Fig. 3) and in Sweden, where Erlinge (1967, 1968) reported more spraints in winter, at the time of the 'rut', than in the summer. In Spain, however, there was a different seasonal pattern, with most spraints counted in summer and fewest in late winter/spring (Lopez-Nieves & Hernando 1984).

Table 2. Details of spraint surveys, numbers of sites counted and dates of visits.

Country	No. of sites counted	Dates of visit	Reference
Albania	31	8–27 May 1985	Prigioni, Bogliani & Barbieri (1986)
Central Algeria	52	24 March–2 April 1984	Macdonald, Mason & De Smet (1985)
Tunisia	75	8–20 July 1982	Macdonald & Mason (1983b)
Central Portugal	90	17–29 August 1980	Macdonald & Mason (1982b)
Yugoslavia	129	22 July–22 August 1982	Liles & Jenkins (1984)
Spain	176	3 June–4 July 1981	Elliot (1983)
Spain, W. Andalusia	354	July 1984	Adrian, Wilden & Delibes (1986)
Southern Italy	188	25 March–14 April 1982	Macdonald & Mason (1983c)
Greece	200	27 March–14 April 1981	Macdonald & Mason (1982a)
Greece, northern	52	28 July–10 August 1983	Macdonald & Mason (1985)
Western France	315	6 Oct.–6 Nov. 1980	Green & Green (1981)
Wales (1)	1030	May 1977–March 1978	Crawford *et al.* (1979)
(2)	1097	May 1984–March 1985	Andrews & Crawford (1986)
Ireland	2373	Jan. 1980–Jan. 1981	Chapman & Chapman (1982)
England	2940	Sept. 1977–Dec. 1979	Lenton *et al.* (1980)
Scotland	4636	Sept. 1977–Sept. 1979	Green & Green (1980)

Seasonal variation in spraint numbers has not been considered in the interpretation of many survey results, however. For example, reviewing the British and Irish otter surveys, Macdonald (1983) mentions the possibility of seasonal fluctuations, but none of the surveys to which she refers was checked for this.

Otters also appear to spraint more frequently in some areas than in others; and within the same area, more spraints were generally found on coasts, fewer on freshwater lochs away from the coasts, and fewer still on streams and rivers (Jenkins & Burrows 1980; Bas, Jenkins & Rothery 1984; Macdonald & Mason 1983a; Veen 1986; our Shetland study). Further, Green & Green (1980) concluded that, in Scotland, there was an altitudinal variation in sprainting, with fewer spraints at higher altitudes.

Given all the variables which affect sprainting, the time and effort required to obtain the data to relate spraint numbers to otter numbers, and the difficulties in interpreting the results from spraint surveys, can any valid estimates be made of the distribution of otters from spraint surveys? In particular, can studies in different areas, or at different seasons, and long- and short-term surveys be compared?

Most of the European and North African surveys were short-term and carried out in summer (Table 2), and many are compared with each other and with the long-term studies of the U.K. Without information at least on seasonality, such comparisons are unjustified.

Long-term studies, especially if repeated at later dates by people of similar experience and at the same time of year, may give acceptable indications of the distributions of spraints over a wide area and possible changes in otter distribution, but changes in the numbers of spraints counted need to be more than 30% before one can assume a 'real' change. Short-term studies are likely to be less reliable and comparisons of one area with another could be meaningless.

Bearing in mind that on a 250 km coastline of Shetland a minimum of over 50 monitoring sites (each containing several spraint sites) was necessary to be sure of detecting changes of 30% or more in spraint numbers, even with detailed knowledge of seasonal fluctuations, can single short-term surveys of countries, based on only a few sites, a high proportion of which may have no spraints, be of any value? Is it, for example, possible to get a sensible indication of the status of the otter in Albania (a country about 40 times the size of Yell Sound) by visiting only 31 sites (Table 2)?

The Shetland study found a significant relationship between numbers of spraints and numbers of otters seen, but this was based on detailed knowledge of the species' sprainting and breeding behaviour. Even then there was no significant relationship over short lengths of coasts or on small islands. More generally, relationships might exist over larger areas, but to date insufficient data are available for a convincing test. Our experience suggests that any such relationship will take a great deal of time and effort to elucidate.

There are other problems associated with counting spraints which can also affect results but are not discussed here. These are largely logistic and environmental, such as choice of monitoring sites (many appear to have been chosen for convenience), the state of vegetation cover of the bank side (a bank surveyed in March may have far less vegetation cover than the same bank in July) and the state of the rivers (a rise in the water level about the time of the survey can cover and/or remove spraints).

The use of questionnaires in monitoring otter numbers has been challenged because they could be selective, may result in a biased sample and give a wrong picture of the status of the otter (Mason & Macdonald 1986). However, from current knowledge, the use of spraint numbers as an index of the status of the otter may be no more reliable unless more detailed studies are undertaken on sprainting behaviour of otters in areas of known otter densities.

Conclusions

Before further spraint surveys are undertaken, there is a need for a more detailed understanding of the sprainting behaviour and habitat preferences of otters.

Extreme caution should therefore be used in the interpretation of the results of spraint surveys until such time as the relationship between otter numbers and spraint numbers has been established; and the probable influence of season and habitat on survey data should be very carefully and explicitly considered in all comparisons between surveys, especially when comparing results from different areas.

The relationship between otter numbers and signs of their presence may be affected by so many variables that, for practical monitoring, the time and effort required to evaluate them may be too costly. Without a full understanding of the relationships, however, spraint surveys are not a valid method for monitoring otters.

Acknowledgements

For their help in Shetland we thank D. Mace and other members of BP staff, in particular those in the Emergency Services Division. For help in the field, P. Dale, C. Cederland, M. Herfst and C. Sasse. For commenting on various drafts of the manuscript and advice, Dr M. G. Richardson, Dr Hans Kruuk, Dr B. W. Staines, Prof. D. Jenkins and K. H. Lakhani.

Many of the data were collected while under contract to the Shetland Oil Terminal Environmental Advisory Group, whose help is gratefully acknowledged.

References

Adrian, M.I., Wilden, W. & Delibes, M. (1986). Otter distribution and agriculture in Southwestern Spain. *Congr. Int. Un. Game Biol.* **17**: 519–26.

Andrews, E. & Crawford, A.K. (1986). *Otter survey of Wales 1984–85*. Vincent Wildlife Trust, London.

Anon. (1969). The otter in Britain. *Oryx* **10**: 16–22.

Anon. (1974). The otter in Britain—a second report. *Oryx* **12**: 429–35.

Arden-Clarke, C.H.G. (1986). Population density, home range size and spatial organization of the Cape clawless otter, *Aonyx capensis*, in a marine habitat. *J. Zool., Lond. (A)* **209**: 201–11.

Baker, J.R., Jones, A.M., Jones, T.P., & Watson, H.C. (1981). Otter *Lutra lutra* L. mortality and marine oil pollution. *Biol. Conserv.* **20**: 311–21.

Bas, N., Jenkins, D. & Rothery, P. (1984). Ecology of otters in northern Scotland, V. The distribution of otter (*Lutra lutra*) faeces in relation to bankside vegetation on the river Dee in summer 1981. *J. appl. Ecol.* **21**: 507–13.

Chanin, P.R.F. & Jefferies, D.J. (1978). The decline of the otter *Lutra lutra* L. in Britain: an analysis of hunting records and discussion of causes. *Biol. J. Linn. Soc.* **10**: 305–28.

Chapman, P.J. & Chapman, L.L. (1982). *Otter survey of Ireland, 1980–81*. Vincent Wildlife Trust, London.

Chehebar, C.E. (1985). A survey of the southern river otter *Lutra provocax* Thomas in Nahuel National Park, Argentina. *Biol. Conserv.* **32**: 299–307.

Conroy, J.W.H. & French, D.D. (Unpublished). *Monitoring otters in Shetland.*

Report to the Shetland Oil Terminal Environmental Advisory Group, Aberdeen, 1985.

Conroy, J.W.H., French, D.D. & Jenkins, D. (In prep.). *Seasonal patterns in the sprainting behaviour of the otter* (Lutra lutra) *in Shetland.*

Crawford, A., Jones, A. & McNulty, J. (1979). *Otter survey of Wales 1977–78*. Vincent Wildlife Trust, London.

Elliot, K.M. (1983). The otter (*Lutra lutra* L.) in Spain. *Mammal Rev.* **13**: 25–34.

Erlinge, S. (1967). Home range of the otter *Lutra lutra* L. in southern Sweden. *Oikos* **18**: 186–209.

Erlinge, S. (1968). Territoriality of the otter *Lutra lutra* L. *Oikos* **19**: 81–98.

Erlinge, S. (1972). The situation of the otter population in Sweden. *Viltrevy, Stockholm* **8**: 379–97.

Erlinge, S. (1980). Die Situation des Fischotters in Schweden. In *Der Fischotter in Europa – Verbreitung, Bedrohung, Erhaltung*: 103–6. (Eds Reuther, C. & Festetics, A.). Aktion Fischotterschutz e.V., Oderhaus und Göttingen.

Green, J. & Green, R. (1980). *Otter survey of Scotland 1977–79*. Vincent Wildlife Trust, London.

Green, J. & Green, R. (1981). The otter (*Lutra lutra* L.) in western France. *Mammal Rev.* **11**: 181–7.

Heggberget, T.M. & Myrberget, S. (1979). Den norske bestand av oter 1971–1977. *Fauna, Oslo* **32**: 89–95.

Heggberget, T.M. & Myrberget, S. (1980). Der norwegische Fischotterbestand 1970–1977. In *Der Fischotter in Europa – Verbreitung, Bedrohung, Erhaltung*: 93–101. (Eds Reuther, C. & Festetics, A.). Aktion Fischotterschutz e.V., Oderhaus und Göttingen.

Jefferies, D.J., Chanin, P.R.F. & Lenton, E.G. (1984). The decline of the otter *Lutra lutra* L. in England and a survey of its present status and distribution. *J. Otter Trust* **1983**: 13–19.

Jenkins, D. & Burrows, G.O. (1980). Ecology of otters in northern Scotland. III. The use of faeces as indicators of otter (*Lutra lutra*) density and distribution. *J. anim. Ecol.* **49**: 755–74.

Jenkins, D. & Conroy, J.W.H. (1982). Methodology for studying habitats used by coastal otters. *A. Rep. Inst. terr. Ecol.* **1981**: 19–23.

Joint Otter Group (1977). *Otters 1977*. Nature Conservancy Council/Society for the Promotion of Nature Conservation, London.

Joint Otter Group (1979). *Otters 1979*. Society for the Promotion of Nature Conservation, Lincoln.

Jonckheere, A.R. (1954). A distribution-free *K*-sample test against ordered alternatives. *Biometrika* **41**: 133–45.

Kruuk, H., Conroy, J.W.H., Glimmerveen, U. & Ouwerkerk, E.J. (1986). The use of spraints to survey populations of otters *Lutra lutra*. *Biol. Conserv.* **35**: 187–94.

Lenton, E.J., Chanin, P.R.F. & Jefferies, D.J. (1980). *Otter survey of England 1977–79*. Nature Conservancy Council, London.

Liles, G. & Jenkins, L. (1984). A field survey for otters (*Lutra lutra*) in Yugoslavia. *J. Zool., Lond.* **203**: 282–4.

Lloyd, J.I. (1962). Where are the otters? *Gamekeeper* **65**: 299–300.

Lopez-Nieves, P. & Hernando, J.A. (1984). Food habits of the otter in the central Sierra Morena (Cordoba, Spain). *Acta theriol.* **29**: 383–401.

Macdonald, S.M. (1983). The status of the otter (*Lutra lutra*) in the British Isles. *Mammal Rev.* **13**: 11–23.

Macdonald, S.M. & Mason, C.F. (1982a). Otters in Greece. *Oryx* **16**: 240–4.

Macdonald, S.M. & Mason, C.F. (1982b). The otter *Lutra lutra* in central Portugal. *Biol. Conserv.* **22**: 207–15.

Macdonald, S.M. & Mason, C.F. (1983a). Some factors influencing the distribution of otters (*Lutra lutra*). *Mammal Rev.* **13**: 1–10.

Macdonald, S.M. & Mason, C.F. (1983b). The otter (*Lutra lutra*) in Tunisia. *Mammal Rev.* **13**: 35–37.

Macdonald, S.M. & Mason, C.F. (1983c). The otter *Lutra lutra* in southern Italy. *Biol. Conserv.* **25**: 95–101.

Macdonald, S.M. & Mason, C.F. (1985). Otters, their habitat and conservation in north-east Greece. *Biol. Conserv.* **31**: 191–210.

Macdonald, S.M., Mason, C.F. & De Smet, K. (1985). The otter (*Lutra lutra*) in north-central Algeria. *Mammalia* **49**: 215–9.

Mason, C.F. & Macdonald, S.M. (1986). *Otters: ecology and conservation.* Cambridge University Press, Cambridge.

Muffling, S. von (1977). *Fischotter in Europa.* Loizenkirchen.

Myrberget, S. & Froiland, O. (1972). Oteren i Norge omkring 1970. *Fauna, Oslo* **25**: 149–59.

Prigioni, C., Bogliani, G. & Barbieri, F. (1986). The otter *Lutra lutra* in Albania. *Biol. Conserv.* **36**: 375–83.

Reuther, C. (1980). Zur Situation des Fischotters in Europa. In *Der Fischotter in Europa – Verbreitung, Bedrohung, Erhaltung*: 71–92. (Eds Reuther, D. & Festetics, A.). Aktion Fischotterschutz e.V., Oderhaus und Göttingen.

Richardson, M.G. (Unpublished). *The environmental effects of the* 'Esso Bernicia' *fuel oil spill, Sullom Voe, Shetland, January 1979.* Unpublished Report of the Nature Conservancy Council, Lerwick, 1979.

van Wijngaarden, A. & van de Peppel, J. (1970). De otter, *Lutra lutra* (L.), in Nederland. *Lutra* **12**: 1–72.

Veen, J. (1985). The distribution of otters and their biotope in the south-west of the Netherlands. *Z. angew. Zool.* **72**: 71–82.

Veen, J. (1986). The distribution of otter faeces (*Lutra lutra* L.) on the coast of Wester Ross, Scotland (1979–1980). *Z. Säugetierk.* **51**: 97–103.

Symp. zool. Soc. Lond. (1987) No. 58: 263–278

Seasonal reproduction, mortality and food of otters (*Lutra lutra* L.) in Shetland

H. KRUUK, J. W. H. CONROY and A. MOORHOUSE

Institute of Terrestrial Ecology, Banchory, Scotland

Synopsis

Otters *Lutra lutra* living in a coastal habitat in Shetland were born mostly in summer (84% between May and August), in contrast to otters in many other parts of Britain, where they are aseasonal. In Shetland 73% of natural mortality occurred in spring, from March till June. This seasonality occurred despite smaller differences in temperature between summer and winter than on mainland Britain, and was associated with large differences in food availability. Biomass of prey species of otters was ten times higher in summer, during the first months of lactation, and seasonal mortality was associated with low numbers of eelpout (*Zoarces viviparus*), rocklings (Gadoidea) and butterfish (*Pholis gunnellus*), and low hunting success. Otter cubs were fed by their mother with larger prey than she ate herself; larger fish were also more abundant in summer. The families coped with the increasing scarcity of fish during the winter at least partly through the increasing proportion of fish caught by the cubs themselves. The mean litter size was relatively small (1.8), possibly related to the timing of gestation during the annual period of food stress.

Introduction

Despite serious concern about the dramatic decline of otter populations throughout Europe (Chanin & Jefferies 1978; Mason & Macdonald 1986), little is known about population mechanisms in the species. Such knowledge is important for understanding the natural limiting factors operating on otter populations, and the way in which pollution affects them (Mason & Macdonald 1986). It is also a prerequisite for the design of conservation management policies. In this study we analyse aspects of the seasonality and success of reproduction, mortality and body condition of otters; these are related to seasonal environmental changes, especially in food.

Most Carnivora living in temperate climates show seasonal reproduction. The otter, however, is an anomaly in that it is reported to give birth at all times of year (Stephens 1957; Harris 1968; Chanin 1985; Mason & Macdonald 1986). It is polyoestrous (Duplaix-Hall 1975), but without a delayed

ZOOLOGICAL SYMPOSIUM No. 58
ISBN 0-19-854006-X

implantation as in the closely related species *Lutra canadensis* (Hamilton & Eadie 1964). Breeding seasons of animals are usually adapted to seasonality in food supply, most food being available at that stage of reproduction when energy demand is highest (Lack 1954). In the case of the otter, this would suggest that fish are either available equally at all times of year, or available in abundance, and therefore not a limiting factor.

It is clearly important for conservation management to distinguish between these possibilities, in order to understand the mechanisms of population limitation. We therefore studied this problem of seasonality in some detail, quantifying differences in reproduction and food supply, and using observations on body condition and mortality as indicators of possible seasonal stresses.

Otters are common along the coasts of Ireland and of north-west Scotland, in the Hebrides, in Orkney and in Shetland (Green & Green 1980; Chapman & Chapman 1982). In all these areas they feed mostly in the sea, and they have access to fish at all times of year. Their food in Shetland has been described by Herfst (1984), and consists mainly of bottom-living fishes, such as eelpout (*Zoarces viviparus*), butterfish (*Pholis gunnellus*), various rocklings (Gadoidea), seascorpion (*Taurulus bubalis*) and flatfish. Cubs are born in holes, 'holts', which are often quite far from water. They remain in these holts for eight to ten weeks, after which they accompany the mother, at least part of the time (Sandegren 1985; pers. obs.), until they are about ten months old. Male otters are not involved in parental care.

Study area and climate

Observations on litter sizes and the relative size of cubs were made along Yell Sound, between the islands Mainland and Yell, Shetland, at all times of the year between 1981 and 1985. Carcasses of otters were collected from all over Shetland.

The area for intensive study, the west coast of the Lunna Ness (60° 25′ Lat. N and 1° 5′ Long. W), was also located along the Yell Sound. This coast, approximately 20 km long, consists mostly of rocky shores below gently sloping peat areas (Kruuk, Conroy, Glimmerveen & Ouwerkerk 1986). There were approximately 120 otter holts in the study area, mostly in peat banks. Only a small proportion was regularly used by families for long periods.

The climate is relatively mild (Berry & Johnston 1980; Anon. 1983). An important characteristic for otters is water temperature, which we calculated from data provided by the Deutsches Hydrographisches Institut, Hamburg, by intrapolation. In Fig. 1 these data are compared with observations from representative inland waters in England; the difference between extremes in Shetland (approximately 5° C) was much smaller than for Windermere or the River Severn (both about 16° C). Winter temperatures of

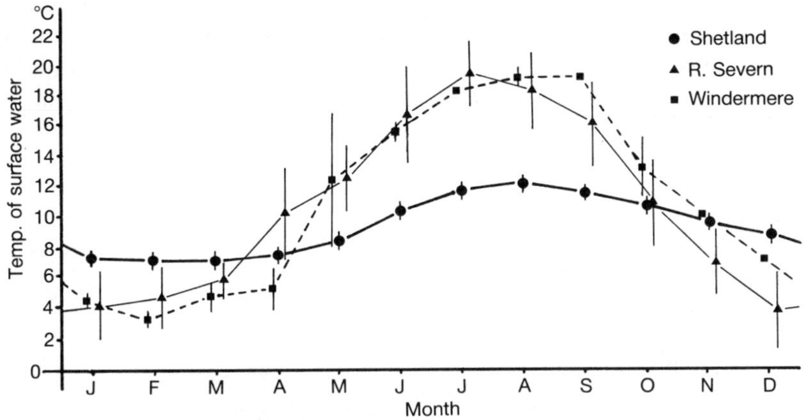

Fig. 1. Water surface temperatures in Shetland in different months, compared with inland waters in England: Windermere and the River Severn at Tewkesbury. Means ± *s.d.* Data for Shetland from Deutsches Hydrographisches Institut, Hamburg, for 1982–1985; for Windermere from Heron (1961) for 1958–1960, and for the River Severn from Severn-Trent Water Authority, Birmingham, for 1980–1985.

Shetland waters were higher than those in mainland Britain. Air temperatures show similar differences, but Shetland was windier at all times of the year (Chandler & Gregory 1976).

Methods

Most of the observations on litter size in Shetland were made while collecting data on the distribution of otter spraints (faeces) along the coasts. For each otter family seen, the number of cubs was counted and their size was classified relative to that of the adult female. We distinguished four size classes: Class 1, < 0.5 female size; Class 2, 0.5–0.75 female size; Class 3, > 0.75 female size yet recognizably smaller; Class 4, same size as female or larger. In general, there was good agreement between observers on size classification. We assumed that cubs of Class 1 would be 3–4 months old; this was based on observations of known-age cubs in captivity and the age at which they first emerge from the holt (Sandegren 1985), and observations of known families in the field. The month in which cubs were born was estimated from observations of Class 1 cubs. All our observations, therefore, refer to cubs which had survived until they left the natal holt.

Dead otters were deep-frozen as soon as possible, and the cause of death was established from post-mortem analysis or from circumstantial evidence (e.g. road casualties). It was assumed that the probability of finding a dead otter was the same throughout the year, as seasonal vegetation differences were small in Shetland.

Dead otters, and animals live-trapped for intensive study, were weighed and the length of body was measured as the straight line distance between the snout and the tip of the tail. Using data only from live otters and those killed on roads, the mean weight W (in kg) was established as a function of mean length L (in m):

$$W = a.L^n$$

For 25 females $a = 5.02$ and $n = 2.33$; for 17 males $a = 5.87$ and $n = 2.39$.

The index of body condition K was calculated for individual otters, following Le Cren (1951):

$$K = W / (a.L^n)$$

Thus, for a female otter of 4.95 kg and 1.03 m long,

$$K = 4.95 / (5.02 \times 1.03^{2.33}) = 0.92.$$

Observations of feeding otters were made in daytime, with binoculars (7 × 50) and telescope (15–45 ×). Whenever possible, the species and length of prey were determined. To check the accuracy of this, we tested each other on our ability to identify species of fish and estimate their length. A dead specimen was shown to an observer at 70 m distance (which is a normal otter observation distance), for a period of 5 s, the fish being handled to imitate the manipulation of prey by an otter. The observer watched through binoculars, noting species and length. Rocklings were not identified to species. In 75 trials, in which 12 species were used, the species was identified correctly in 84%, whilst length was underestimated by a mean of 9.7%, with negligible differences between observers. In the natural situation, prey identification was somewhat easier than in these tests, because the movements of some fish species were characteristic, and the colours in live fish were brighter than in dead ones.

Fish populations were sampled with cylindrical traps made of stiff, plastic gauze ('Netlon') with a 7 mm mesh size, 80 cm long and 27 cm in diameter. One end of the cylinder was blocked off, the other fitted with a funnel-shaped entrance, 7 cm in diameter. These traps, attached to a string, were placed in water 1–2 m deep below low tide, at various places in the intensive study area, and left out for 24 h at a time. Trapping efficiency for different species was estimated by comparing trap catches with exhaustive searches of a 20 m^2 area around the trap site, either using a dry suit and diving equipment, or when the area was exposed during very low tides.

Results

Birth seasons, in Shetland and elsewhere

Figure 2 shows the estimated month of birth for 34 litters observed in Shetland. Of these, 29 (85%) were born during the four summer months, May

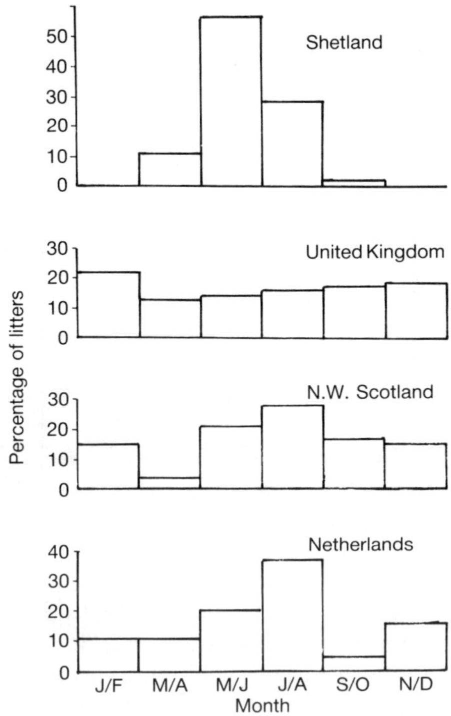

Fig. 2. Otter litters born at different times of year, for different areas. Data for UK from Harris (1968), for coastal otters in north-west Scotland from J. Rowbottom (pers. comm.), and for the Netherlands from Wijngaarden & Peppel (1970). Numbers of litters per month, for May–August compared with rest of year (*d.f.* = 1): Shetland, $n = 34$, $\chi^2 = 37.3$, $p < 0.001$; UK, $n = 168$, $\chi^2 = 0.2$, $p > 0.5$; north-west Scotland, $n = 47$, $\chi^2 = 5.1$, $p < 0.05$; Netherlands, $n = 19$, $\chi^2 = 5.2$, $p < 0.05$.

to August; more than half were born in May–June, and none in winter. The seasonal trend is highly significant. Also in coastal populations in north-west Scotland, and in inland waters in the Netherlands, significantly more cubs were born in summer (J. Rowbottom pers. comm.; Wijngaarden & Peppel 1970). However, there was no such trend in the large sample from inland Britain (Harris 1968; Fig. 2). The ratio of numbers of cubs born per summer month over numbers born per month in the rest of the year was 10.8 for Shetland, 1.9 for north-west Scotland, 2.7 for the Netherlands, but only 0.9 for inland Britain.

There are some data from other areas; in Sweden, inland otters are born in spring (Erlinge 1967), in Russia in spring and summer (Danilov & Tumanov 1975), and in West Germany mostly in spring (Reuther 1980). However, coastal otters in Norway appear to be non-seasonal in their breeding (T. M. Heggberget pers. comm.). For the north-east of Scotland data are few

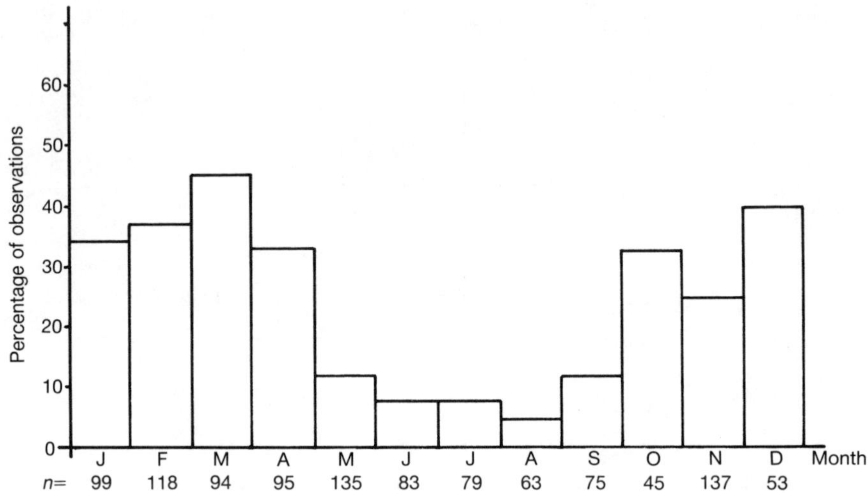

Fig. 3. Frequency of sightings of otter families, as a proportion of all observations, in 1984 and 1985, Lunna Ness. $\chi^2 = 100.3$, $d.f. = 11$, $p < 0.001$.

and conflicting (Jenkins 1980; Green, Green & Jefferies 1984). In captivity, otters are born during most months (Reuther 1985), but seasonal patterns are not meaningful, because matings could have been arranged.

Figure 3 shows the proportion of otter sightings in which a family group was seen, in the intensive study area on Shetland. Because groups are more conspicuous than single otters, this figure does not show the true proportion of families in the population. It does, however, show the seasonal trend, and indicates the approximate timing of births, the period when cubs are still in the holt (summer), the time of free-roaming family groups (from September onwards), and the dispersal of the cubs (April and May). About 15 different families were involved.

Between 1983 and 1986 otters were observed mating on seven occasions. All copulations occurred during March, April or May, and on three occasions cubs of the last litter were still present. In 1985 and 1986 there were five families of which the breeding performance from the previous year was also known; four had a litter in the previous breeding season, one of which had failed.

To summarize, the available data show that in many places, and particularly in Shetland, otters breed in spring and summer. There are areas, however, where cubs may be born at any time of year, e.g. the inland waters of Britain, and the coasts of Norway.

Litter sizes

The mean litter size in Shetland was 1.78 ± 0.57 *s.d.* (n=80; 30% of litters contained only one cub, 62% two cubs, and 7% three cubs). Values were

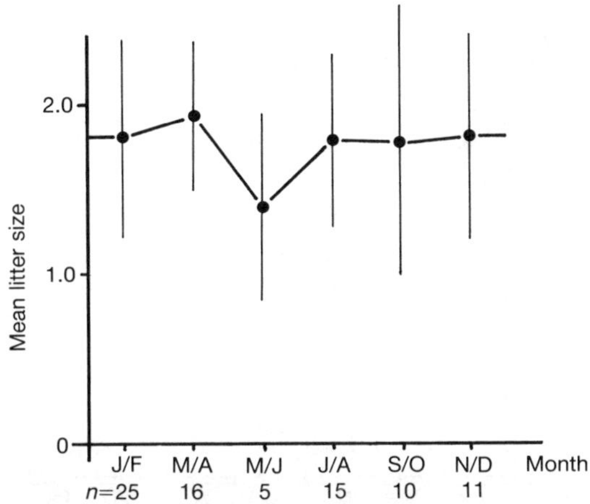

Fig. 4. Mean numbers of cubs per family, ± *s.d.*, at different times of year. Kruskal–Wallis analysis of variance: $H = 4.75$, $d.f. = 5$, $p > 0.3$.

similar for our intensive study area, with a mean litter size of 1.80 (n=11). In coastal areas in north-west Scotland J. Rowbottom (pers. comm.) found 1.55 cubs per litter (n=47), and litters of coastal otters in Norway had 1.95 cubs (T. Heggberget pers. comm.). Litter sizes in marine habitats were smaller than in most inland areas; in the Netherlands litter size was 2.8 ± 0.8 (n=48) (Wijngaarden & Peppel 1970), in East Germany 2.3 (n=70) (Stubbe 1980), in Poland 2.4 (n=41) (Wlodek 1980), and in U. K. inland habitats 2.5 (n= 160) (Mason & Macdonald 1986).

All our observations were of cubs accompanying their mother, and mostly from September onwards. In Shetland the number of cubs per family changed little during the season, except for the decrease at the time of cub dispersal, around April (Fig. 4). This suggests that cub mortality was low as long as the family stayed together, in winter. However, our method would not detect the loss of whole families, nor would it register mortality of cubs either before they became mobile, or after they dispersed. The loss of whole families was probably rare; among 11 families in the intensive study area, in the period between August and March it happened only once (to a single-cub family).

Condition and mortality

The index of condition, K, was determined for 27 live and 36 dead otters, at different times of year (Fig. 5). Condition varied significantly from 1.05 in November–December, to 0.77 in May–June, so otters were significantly lighter in summer. The same differences were found using data only from

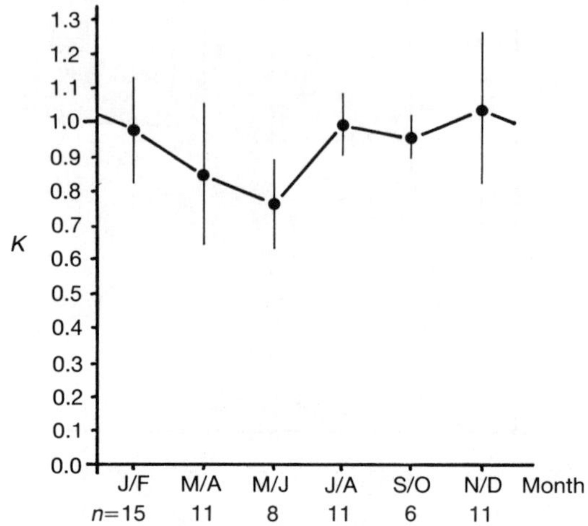

Fig. 5. Index of body-condition *K*, of otters caught alive or found freshly dead on Shetland. Kruskal–Wallis analysis of variance: $H = 11.6$, $d.f. = 2$, $p < 0.01$.

live otters or those killed by accidents, but these were statistically not significant. Preliminary analyses suggested that the differences in condition cannot be explained by differences in age-structure in the samples from summer and winter. The seasonality of condition was the same in males and females.

The mortality amongst otters in Shetland also varied significantly with the season (Fig. 6). Violent deaths, mostly traffic accidents, happened throughout the year, with perhaps slightly fewer in May–June, but this was not significant. However, non-violent mortality, i.e. death from disease and/or starvation, was highly concentrated in spring, with 73% occurring between March and June.

Diet and food supply

In the intensive study area 833 prey items caught by otters were identified (Fig. 7). In winter and summer more than three quarters of the prey caught were bottom-dwelling, eel-like fishes (eelpout, rockling, butterfish). There was some variation in the species composition of the diets in summer and in winter, but the only substantial and significant difference was the change from eelpout in summer to rocklings in winter.

Fish traps, in the centre of our study area, caught the same species of fish as the otters did; 79% were eelpout, rockling or butterfish. The numbers of prey caught per trap in the summer were much higher than those in winter (Fig. 8); however, in winter fish were heavier, mostly because more rock-

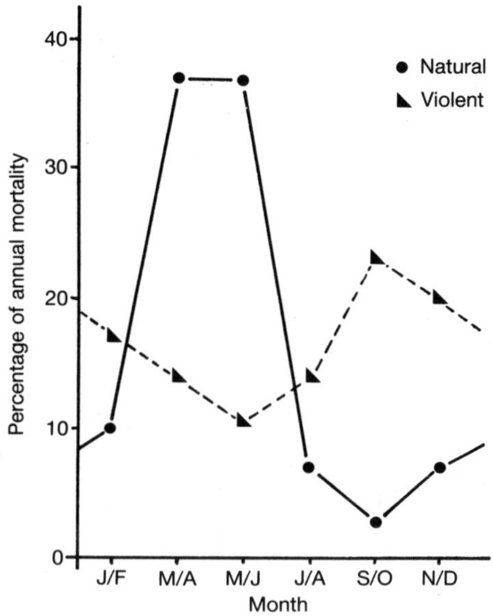

Fig. 6. Mortality of Shetland otters in different months. Natural deaths: $n = 30$, $\chi^2 = 22.0$, $d.f. = 5$, $p < 0.001$. Violent deaths (mostly road-kills): $n = 35$, $\chi^2 = 1.87$, $d.f. = 5$, $p < 0.9$.

lings were caught then, and fewer eelpout. These differences were highly significant.

The exact, quantitative interpretation of the numbers of fish caught in the traps was difficult, because of the changes in species composition, and because the probability of being caught varied between species. These probabilities were calculated from comparisons between trap-catches and direct observations of fish densities around the trap (unpubl. data). The application of these correction factors did not substantially alter the conclusions, however (Fig. 9); total joint biomass of eelpout, rockling, and butterfish was more than ten times higher in July–August than in March–April. The catches of various species did not fluctuate together (Kendall Coefficient of Concordance: $W = 0.50$, $\chi^2 = 19.4$, $d.f. = 13$, $p < 0.2$). In winter, for instance, there were more rocklings, in summer more eelpout; presumably, therefore, there was no overall, seasonal trap effect. The traps did not catch any fish with a diameter larger than 7 cm.

Feeding observations

An otter which was fishing would usually be swimming along the surface, diving to the seabed at depths of up to 14 m, though mostly in water less than 5 m deep (Kruuk, Glimmerveen & Ouwerkerk 1985). It would come up again after up to 70 s, and, if successful, with a fish in its mouth. There

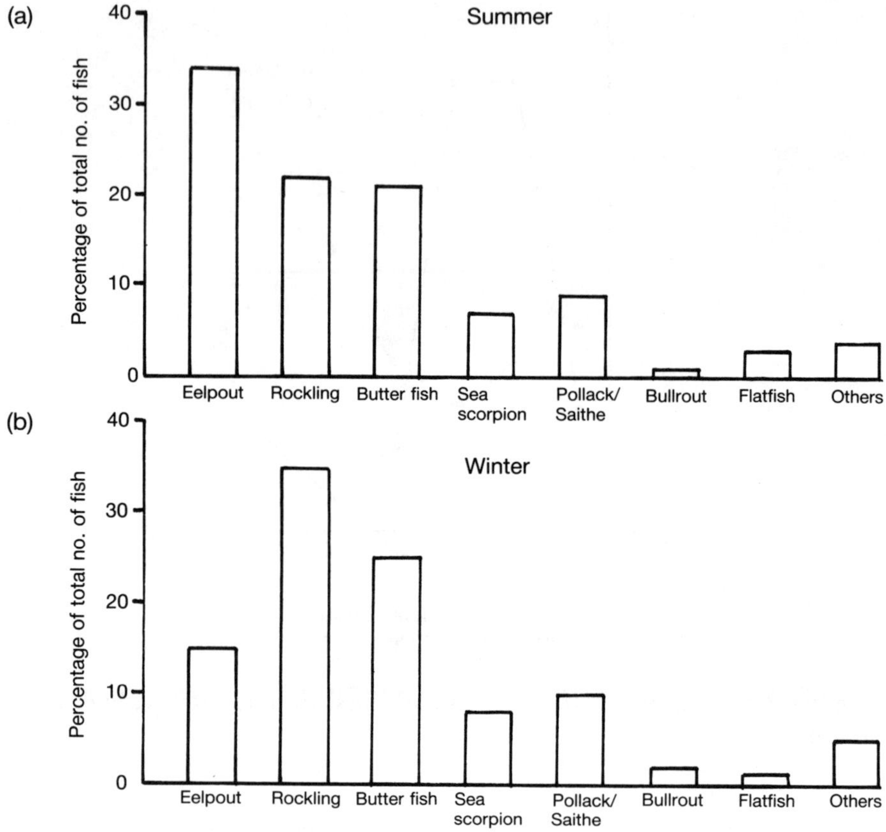

Fig. 7. Food of otters in summer and winter; direct observations.
(a) Fish eaten from May to August, 1984 and 1985 (n = 565).
(b) Fish eaten from November to February, 1984, 1985 and 1986 (n=268). Bullrout = *Myoxocephalus scorpius*, pollack = *Pollachius pollachius*, saithe = *Pollachius virens*.
Eelpout/rockling/butterfish/others, winter versus summer: $\chi^2 = 40.0$, $d.f. = 3$, $p < 0.001$.

was circumstantial evidence (from close observations of animals in the wild, and from otters in captivity) that otters did not swallow fish under water. Most fish were eaten on the surface, but larger prey was taken ashore.

The diving success rates were higher in summer (May–September 36% successful, n = 1181) than in winter (November–March, 25% successful, $n = 648$, $\chi^2 = 22.0$, $d.f. = 1$, $p < 0.001$).

Female otters usually fed their cubs on land, but if the prey was small the cubs might take it from their mother in the water. Females often travelled 100 m or more to take a fish to their cubs, especially when the cubs were still small. In this way one single prey item could take 5 min or more to process.

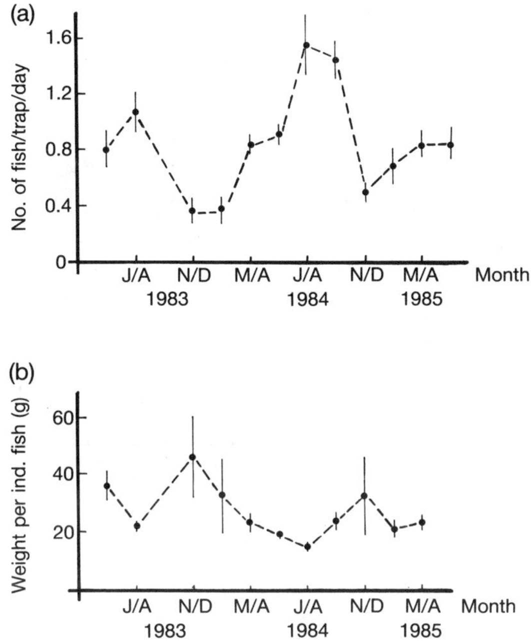

Fig. 8. Numbers and weights of fish caught in traps, at different times of year. Total number of fish caught: n = 1,748. (a) Mean uncorrected numbers, ± *s.d.* (b) Weight per fish in the traps, ± *s.d.*

Weaning occurred over a long period, and during several months in autumn and winter cubs gradually caught more prey for themselves (e.g. in January–February, 19% of their prey, n = 110).

We compared the length of fish taken by the female to the cubs with that of fish eaten by herself (Fig. 10). The median estimated weight per fish taken to cubs was more than six times heavier than that eaten by the female. The size of fish taken by single female otters during this period was similar to that of fish taken and eaten by the otter mothers ($n=68$; $\chi^2 = 6.8$, $d.f.=3$, $p>0.05$). The size of prey taken in summer by single otters was similar to that taken in winter.

Discussion

The density of fish along the Shetland shores fluctuated strikingly with the time of year; the biomass of fish which otters usually ate (eelpout, rocklings and butterfish) was about ten times higher in July–August than during winter and early spring (January–April). Presumably, in winter these species moved into deeper water, where otters did not take them. The otters' fishing

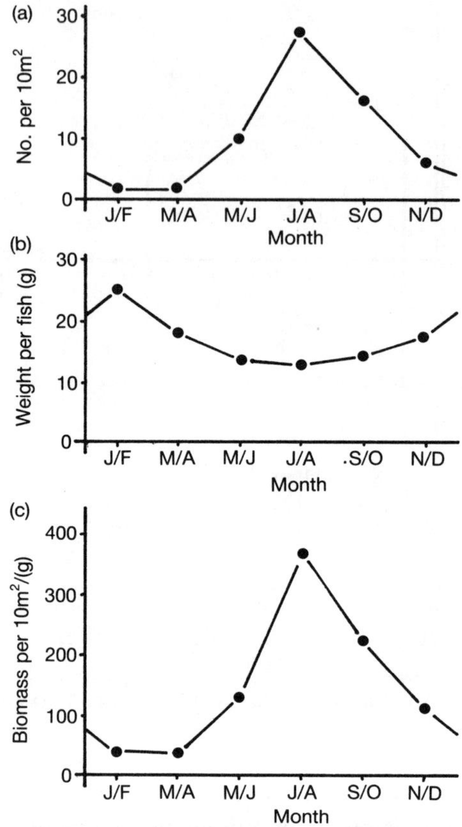

Fig. 9. Densities and weights of eelpout, butterfish and rocklings in the study area, calculated by applying species-specific correction factors to trap-data, 1984–1986. (a) Estimated numbers of fish per 10 m^2. (b) Mean weight per fish in the population. (c) Total biomass of fish per 10 m^2.

success was considerably lower in winter; at the same time the water was colder (6° C, compared with 12° C in summer), and this would increase the energetic costs of foraging. Thus, conditions for foraging were considerably harsher during the first months of the year.

It was likely that the observed high mortality of otters in spring, March–June, was directly related to this pattern of feeding conditions. Otters which died a violent death, usually on roads, were found somewhat more often during the time of year that daylight was short; road kills usually occurred at night, when animals were dazzled. However, this difference between seasons was not significant. But animals which died naturally, of various proximate causes, were found significantly more often during or immediately after the period of worst foraging conditions. Also, the otters' body

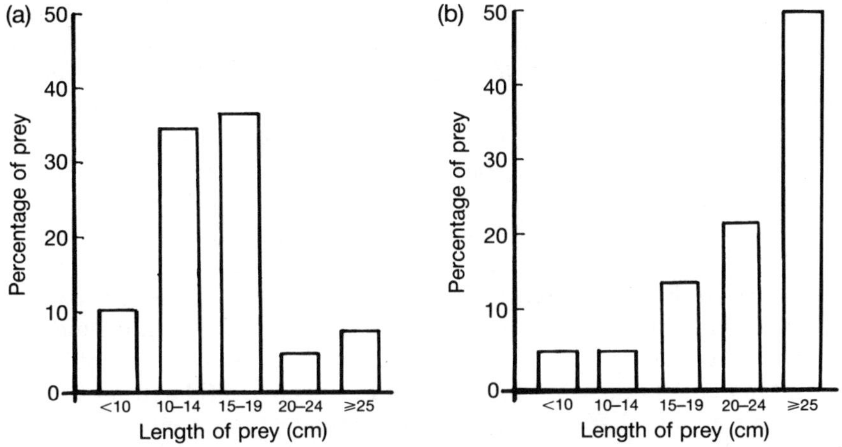

Fig. 10. Size distribution of fish caught by otter families. December 1985–February 1986: (a) caught and eaten by female (n = 103); (b) caught by female, eaten by cubs (n = 22). $\chi^2 = 49.0$, $d.f. = 2, p < 0.001$.

condition, a function of weight over length, was lowest during that time. There was strong evidence, therefore, of seasonal environmental stress on the otter population, caused by shortage of food.

The timing of reproduction may also be geared to feeding conditions. Over 80% of otters in Shetland had their litters between May and August, i.e. just before the fish biomass reached its peak. Thus, food availability was high during the months when cubs were fed exclusively through lactation (the time of maximum energetic requirements in mammalian reproduction: Randolph, Randolph, Mattingly & Foster 1977; Widdowson 1981; Loudon & Racey 1987). When older, cubs were fed mostly with fishes which were significantly larger than those eaten by the mother herself (as in some birds: Royama 1966), but although such large fishes were *relatively* more common in winter, in absolute terms their abundance was higher in summer. During the autumn and winter cubs caught increasingly more fish for themselves, and this should affect the energy expenditure by the mother. Also, as cubs moved around with the female more and more from the age of two to three months onwards, the mother's effort of carrying food to the cubs should decrease. Thus, the female would expend less energy in providing for her cubs during the later months of parental care. The families broke up (females abandoning cubs) during the period of greatest food stress, March–April, and soon after that the newly independent otters would again find an increase in the biomass of fish.

Litter sizes were small in Shetland, averaging 1.8 cubs per litter, compared with 2.5 or more in some areas elsewhere. It is possible that this was a consequence of gestation during the annual period of stress. But as we could only assess litter size after cubs had emerged from their natal holts, at two months old or older, there could have been early mortality which we did not detect. We could not distinguish between these possible explanations for the small litter size, but when more data become available it will be instructive to analyse for possible correlations between cub numbers and fish biomass during early or late gestation, or during early lactation. It is possible that events before the time that cubs emerge above ground are critical in deciding reproductive success, as cub mortality was low later on.

Apart from the functional significance of the different seasonal patterns of reproduction in various areas, the question arose as to what the mechanism was: how was synchronization achieved? *Lutra lutra* does not have a delayed implantation, as has for instance *L. canadensis* (Hamilton & Eadie 1964); this implies that in our species the mating period itself was synchronized. Shetland otters could be seasonally polyoestrous. The seasonality, and lack thereof, at different latitudes suggested that there was no species-specific photoperiod response, although it was possible that populations in different areas had evolved a different timing of their mating season in response to particular photoperiod changes. Alternatively, the observed variation could be explained by a system in which, in all otter populations, oestrus occurred during the annual period of environmental stress.

Whatever the physiological mechanism of synchronization between reproduction and food availability, the relationship was highly significant and therefore may be very important. The absence of seasonal breeding in other areas, despite much greater seasonal differences in temperature, suggests that either there is no substantial seasonal difference in food availability, or that the advantages associated with a seasonal food abundance are offset in some other way. In most fresh water the fish would not be able to move out of otters' reach during the winter, as they did in the sea, so this possibly explains the lack of seasonality in inland Britain. In countries with a more continental climate, ice formation in winter might be a significant factor determining access to food. Clearly, in the light of our observations on mortality and reproduction in Shetland, there is a considerable need for data on food availability for otters in fresh water.

Acknowledgements

We are very grateful to the World Wildlife Fund and Heinz Ltd. for financial support, to J. Rowbottom, T. Heggberget, the Deutsches Hydrographisches Institut and the Severn-Trent Water Authority for allowing us to use unpublished data, to P. Dale, B. Nolet, and D. Wansink for help with the field work, to BP for logistical support and to Dr M. Gorman, Prof. P. Racey and Dr B. Staines for critical comments on the manuscript.

References

Anon. (1983). *Shetland in statistics*. Shetland Islands Council, Lerwick.

Berry, R.J. & Johnston, J.L. (1980). *The natural history of Shetland*. Collins, London.

Chandler, T.J. & Gregory, S. (Eds). (1976). *The climate of the British Isles*. Longman, London & New York.

Chanin, P.R.F. (1985). *The natural history of otters*. Croom Helm, London.

Chanin, P.R.F. & Jefferies, D.J. (1978). The decline of the otter *Lutra lutra* L. in Britain: an analysis of hunting records and discussion of causes. *Biol. J. Linn. Soc.* **10**: 305–28.

Chapman, P.J. & Chapman, L.L. (1982). *Otter survey of Ireland, 1980–81*. Vincent Wildlife Trust, London.

Danilov, P.I. & Tumanov, I.L. (1975). The reproductive cycles of some Mustelidae species. *Byull. Mosk. O-va Ispyt. Oto. Biol.* **80**: 35–47.

Duplaix-Hall, N. (1975). River otters in captivity: a review. In *Breeding endangered species in captivity*: 315–27. (Ed. Martin, R.D.). Academic Press, London.

Erlinge, S. (1967). Home range of the otter *Lutra lutra* L. in southern Sweden. *Oikos* **18**: 186–209.

Green, J. & Green, R. (1980). *Otter survey of Scotland 1977–79*. Vincent Wildlife Trust, London.

Green, J., Green, R. & Jefferies, D.J. (1984). A radio-tracking survey of otters *Lutra lutra* on a Perthshire river system. *Lutra* **27**: 85–145.

Hamilton, W.J. & Eadie, W.R. (1964). Reproduction in the otter, *Lutra canadensis*. *J. Mammal.* **45**: 242–52.

Harris, C.J. (1968). *Otters: a study of the recent Lutrinae*. Weidenfeld & Nicholson, London.

Herfst, M.S. (1984). Habitat and food of the otter *Lutra lutra* in Shetland. *Lutra* **27**: 57–70.

Heron, J. (1961). The seasonal variation of phosphate, silicate and nitrate in waters of the English Lake District. *Limnol. Oceanogr.* **6**: 338–46.

Jenkins, D. (1980). Ecology of otters in northern Scotland. I. Otter (*Lutra lutra*) breeding and dispersion in mid-Deeside, Aberdeenshire in 1974–79. *J. Anim. Ecol.* **49**: 713–35.

Kruuk, H., Conroy, J.W.H., Glimmerveen, U. & Ouwerkerk, E.J. (1986). The use of spraints to survey populations of otters *Lutra lutra*. *Biol. Conserv.* **35**: 187–94.

Kruuk, H., Glimmerveen, U. & Ouwerkerk, E.J. (1985). The effects of depth on otter foraging in the sea. *A. Rep. Inst. terr. Ecol.* **1984**: 112–5.

Lack, D. (1954). *The natural regulation of animal numbers*. Clarendon Press, Oxford.

Le Cren, E.D. (1951). The length–weight relationship and seasonal cycle in gonad weight and condition in the perch (*Perca fluviatilis*). *J. anim. Ecol.* **20**: 201–19.

Loudon, A. & Racey, P. (Eds). (1987). *Reproductive energetics in mammals. Symp. zool. Soc. Lond.* No. 57: 1–371.

Mason, C.F. & Macdonald, S.M. (1986). *Otters: ecology and conservation*. Cambridge University Press, Cambridge.

Randolph, P.A., Randolph, J.C., Mattingly, K. & Foster, M.M. (1977). Energy costs of reproduction in the cotton rat, *Sigmodon hispidus*. *Ecology* **58**: 31–45.

Reuther, C. (1980). Zur Situation des Fischotters in Europa. In *Der Fischotter in Europa—Verbreitung, Bedrohung, Erhaltung*: 71–92. (Eds Reuther, C. & Festetics, A.). Aktion Fischotterschutz e.V., Oderhaus & Göttingen.

Reuther, C. (1985). Otters in captivity—first results of an enquiry. *J. Otter Trust* **1984**: 45–56.

Royama, T. (1966). Factors governing feeding rate, food requirement and brood size of nestling great tits *Parus major*. *Ibis* **108**: 313–47.

Sandegren, F. (1985). Project Otter—Sweden. *J. Otter Trust* **1984**: 41–4.

Stephens, M.N. (1957). *The natural history of the otter. A report to the Otter Committee.* Universities Federation for Animal Welfare, London.

Stubbe, M. (1980). Die Situation des Fischotters in der DDR. In *Der Fischotter in Europa—Verbreitung, Bedrohung, Erhaltung*: 179–82. (Eds Reuther, C. & Festetics, A.). Aktion Fischotterschutz e.V., Oderhaus & Göttingen.

Widdowson, E.M. (1981). The role of nutrition in mammalian reproduction. In *Environmental factors in mammal reproduction*: 145–59. (Eds Gilmore, D. & Cook, B.). Macmillan Publ., London.

Wijngaarden, A. van & Peppel, J. van den (1970). De otter, *Lutra lutra* (L.) in Nederland. *Lutra* **12**: 3–70.

Wlodek, K. (1980). Der Fischotter in der Provinz Pomorze Zachodnie (West-Pommern) in Polen. In *Der Fischotter in Europa—Verbreitung, Bedrohung, Erhaltung*: 187–94. (Eds Reuther, C. & Festetics, A.). Aktion Fischotterschutz e.V., Oderhaus & Göttingen.

Symp. zool. Soc. Lond. (1987) No. 58: 279–294

Badger population dynamics in a high-density area

C. L. CHEESEMAN[1], J. W. WILESMITH[2], J. RYAN[2] and P. J. MALLINSON[1]

[1] *Ministry of Agriculture, Fisheries and Food*
Worplesdon Laboratory
Tangley Place
Worplesdon
Guildford
Surrey GU3 3LQ, U.K.

[2] *Ministry of Agriculture, Fisheries and Food*
Epidemiology Unit
Central Veterinary Laboratory
New Haw
Weybridge
Surrey KT15 3NB, U.K.

Synopsis

The study area for this investigation was in the Cotswold escarpment region of Gloucestershire. Twenty-one social groups of badgers occupied an area of 7.2 km^2. An annual capture–recapture programme from 1977 to 1985 provided data for the study of the dynamics of this population of badgers. The total number of adult badgers present did not vary greatly, ranging from 75 in 1980 to 103 in 1982 and 1983. This represented a density of between 10.4 and 14.3 adult badgers per km^2. The number of adult badgers present in individual social groups showed a greater variation than the population as a whole.

An estimated average of 59% of cubs died in their first year of life. In the adult population males exhibited a higher annual mortality rate (average 34.6%) than females (average 24.0%), which resulted in an adult sex ratio biased towards females. The number of cubs caught each year was inversely proportional to the number of adults caught.

Data on the lifespan of individual badgers of known age showed that very few animals exceeded the age of six years. There was evidence that females which bred had a lower expected survival rate than females which did not breed. The proportion of females which bred was higher in the four-to seven-year-old age classes than in the two- to three-year-old age classes. A single one-year-old female was observed to be lactating, providing evidence that occasionally females may breed in their first year.

ZOOLOGICAL SYMPOSIUM No. 58
ISBN 0-19-854006-X

Breeding females showed a higher body weight than non-breeding females in the quarter October–December, which is prior to the main gestation period, and lower body weights in the quarter April–June, which corresponds to the lactation period.

Introduction

Previous studies on the biology of the European badger (*Meles meles*) have concentrated mainly on the social behaviour, diet and reproductive biology of the species (for a literature review see Neal 1986). Kruuk & Parish (1982) investigated the relationship between the group size, territory size and population density of badgers and the biomass of their main food sources. They concluded that badger territory size is determined by the distribution of suitable feeding areas, while social group size is correlated with the biomass of food available. Some work has been published on badger density in different habitats (Kruuk 1978; Cheeseman, Jones, Gallagher & Mallinson 1981; Cheeseman, Little *et al.* 1985) but these studies only present population estimates taken at one point in time.

The first attempt to examine the dynamics of badger populations was by Anderson & Trewhella (1985). These authors developed mathematical models to analyse breeding rates, mortality rates and possible cyclic fluctuations in abundance in areas of different badger density. This was designed to understand the processes that control the dynamics of badger abundance and the role of the badger in the epidemiology of bovine tuberculosis. The authors synthesized all the available information, including that derived from the present study, on the population biology of the badger. In their conclusions they drew attention (*inter alia*) to the need for long-term studies monitoring adult badger and cub abundance in areas of good, medium and poor badger habitat.

The study described in this paper is directed towards this end, with the limitation that it is taking place in one area of high-quality badger habitat. The work forms part of a programme of research into the problem of bovine tuberculosis in badgers and this paper represents an interim account of a long-term study. Data covering a six-year period are presented on various aspects of the dynamics of this badger population. The paper in this volume by Harris & Cresswell (p. 295) deals with the dynamics of a badger population in suburban Bristol. In order to facilitate comparison between these two studies the methodology has been standardized as far as possible.

Methods

The study area for this investigation lies in the Cotswold escarpment region of Gloucestershire. Habitat details are broadly described in Cheeseman,

Jones *et al.* (1981) and Cheeseman, Little *et al.* (1985). Farming in the area is mixed and, apart from the rotation of some pasture and arable fields which affected local habitat composition, the overall proportions of pasture, arable land, woodland and scrub remained approximately the same throughout the period of study. Management of the woodland by a forestry company has resulted in fairly constant proportions of mature and immature woodland. Data were gathered from 21 social groups of badgers from 1977 to 1985 inclusive. During the period of study the only interference to the badger population which took place was the complete removal of one social group (the 'Beech' group) in July 1977. Otherwise, to our knowledge, there was no interference to the badger population by man. The 21 social groups were located contiguously and occupied an area of 7.2 km^2.

Social group territories

The territorial boundaries of social groups were established each year using the bait-marking method described by Kruuk (1978). This was carried out during the period of maximum territorial activity from February to April. The information obtained from bait-marking was supplemented by data from individual badgers which were radio-tracked (for methods see Cheeseman & Mallinson 1980).

Capture of badgers

Badgers were captured using cage traps, snares or hand-held nets (Cheeseman & Mallinson 1980). Traps were normally pre-baited for up to 10 days. This resulted in a high capture rate over a short period of time which kept disturbance of social groups to a minimum. Thirteen of the 21 social groups were subjected to regular quarterly catch-ups as part of an experiment to monitor the extent of bovine tuberculosis infection in the population (Cheeseman, Wilesmith, Stuart & Mallinson in press). In the remaining eight groups, catching took place mainly during the months May–October. Experience has shown that catching success is highest during this period of the year. The summer months are also the period when cubs of the year have been weaned but are still resident in their natal social groups. As far as possible a similar catching effort was maintained at each social group, the objective being to catch all or most of the badgers present in each group each year. Snares or nets were employed as a capture method when badgers were suspected of being trap-shy.

Badgers were marked with ear tags and a tattoo (Cheeseman & Harris 1982). The year of birth, if evident, and sex were recorded at the first capture of a badger. Body weight and the location where caught were recorded for each badger at every capture. Evidence of breeding, by abdominal palpation for evidence of pregnancy and clinical examination of mammary glands for signs of lactation, was also recorded for females.

Population estimates

Population estimates were obtained from the capture records by the method of direct enumeration. This includes all animals caught on a particular occasion, plus those caught and marked previously that were not caught that session but reappeared in a later capture session. In the case of a badger first caught as an adult, that badger was assumed to have been present in the same social group the previous year. It was therefore included in the total population estimate but not assigned to either the adult or cub categories. The method of direct enumeration assumes little or no movement in the population, which was the case here (C. L. Cheeseman unpublished data).

Age structure of population and estimation of mortality rates

Numbers of badgers in each 12-month age class for each year from 1980 to 1985 were determined from the records of captures of known-age badgers since 1977. Known age badgers were those first caught in their calendar year of birth and identified as cubs on the basis of their size and weight. This was sometimes difficult with animals first captured late in the year when cubs and adults could not be reliably separated. In such cases radiography was employed to check whether the epiphysis of the tibia was open (cubs) or closed (adults) (Ahnlund 1976).

An individual's lifespan was taken to be from the year of birth to the year of death. In the case of animals not found dead, the year of death was assumed to be the year in which it was last caught (it was rare for individual badgers to avoid capture for more than one year). Sex-specific mortality rates were estimated from these data on known-age badgers for each year from 1980 to 1984 for both cubs and adults.

The estimate of cub mortality derived in this way was supplemented by calculating a pre-capture mortality rate. The method of calculating this rate was as follows. The observed proportion of females, caught during the months February to August in each year, 1980 to 1984, for which there was evidence that they had bred, was applied to the direct enumeration estimate of the total number of females, in their second year or older, known to be alive in the year—the second year has been established as the age of sexual maturity for the female badger (Neal & Harrison 1958; Ahnlund 1980; Wandeler & Graf 1982). It was assumed that each of these females produced an average litter size of 2.7 cubs, the average litter size for badgers in Europe derived from published and unpublished studies (Anderson & Trewhella 1985). This gave an estimate of the number of cubs born in each year and provided the denominator for estimating the annual pre-capture mortality rate. The difference between the estimated number of cubs born and the number of cubs caught in the year provided the numerator for this rate.

Life tables and survivorship curves

Life tables were constructed for known-age male and female badgers using the mean pre-capture mortality rate (assumed to be the same for male and female cubs) for 1980–84 and the mean age- and sex-specific mortality rates for 1980–84. A survivorship curve was derived from these data. Standard methodology was used for the construction of these life tables (Pollard, Yusuf & Pollard 1984).

Breeding potential and survival of adult females

Capture records of female badgers which were first caught as cubs and whose reproductive status was known in subsequent years were used to compare the survival rate of breeding females with that of non-breeding females for three age groups: two to three, three to four and four to five years of age.

Seasonal variation in body weight

Body weights of males and of breeding and non-breeding females of 18 months of age or over were used to compare seasonal variations in weight between these three categories. A female which bred in a particular year was classed as a breeding female for the 12-month period from the previous July to June of that year. For comparison the same 12-month period was used for non-breeding females and males. Data were aggregated for the years 1977 to 1985 and mean body weights, for each of the three categories of badger, were calculated for each quarter-year period. If more than one weight was recorded for an individual during one quarter of a year, the first weight recorded was used in the calculation of the mean. Ninety-five percent confidence intervals were calculated for each of the means.

Results

Social group territories

Analysis of the information from bait-marking, combined with capture–recapture data, has shown that territorial boundaries of social groups were stable during the period of study and there was only a limited amount of movement of badgers between social groups (C. L. Cheeseman unpublished data). The territory occupied by the 'Beech' group (removed in 1977) was recolonized and the territorial boundary of the new social group remained in the same position as it was prior to the removal of the original group.

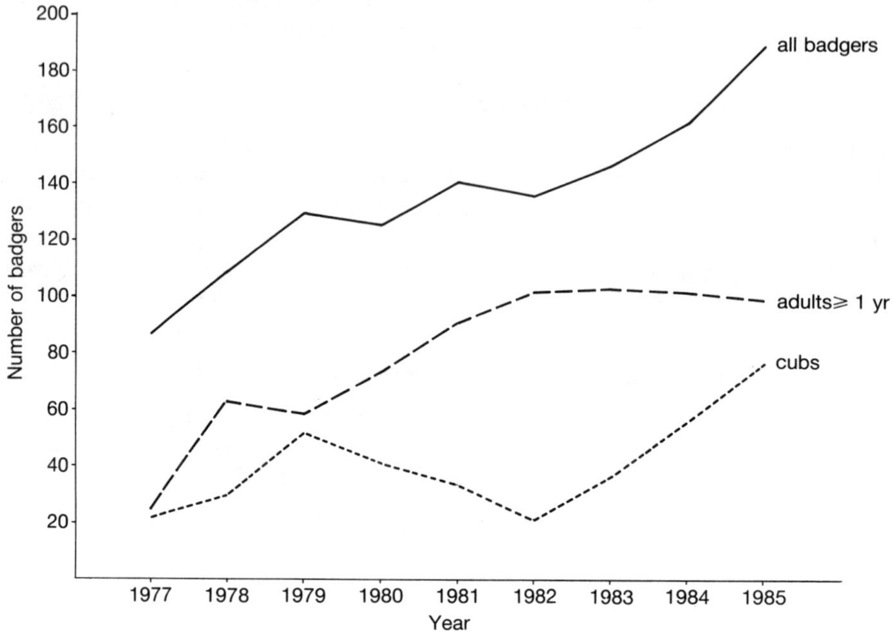

Fig. 1. Total numbers of badgers present each year in the 21 social groups, estimated by direct enumeration.

Population density

Estimates of the total number of badgers present each year in the 21 social groups are shown in Fig. 1. This includes separate estimates of the number of adults and cubs. It should be noted that because of the inclusion of badgers first caught as adults in the estimate for the year prior to their first capture (see Methods), the sum of the estimates of adults and cubs in one year does not necessarily equal the estimate of the total number of badgers present that year.

Over the period from 1979 to 1985 the total number of badgers present ranged from 126 to 189, with a tendency to increase steadily over the period of study (estimates for 1977 and 1978 are considered unreliable and are therefore excluded—see Discussion). During the same period the number of adults ranged from 59 to 103, and the number of cubs from 22 to 77. It is of interest to note that the years when relatively low numbers of cubs were caught coincided with periods when the highest number of adult badgers were recorded. These estimates gave densities of between 10.3 and 14.3 adult badgers per km^2, and overall badger densities of between 17.5 and 26.3 per km^2.

There was greater variation in the numbers of badgers present in individ-

Table 1. Number of adult badgers and cubs present in individual social groups. Mean and range for the years 1980–1985.

Name of social group	Adults Mean	(Range)	Cubs Mean	(Range)
Arthurs	6.67	(4–10)	4.67	(1–8)
Beech	4.33	(3–5)	1.33	(0–5)
Cedar	1.67	(0–4)	0.83	(0–2)
Cole Park	5.00	(3–7)	2.00	(1–4)
Colliers Wood	4.67	(2–8)	1.67	(0–4)
Hedge	4.67	(3–6)	1.67	(0–3)
Honeywell	1.83	(0–5)	0.67	(0–2)
Inchbrook	0.83	(0–1)	0.67	(0–2)
Jacks Mirey	3.67	(0–7)	2.00	(0–3)
Junction	5.16	(4–6)	2.00	(0–4)
Kennel	5.16	(2–9)	3.16	(0–7)
Larch	6.16	(3–8)	2.00	(0–4)
Nettle	1.67	(1–2)	1.50	(0–4)
Old Oak	5.67	(3–8)	2.00	(0–4)
Pegler's	4.50	(0–10)	4.00	(0–11)
Septic Tank	6.50	(3–8)	2.50	(0–4)
Top	8.33	(6–10)	2.33	(0–5)
West	6.50	(5–7)	3.67	(2–5)
Windsor Edge	3.00	(0–6)	2.16	(0–5)
Wood Farm	1.33	(0–2)	0.33	(0–1)
Yew	10.00	(8–12)	3.83	(0–9)

ual social groups than in the population as a whole. Table 1 shows the mean and range of the number of badgers in each of the 21 groups over the years 1980–1985 inclusive. The ranges of the estimated number of badgers present in individual social groups varied considerably, both between groups and within individual groups over time. One extreme may be seen in 'Pegler's' group, where the number of adults ranged from zero to 10 individuals. The maximum of 10 adults occurred in 1985 when the maximum of 11 cubs was also recorded in that group. Using data available up to 2 October 1986 which include badgers first caught as adults in 1986, the direct enumeration estimate of the number of badgers present in 'Pegler's' group in 1985 is 23. It is evident, from the fact that the overall density of badgers in the population did not vary greatly, that years of maximum and minimum group size in any one social group did not necessarily coincide with such years in other groups. The number of cubs recorded each year in individual social groups showed a similar variation.

Age structure

Figure 2 shows the age structure of known-age badgers in the population from 1980 to 1985 inclusive. It should be noted that during the first few years of study there were no animals in the older age groups because of the

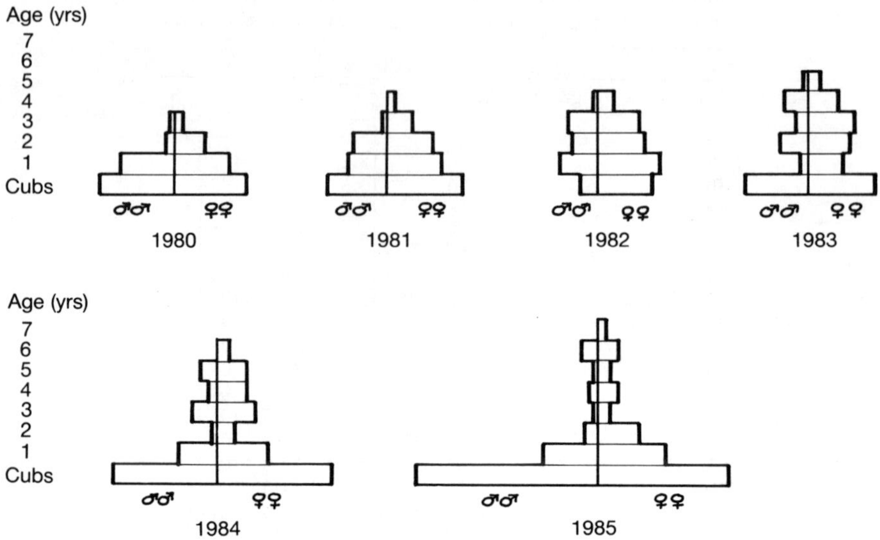

Fig. 2. Age structure of known-age badgers in the study area for the years 1980–1985 inclusive.

time necessary to establish a known-age population (in the absence of a reliable method of ageing badgers other than cubs). Even by 1985, when the proportion of known-age badgers was 87.1% compared with 57.5% in 1980, there were very few animals in the six- to seven- and seven- to eight-year-old age groups. The oldest known badger in this population was a female first caught in 1977 when it was lactating and therefore probably at least in its second year. This female was last caught in November 1986 when it would have been at least 11 years old.

Mortality rates

The estimated mortality rates of known-age adult badgers and cubs are shown in Tables 2 and 3 respectively. Means were calculated for the years 1980–1984 when there were sufficient known-age animals in the population to give reliable estimates (estimates for 1985 were not included because data to the end of 1986 are needed for their calculation). The average mortality rate of adult males was greater than that of adult females, although this difference was not significant ($\chi^2 = 3.58, p>0.05$). For cubs there was a significant difference between the rates of males and females ($\chi^2 = 4.05$, $p<0.05$). In neither adults nor cubs were the sample sizes large enough to determine whether there were significant differences in the mortality rates observed between individual years.

Table 2. Apparent mortality rates for known-age badgers: adults.

Year	Apparent mortality % (*N*)					
	♂♂		♀♀		Total	
1978	80.0	(5)	25.0	(4)	55.6	(9)
1979	50.0	(6)	18.2	(11)	29.4	(17)
1980	43.8	(16)	13.6	(22)	26.3	(38)
1981	22.2	(18)	21.8	(32)	22.0	(50)
1982	26.0	(23)	22.5	(40)	23.8	(63)
1983	31.5	(19)	23.0	(39)	25.8	(58)
1984	50.0	(22)	33.3	(42)	39.0	(64)
1980–1984 mean	34.6	(98)	24.0	(175)	27.8	(273)

Table 3. Apparent mortality rates for known-age badgers: cubs.

Year	Apparent mortality % (*N*)					
	♂♂		♀♀		Total	
1977	54.5	(11)	60.0	(10)	57.1	(21)
1978	68.8	(16)	38.5	(13)	55.1	(29)
1979	48.0	(25)	50.0	(26)	49.0	(51)
1980	50.0	(18)	23.5	(17)	37.1	(35)
1981	35.7	(14)	16.7	(18)	25.0	(32)
1982	50.0	(4)	38.4	(13)	41.1	(17)
1983	40.0	(15)	25.0	(16)	32.3	(31)
1984	48.0	(25)	40.7	(27)	44.2	(52)
1980–1984 mean	44.7	(76)	29.6	(91)	36.5	(167)

The estimated pre-capture mortality rates of cubs for the years 1980–1984 are shown in Table 4. The average rate for the years 1980–1984 was 36.6%, i.e. this proportion of cubs was estimated to have died between birth and reaching the age when cubs were first available for capture. The average mortality rates for cubs and adults were used in the construction of life tables which are shown in Table 5 for males, females and all badgers. When the estimated pre-capture mortality rate of cubs is aggregated with the mortality rate for the remainder of the first year of life, an average of 406 out of 1000 individuals (both sexes combined) survived to one year of age. This represents a mortality of 59.4%. The higher mortality rate of male badgers in each of the age classes is evident from inspection of the life tables. This is shown in the survivorship curves for males, females and the total badger population which were derived from these data (Fig. 3).

Breeding potential and survival of adult females

The proportion of known-age females which bred in each age group is shown in Table 6. The proportion of females breeding in each age group

Table 4. Estimated pre-capture mortality rates of cubs 1980–1984; for methods of calculation see text.

Year	Observed proportion of females which bred	Total number of females	Estimate of number of females which bred	Total number of cubs caught	Estimate number of cubs born	Estimated pre-capture mortality %
1980	9/16	39	22	39	59	33.9
1981	12/22	46	25	34	68	50.0
1982	8/32	45	11	22	30	26.7
1983	18/46	56	22	37	59	37.3
1984	24/39	51	31	55	85	35.3
1981–1984 mean						36.6

Table 5. Life table for known-age badgers.

Males				Females			Total		
x	lx	dx	$1000qx$	lx	dx	$1000qx$	lx	dx	$1000qx$
0	1000	366	366	1000	366	366	1000	366	366
0*	634	275	434	634	188	297	634	288	359
1	358	142	395	446	110	246	406	125	308
2	217	54	250	336	31	93	281	42	149
3	163	63	389	305	88	289	239	77	321
4	99	22	222	217	65	300	162	45	276
5	77	0	0	152	30	200	117	15	125

x = age (years)
lx = number surviving at beginning of age interval
dx = number dying within age interval
$1000qx$ = mortality rate (number dying per 1000 alive at beginning of age interval)
0* = estimated pre-capture mortality from Table 4

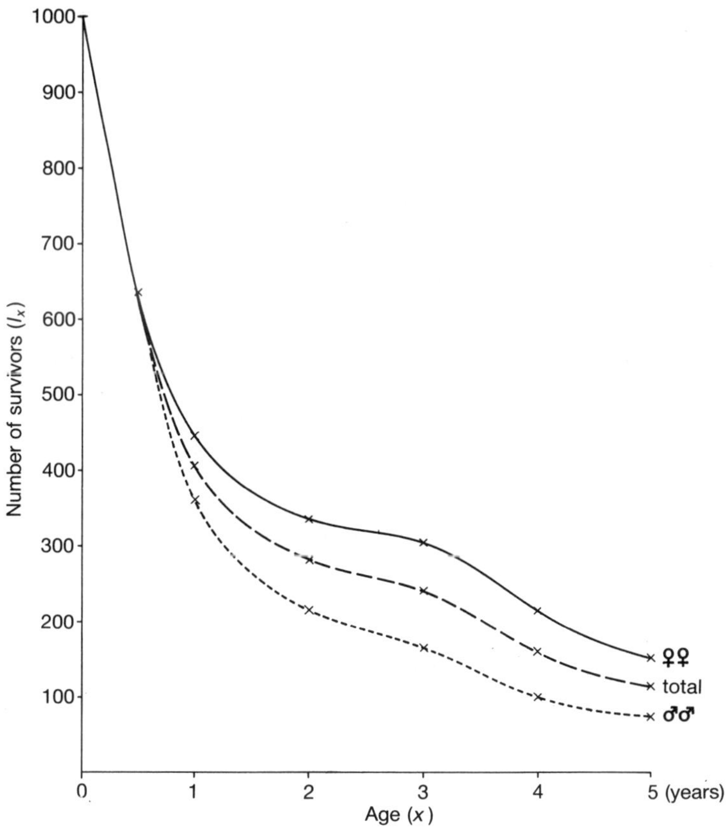

Fig. 3. Survivorship curves for males, females and the total badger population, derived from the data shown in Table 5.

Table 6. Proportion of known-age females which bred.

Age class (years)	Proportion which bred %	(*N*)
1–2	1.3	(80)
2–3	27.7	(47)
3–4	36.4	(33)
4–5	66.7	(21)
5–6	69.2	(13)
6–7	60.0	(5)
7–8	50.0	(2)

Table 7. Survival rates of known-age breeding and non-breeding females for those age categories for which sufficient data are available.

Age class (years)	Survival rate of breeding ♀♀ %	(*N*)	Survival rate of non-breeding ♀♀ %	(*N*)
2–3	62.5	(8)	85.2	(27)
3–4	45.5	(11)	63.2	(19)
4–5	44.4	(9)	71.4	(7)

increased with increasing age, reaching a maximum in the five- to six-year-old age group. Sample sizes in the six- to seven- and seven- to eight-year-old age groups were small but indicated that some females in these age groups were still breeding.

The survival rate of known-age females which showed evidence of breeding was lower than that of females which did not breed in each of the age groups for which there were sufficient data to make comparison (Table 7).

Seasonal variation in body weight

Figure 4 depicts the seasonal variation in body weight of males, non-breeding females and breeding females. Peak body weights were observed in the October to December quarter and lowest body weights from April to September. This pattern was evident for all three classes. However, in the April to June quarter there is a significant difference in the mean weights of these three classes of badgers with males heaviest and breeding females lightest. Also, in the July to September quarter males were significantly heavier than the two classes of females. Of the three classes, breeding females exhibited the most variation in body weight; their mean body weight was higher than that of the other two classes in the October to December quarter. Although only data aggregated over the period 1977 to 1985 are shown, this general pattern of seasonal variation was evident in the majority of years.

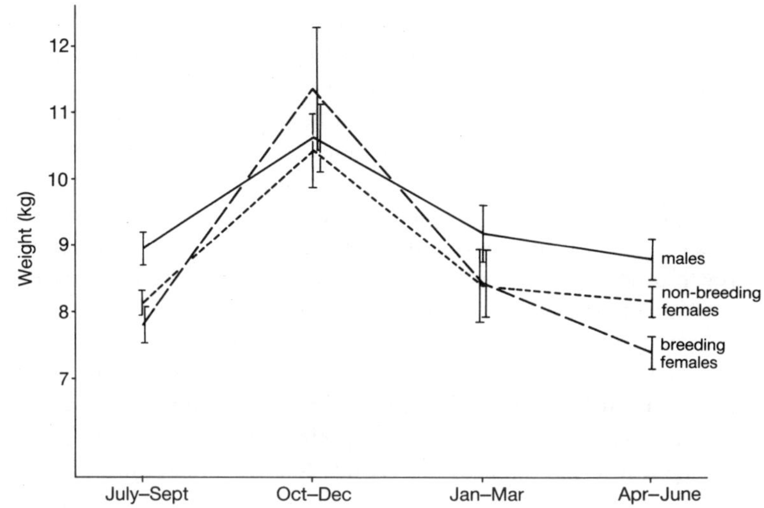

Sample sizes	July–Sept	Oct–Dec	Jan–Mar	Apr–June
males	134	54	43	67
breeding females	83	15	29	60
non-breeding females	131	38	32	86

Fig. 4. Seasonal variation in body weight (mean ± 95% confidence limits) for animals caught 1977–1985 inclusive.

Discussion

The density estimates presented here represent the maximum number of badgers known to be present during the course of each calendar year and not the actual number present at a particular point in time. Despite this, the maximum adult badger density of 14.3/km^2 was lower than that recorded in a removal experiment in the same region of Gloucestershire in 1979, when an exceptionally high density of 19.4 adults/km^2 was recorded (Cheeseman, Jones *et al.* 1981).

With the use of mathematical models Anderson & Trewhella (1985) suggested that in good habitats within Britain badger populations are likely to be stable, with perhaps small-scale fluctuations in adult numbers which in practice would be difficult to detect (given the sampling error inherent in population census techniques). Their model did, however, predict detectable oscillations in cub density.

In the interpretation of the density estimates in the present study allowance must be made for the fact that the ability to catch badgers efficiently was being developed during the early phase of work and the proportion of marked animals in the population was increasing. The estimates for 1977 and 1978 are therefore not considered reliable.

Although we have only a limited amount of data available at this interim stage of our study, it is of interest that the trends which are emerging tend to support Anderson & Trewhella's model predictions.

The stability of the population as a whole was contrasted by large fluctuations in individual group size. Some extremely large group sizes were recorded. The total of 23 badgers in Pegler's group in 1985 was the highest recorded and although this may be exceptional, one other group exceeded 20 individuals and several groups of over 12 individuals were recorded. Kruuk & Parish (1982) suggested that group size is determined by the biomass of prey within the territory. It might be assumed that any variation in prey biomass would be similar throughout the study area. However, the variations in the number of badgers present in individual social groups were not synchronous, i.e. in years when some groups were at a maximum, others were at a minimum. Given that territorial boundaries were stable during the period of study, and that no major habitat changes took place, the reason for the observed variation in group sizes is not clear within the context of the hypothesis put forward by Kruuk & Parish (1982).

The survivorship curve shows a slight reduction in the mortality rate in the two- to three-year-old age group. The reason for this is not apparent. However, the higher mortality rate of males, resulting in a preponderance of females in the adult population, would appear to be a typical demographic feature of badger populations (Cheeseman, Little *et al.* 1985; Pritchard *et al.* 1986; Harris & Cresswell, this volume).

There is limited published information on fecundity in the European badger. In a sample of 230 adult female badgers examined in Switzerland, Wandeler & Graf (1982) found that a large proportion of those examined in spring showed fresh placental scars, but bore no signs of suckling offspring. The authors considered that this indicated a high perinatal mortality. In the present study the finding that an estimated 36.6% of cubs die before they emerge and are available for capture supports the observations of the previous authors. The estimate of pre-capture mortality was based on an estimate of the number of females which bred from clinical evidence of pregnancy and/or lactation. It is likely that this was an underestimate because of the difficulty in making a firm judgement of whether an animal was pregnant, lactating, or had lactated recently. Therefore the true pre-capture mortality rate, and thus the proportion of cubs dying in their first year, was probably higher than that estimated.

The crucial question, of course, is at what point in the female badger reproductive cycle does any possible mechanism for the regulation of the birth rate operate? In a sample of over 700 badgers from south-west England taken from 1973 to 1977, R. Page (pers. comm.) found that the maximum proportion of adult females carrying blastocysts occurred in August (90–95%). The maximum proportion of females with placental scars was observed in April and May when between 55 and 60% of females carried scars. The inference from these data is that although most adult females carry blastocysts, not all of them produce litters and it seems likely that in the unsuccessful females the blastocysts are not implanted.

Our finding that there is a weight difference between breeding and non-breeding females in the October to December period adds further evidence to this story. Given that our sample size is still limited, it might be postulated that the heaviest females, and therefore those which are fittest, have the highest probability of implanting blastocysts. A study of hormone levels in female badgers during the reproductive cycle is currently in progress and may shed more light on this aspect of badger breeding biology.

Given that the mean litter size for badgers is 2.7 (Anderson & Trewhella 1985) the maxima of nine and 11 cubs recorded in the 'Yew' and 'Pegler's' groups respectively suggest that two or possibly three litters were produced in the years in question. It was quite common to observe more than one lactating female in a single group during one breeding season.

Ferris (1986) reported evidence of a yearling female badger successfully rearing a litter of cubs. All other published work concerning the reproductive biology of badgers gives the second year as the age of sexual maturity for females (Neal & Harrison 1958; Ahnlund 1980; Wandeler & Graf 1982). Our discovery of one yearling female which was lactating demonstrates that occasionally females of this age may breed.

The better survival rate of non-breeding females probably has some biological significance, although small sample sizes precluded testing for statistical significance. The most likely explanation for this is the greater physiological demands which pregnancy and lactation make on breeding females, leading to a higher mortality rate in these individuals.

As yet there is no evidence to suggest that the presence of tuberculosis infection in this badger population has had any effect on the dynamic aspects described (Cheeseman, Wilesmith *et al.* in press). It is too early to draw firm conclusions, but during the six-year period of study this high-density badger population displayed no major fluctuations in density. The regulation of density appears to be through limitation of the number of cubs born and high cub mortality, though the processes which control this regulation are not clear at present.

Acknowledgements

We gratefully acknowledge the co-operation of farmers, landowners and members of the public who have given access to their property in order for the field work to be carried out. Our thanks to the field workers themselves for their expert technical assistance, to colleagues in MAFF who provided helpful comments on a draft of this paper and to secretarial staff for typing the manuscript. The project is funded by the Ministry of Agriculture, Fisheries and Food.

References

Ahnlund, H. (1976). Age determination in the european badger, *Meles meles* L. *Z. Säugetierk.* **41**: 119–25.

Ahnlund, H. (1980). Sexual maturity and breeding season of the badger, *Meles meles* in Sweden. *J. Zool., Lond.* **190**: 77–95.

Anderson, R.M. & Trewhella, W. (1985). Population dynamics of the badger (*Meles meles*) and the epidemiology of bovine tuberculosis (*Mycobacterium bovis*). *Phil. Trans. R. Soc. Lond. (B)* **310**: 327–81.

Cheeseman, C.L. & Harris, S. (1982). Methods of marking Badgers (*Meles meles*). *J. Zool., Lond.* **197**: 289–92.

Cheeseman, C.L., Jones, G.W., Gallagher, J. & Mallinson, P.J. (1981). The population structure, density and prevalence of tuberculosis (*Mycobacterium bovis*) in badgers (*Meles meles*) from four areas in south-west England. *J. appl. Ecol.* **18**: 795–804.

Cheeseman, C.L., Little, T.W.A., Mallinson, P.J., Page, R.J.C., Wilesmith, J.W. & Pritchard, D.G. (1985). Population ecology and prevalence of tuberculosis in badgers in an area of Staffordshire. *Mammal Rev.* **15**: 125–35.

Cheeseman, C.L. & Mallinson, P.J. (1980). Radio tracking in the study of bovine tuberculosis in badgers. In *A Handbook on biotelemetry and radio tracking*: 649–56. (Eds Amlaner, C.J. & Macdonald, D.W.). Pergamon Press, Oxford.

Cheeseman, C.L., Wilesmith, J.W., Stuart, F.A. & Mallinson, P.J. (In press) The dynamics of tuberculosis in a naturally infected badger population. *Mammal Rev.*

Ferris, C. (1986). Mating and early maturity of badgers in Kent. *J. Zool., Lond. (A)* **209**: 282.

Kruuk, H. (1978). Spatial organization and territorial behaviour of the European badger *Meles meles*. *J. Zool., Lond.* **184**: 1–19.

Kruuk, H. & Parish, T. (1982). Factors affecting population density, group size and territory size of the European badger, *Meles meles*. *J. Zool., Lond.* **196**: 31–9.

Neal, E.G. (1986). *The natural history of badgers*. Croom Helm, London, England.

Neal, E.G. & Harrison, R.J. (1958). Reproduction in the European badger (*Meles meles* L.). *Trans. zool. Soc. Lond.* **29**: 67–130.

Pollard, A.H., Yusuf, F. & Pollard, G.N. (1984). *Demographic techniques*. Pergamon Press, Oxford, England.

Pritchard, D.G., Stuart, F.A., Wilesmith, J.W., Cheeseman, C.L., Brewer, J.I., Bode, R. & Sayers, P.E. (1986). Tuberculosis in East Sussex III. Comparison of post-mortem and clinical methods for the diagnosis of tuberculosis in badgers. *J. Hyg., Camb.* **97**: 27–36.

Wandeler, A.I. & Graf, M. (1982). The reproductive cycle of female badgers (*Meles meles* L.) in Switzerland. *Revue suisse Zool.* **89**: 1009–16. (In German, English summary.)

Symp. zool. Soc. Lond. (1987) No. 58: 295–311

Dynamics of a suburban badger (*Meles meles*) population

Stephen HARRIS and
W. J. CRESSWELL

Department of Zoology
University of Bristol
Woodland Road
Bristol BS8 1UG

Synopsis

The study area was in the north-west suburbs of Bristol. Fifteen social groups of badgers occupied an area of about 8 km^2, although not all the social groups were present throughout the study. An annual capture–mark–recapture programme from 1978 to 1985 provided data for the study of the dynamics of this population of badgers. The total number of adults ranged from 35 in 1984 to 60 in 1981. This represented a density of between 4.4 and 7.5 adult badgers per km^2; this figure included those parts of the study area never utilized by badgers, and locally densities were much higher.

To construct life tables, the number of breeding females and mean litter size were used to estimate cub production each year. Pre-capture mortality of young cubs was calculated to be 38.4%, and 63.9% of cubs died in their first year. In the adult population, males exhibited a higher annual mortality rate (average 40.3%) than females (average 28.1%), which resulted in the adult sex ratio being significantly biased towards females. This sex difference in mortality was most pronounced in animals over three years old. The loss of adult males may be due, in part, to increased rates of group change and long-distance movements, which increase the chance of mortality in older boars.

Simple models were constructed to examine the processes involved in the dynamics of this badger population. The results are compared with those from a higher-density badger population, and the possible influences of density and social organization on badger population dynamics are discussed.

Introduction

In a recent paper, Anderson & Trewhella (1985) reviewed the available information on badger (*Meles meles*) population dynamics. In this they showed that there was a paucity of data, and that few studies have been able to provide precise information on the age structure of the older segments of

ZOOLOGICAL SYMPOSIUM No. 58
ISBN 0-19-854006-X

badger populations. They concluded that the current knowledge of the population biology of badgers is somewhat limited.

Ahnlund (1976) compared methods of age determination in a sample of badgers from Sweden. He demonstrated that there were clear annuli in the teeth, especially in the third upper incisor, and by this means he was able to determine the age structure of his sample. Such an approach is not, however, universally applicable. In a sample of badgers collected from south-west England (including the area of the present study), S. Fargher & P. Morris (unpublished) found that incremental lines were unreliable for determining the age of badgers, although younger animals could be recognized by the stage of epiphyseal closure. Also, post-mortem material can be misleading with respect to the age and/or sex composition of the sample, since badgers show different activity patterns according to age, sex and season (Anderson & Trewhella 1985). Care must, therefore, be taken when interpreting post-mortem material. Harris & Smith (1987) found that large samples of suburban foxes (*Vulpes vulpes*), collected from a variety of sources, would provide a representative sample for population analysis, but it would be difficult to obtain a comparable sample from a single population of badgers.

In view of the problems with both obtaining sufficient post-mortem material and then accurately ageing the sample, we have used a long-term trapping study to analyse the population dynamics of badgers living in suburban Bristol. This approach also has its limitations: it requires a long period to mark a substantial proportion of the badger population and for known-age animals (i.e. those marked as cubs or yearlings) to enter the older age classes. Therefore, in a trapping study such as this, sample sizes will of necessity be small. Yet despite these limitations, the approach does have its advantages: individual animals can be studied throughout their life and their reproductive success, any group changes and social status monitored. This paper and the one by Cheeseman, Wilesmith, Ryan & Mallinson (this volume p. 279) are the first two such long-term badger trapping studies. They should be viewed as companion papers: wherever possible the methodology and analyses have been standardized to facilitate direct comparisons between the studies, which have examined two badger populations of very different social organization, living at different densities and in contrasting habitats.

Methods

The study area and field techniques

The present study included all the social groups described by Harris (1982), plus one social group to the south and two to the north. In total there were 15 social groups, although not all were present throughout the study; some of the groups described by Harris (1982) had subsequently split into two, and some new groups had become established on the central part of the area. In total these 15 groups occupied an area of approximately 8 km^2, although this included areas never visited by badgers. The social and territorial organization of this population of badgers is described by Cresswell & Harris (in press), who also illustrate the distribution of the setts, with the ranges of the social groups active in 1985. In addition, setts were trapped north of the main study area to the river Severn (a distance of 6 km), and cage-trapping was carried out in the suburbs of Bristol to the east and south of the study area in response to reported badger sightings. This area contained no resident badger populations (Harris 1984), but itinerant or solitary animals did occasionally move into the area. The western boundary of the study area was the tidal stretches of the river Avon, and this was assumed to be an effective barrier to badger movements, although one animal was known to have crossed the river (S. Harris & W. J. Cresswell unpublished data).

Badgers were caught each year from 1978 to 1985 inclusive, in cage-traps baited with peanuts, in snares or with hand-held nets while they were foraging at night in open spaces (Cheeseman & Mallinson 1980; Cresswell & Harris in press). Cage-trapping was limited in 1978, but thereafter commenced in late June or July each year, and continued until November or December. Some trapping was undertaken at other times of the year to catch specific animals or to replace radio-collars. The trapping effort and success each year were variable, depending on weather conditions (for instance, badgers were easier to catch in hot dry summers), human disturbance and co-operation from local householders. Snares and nets were only used to supplement the trapping programme in areas where it was difficult to trap, for instance because of human disturbance. The anaesthesia and marking techniques are described by Cheeseman & Harris (1982) and Harris (1982).

Population estimates

Population estimates were based on capture records and calculated by the method of direct enumeration, which is the minimum number of animals known to be alive each year. By this method, the estimate for a given year included all those badgers caught that year, plus those already marked which were not caught that year but were recaptured in a subsequent year. Badgers were aged principally on the basis of wear to the upper rear molar; by reference to material of known age, unmarked animals could be reliably aged as cubs, yearlings (animals in their second year) or animals in their third year or older. In this paper the term cub applies to all animals in their first year and adult refers to animals which are in their second year or older.

All animals enter the next year class on January 1st each year. In the case of a badger first caught as a yearling, the previous year's estimate of cub numbers was retrospectively increased. Similarly, an unmarked animal identified as three years of age or older was included as an adult of unknown age in the population estimate for the previous year. In each case the animal was assumed to have been present in the same social group during the previous year.

Estimation of group size by radio-isotope labelling of faeces

To validate the population estimates obtained by direct enumeration, and to determine whether any animals in the population had become 'trap-shy', four social groups were chosen for a radio-isotope labelling experiment. Within these four groups, selected individuals were marked with one of three radioactive isotopes, the output of which was monitored in a series of faeces samples in an attempt to estimate group size. Intramuscular injection was used to administer 100 microcurie equivalents of the isotopes ^{65}Zn, ^{54}Mn and ^{134}Cs. The animals in each social group were labelled with the same isotope, in such a way that no groups with contiguous ranges were given the same isotope, and faeces samples were only collected from dung pits close to the main sett. This precluded the possibility of cross-contamination of samples between adjacent groups. As far as possible an equal number of yearling and older animals of either sex were marked. Two individuals were marked from the smaller social groups, three from the larger. Faeces were recovered between April and July 1985 and samples were collected until the cumulative percentage of labelled faeces (when plotted against the total number of faeces samples) levelled out to a constant figure.

Age structure and mortality rates

All subsequent analyses were based on the capture records of known-age animals. Badgers were said to be of known age if first captured in their calendar year of birth or as yearlings. The number of badgers in each year class for the years 1980 to 1985 was calculated from capture records based on known-age animals marked since 1978.

Age- and sex-specific mortality rates were estimated for each year from 1980 to 1984. When an animal disappeared from the population and its corpse was not recovered, it was assumed to have died in the year in which it was last caught. An animal not caught for a year was assumed to have died.

In an attempt to account for mortality factors acting before the cubs were available for capture, an estimate of the pre-capture mortality rate was calculated as follows. Each year adult females were assigned to one of three breeding categories. First, those known to have bred. This category included animals with evidence of lactation or palpable embryos. Secondly, those likely to have bred. This category included animals known not to have bred in the previous year but recaptured late in the following year with evidence of having produced cubs, and any animal which was the only adult female in a group from which cubs had been caught but no breeding female identified. Lastly, those believed not to have produced cubs that year, or never to have bred. The latter animals had small and inconspicuous nipples, which was taken to imply that they had never lactated; the difference in nipple size between non-breeding and breeding animals for a variety of mustelids, including *Taxidea taxus*, is described by Petrides (1950). The number of females falling into the first two categories was expressed as a proportion of the number of adult females caught that year. This proportion was then used to calculate the number of breeding females that year from the total population estimate for sows two years of age or older, as determined by direct enumeration. It was assumed that each of these females produced an average litter of 2.7, which was the mean number of blastocysts, foetuses and placental scars recorded by a number of studies (Anderson & Trewhella 1985). This gave an estimate of the number of cubs born each year. By subtracting the number of cubs actually caught each year an estimate of pre-capture mortality was obtained.

Information on the reproductive state of known-age females (as calculated above) was also used to investigate the proportion of female badgers which bred in each age class.

Life tables and survivorship curves

Life tables were constructed using figures derived from the mean pre-capture mortality rate for the years 1980 to 1984 and the mean yearly age- and sex-specific mortality rates for the same period. Survivorship curves for male and female badgers were derived from these data.

Modelling the population

As a further test of the accuracy of our data, a simple population model, based on our calculated population parameters, was constructed as follows. Initial numbers were set at the average for each age class recorded during 1980 to 1984. Initial cub production was generated by using the average number of breeding females for 1980 to 1984, a cub sex ratio of 50:50 and a mean litter size of 2.7. The mortality rate in each age group was based on the age- and sex-specific figures calculated in the life tables presented here. The programme then continued to increment the age of all badgers by one year, with the maxima for males and females decided by the final mortality rates from the life tables.

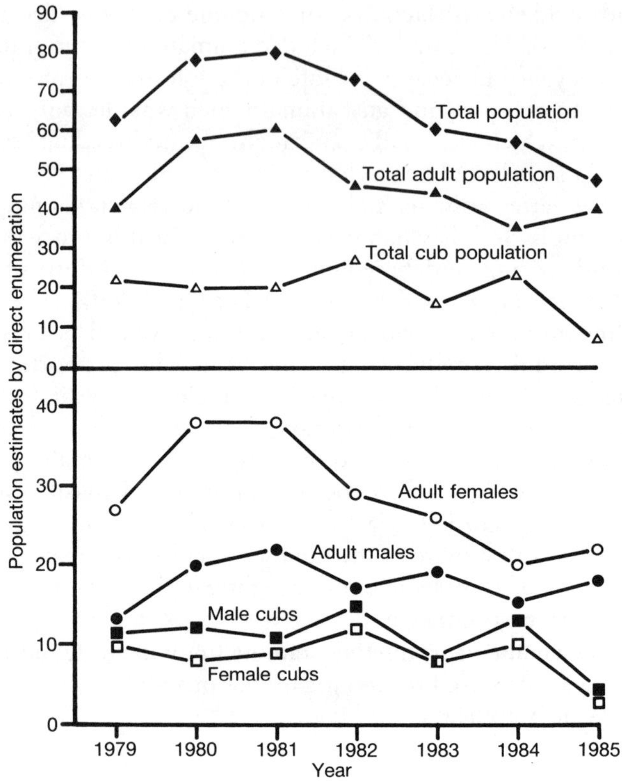

Fig. 1. Badger population estimates for the study area from 1979 to 1985. Above: total population estimate, solid diamonds; total adult population, solid triangles; total cub population, open triangles. Below: adult males, solid circles; adult females, open circles; male cubs, solid squares; female cubs, open squares.

Results

Population estimates

Estimates of the total number of badgers present on the study area each year from 1979 to 1985 are shown in Fig. 1. Separate estimates of the number of adults and cubs of each sex are also indicated. With the direct enumeration technique, calculations for the first year of the study, before the entire population had been marked, are likely to be slight underestimates. Similarly, the estimates for 1985 will be slightly inflated by the 1986 trapping results, in which any animals missed during the 1985 season will be added to the population estimate. The calculations of sex- and age-specific mortality rates presented below are based therefore on the years 1980 to 1984 only. Over the period 1979 to 1985 the total number of badgers present ranged from 80 in 1981 to 47 in 1985; the population exhibited a steady decrease

Table 1. Badger social group sizes for the years 1980 to 1985. Not all the groups were present throughout the study period, and the means are based on all years and not only the years when badgers were present.

Group	Adults		Cubs	
	Mean	(Range)	Mean	(Range)
New Cote	2.83	(2–4)	0.83	(0–3)
Hiatt Baker	2.00	(1–4)	1.17	(0–3)
Horsefield	3.67	(2–5)	2.17	(0–4)
Coombe Bridge Avenue	2.33	(1–3)	1.67	(0–5)
Canford Lane	1.33	(0–2)	0.33	(0–1)
Penpole	2.50	(2–4)	3.50	(0–6)
Druid Stoke Avenue	2.33	(0–7)	1.00	(0–4)
Glen Avon	8.33	(5–13)	2.17	(1–3)
Bishop's Knoll	2.50	(1–5)	0.33	(0–1)
Bridge Valley Road	1.00	(0–3)	0.33	(0–2)
Sharlands	2.33	(1–4)	1.33	(0–3)
Orchard House	4.17	(2–6)	1.33	(0–3)
Stoke Park Road	2.67	(2–4)	2.33	(1–4)
Elmfield Deaf School	0.67	(0–2)	0.33	(0–1)
10–12 Old Sneed Road	0.50	(0–2)	0	0

from 1981 onwards. The number of adults tended to mirror the fluctuations seen in the total population and ranged from 60 in 1981 to 35 in 1984, whereas there seemed to be independent fluctuations in the number of cubs present.

There was no significant difference ($\chi^2 = 0.20$, $P > 0.05$) between the numbers of male and female cubs produced, although there was a slightly greater number of male cubs caught each year; however, there was a significantly greater proportion of females in the adult population ($\chi^2 = 17.83$, $P < 0.001$).

The sizes of the individual social groups were found to fluctuate considerably. The mean and range of the number of badgers in each social group from 1980 to 1985 is shown in Table 1. Only two groups (Glen Avon and Stoke Park Road) were found to produce cubs every year. By comparing different groups in given years, it was found that these fluctuations in group size were not synchronous and no overall pattern could be discerned.

Estimation of group size by radio-isotope labelling of faeces

Discernible counts of radioactivity were recorded from samples throughout the period of collection, up to 16 weeks after the first animal was labelled. As is the case for the Glen Avon group shown in Fig. 2, some oscillations in the cumulative percentage of labelled faeces were observed after the initial period of collection, commensurate with an influx of additional faeces produced by the cubs born in 1985.

The group size estimates were produced by the following simple equation: $G = I \times N/n$, where G = group size, I = number of individuals

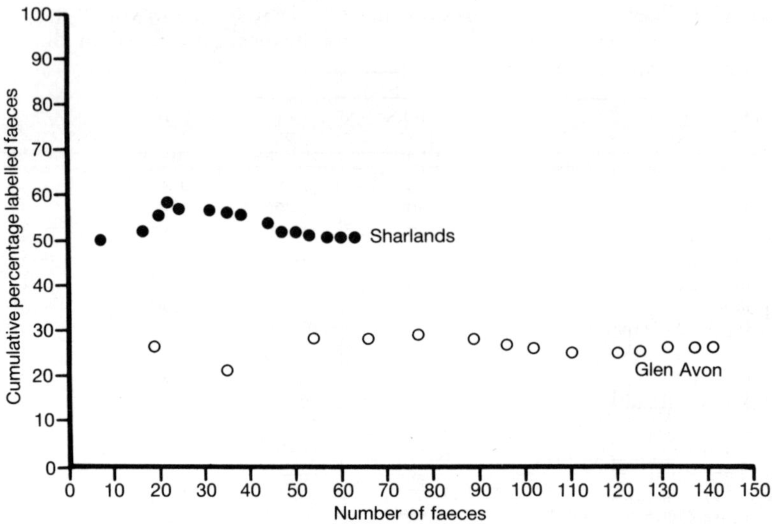

Fig. 2. Results of the radio-isotope labelling experiment for two badger social groups—'Sharlands', solid circles, and 'Glen Avon', open circles. Faeces were collected until a constant cumulative percentage of labelled faeces was obtained. The decline in the cumulative percentage of labelled faeces during the course of sampling was due to the emergence of cubs above ground and their contribution to the faeces pool.

Table 2. Population estimates for four badger social groups in 1985, comparing the estimates obtained by direct enumeration and by the isotope labelling experiment. The figures show the estimates before and after the cubs became active above ground.

Group	Estimate before cub recruitment		Estimate after cub recruitment	
	Isotope labelling	Direct enumeration	Isotope labelling	Direct enumeration
Glen Avon	9–10	9	11–12	11
Sharlands	4	4	4	4
Horsefield	5–6	5	7–8	8
Stoke Park Road	4	4	5–6	5

labelled, N = total number of faeces collected, n = number of labelled faeces (after Kruuk, Gorman & Parish 1980). In each case a remarkably close agreement was found between the group size estimated by isotope labelling and that produced by the direct enumeration technique (Table 2), both before and after cub recruitment.

Age structure and mortality rates

Figure 3 shows the age structure of known-age badgers in the population from 1980 to 1985, with an indication of the number of animals of

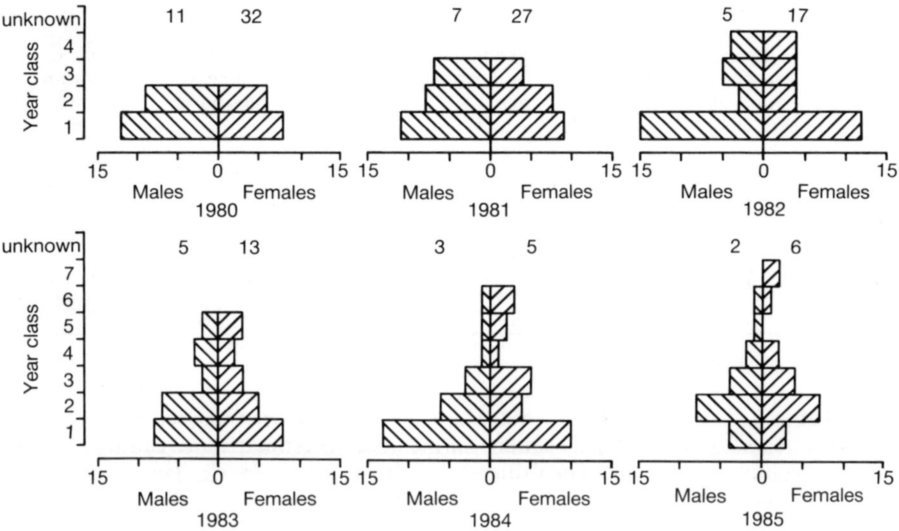

Fig. 3. Age structure of the known-age badger population for the years 1980 to 1985; the numbers of animals of unknown age each year are indicated by the figures above each histogram. The poor survival of cubs in some years (e.g. 1981) can be seen, as can the subsequent imbalance produced in the age structure of the adult population.

Table 3. Apparent mortality rates for known-age badgers based on trapping returns. The figures are percentages, with the sample size in brackets.

Year	Apparent mortality rate					
	Cubs			Adults		
	Males	Females	Total	Males	Females	Total
1979	25.0(12)	40.0(10)	31.8(22)	—	—	—
1980	25.0(12)	12.5(8)	20.0(20)	33.3(9)	33.3(6)	33.3(15)
1981	72.7(11)	55.5(9)	65.0(20)	40.0(15)	27.3(11)	34.6(26)
1982	53.3(15)	58.3(12)	55.6(27)	41.7(12)	33.3(12)	37.5(24)
1983	25.0(8)	50.0(8)	37.5(16)	57.1(14)	15.4(13)	37.0(27)
1984	38.5(13)	30.0(10)	34.8(23)	25.0(12)	33.3(15)	29.6(27)
Mean for 1980 to 1984	40.8(71)	42.1(57)	41.4(128)	40.3(62)	28.1(57)	34.5(119)

unknown age present each year. As was found for the total population, there were significantly more adult females than adult males in the sample of unknown-age animals ($\chi^2 = 5.16$, $P < 0.05$).

The estimated mortality rates of known-age adults and cubs for the years 1979 to 1984 are shown in Table 3. The average apparent mortality rates (1980 to 1984) for male and female cubs were approximately equal—

Table 4. Calculation of pre-capture mortality for the years 1980 to 1984. For a full explanation see text.

Year	Observed proportion of females which bred	Estimated total number of females	Estimated number of females which bred	Total number of cubs caught	Estimated number of cubs born	Estimated pre-capture mortality
1980	8/24	38	12	20	32	37.5%
1981	9/29	38	12	20	32	37.5%
1982	13/19	29	20	27	54	50.0%
1983	12/20	26	16	16	43	62.8%
1984	6/13	20	9	23	24	4.2%
						Mean = 38.4%

Table 5. Life table for known-age badgers, based on trapping returns for the years 1980 to 1984. The first year includes the estimated pre-capture mortality rate calculated in Table 4.

Year class	Males			Females		
	lx	*dx*	*qx*	*lx*	*dx*	*qx*
First	1000	635	0.635	1000	643	0.643
Second	365	133	0.364	357	83	0.232
Third	232	96	0.414	274	120	0.438
Fourth	136	85	0.625	154	26	0.169
Fifth	51	26	0.510	128	26	0.203
Sixth	25	25	1.000	102	34	0.333

40.8% and 42.1% respectively. The mortality rate for adults (both sexes combined) for 1980 to 1984 was 34.5%, but male mortality (40.3%) was greater than that for females (28.1%); however, this difference was not significant ($\chi^2 = 1.453$, $P > 0.05$).

The estimated pre-capture mortality rates of cubs for the years 1980 to 1984 are shown in Table 4. The mean annual pre-capture mortality rate was 38.4%, and this figure was used to estimate the number of cubs which had died between birth and when they were first available for capture.

Life tables and survivorship curves

Life tables for known-age male and female badgers are shown in Table 5. Overall some 64% of badgers died within their first year of life. There was little difference between the mortality rates (*qx*) of males and females in the first few years of life. However, the mortality rates of males were considerably higher in the older age classes, i.e. three years of age and above. These different mortality rates in adult males and females are also shown in the survivorship curves (Fig. 4).

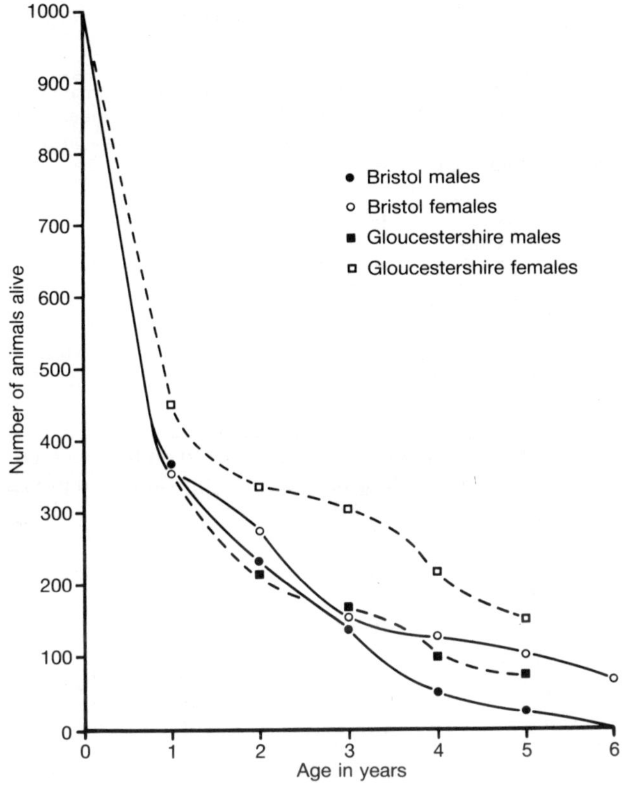

Fig. 4. Survivorship curves for Bristol badgers (solid lines) compared with the results from Cheeseman *et al.* (this volume) for Gloucestershire (broken lines). Bristol males, solid circles, females, open circles; Gloucestershire males, solid squares, females, open squares. The first year mortality includes the estimated rate of pre-capture mortality.

Proportion of females breeding

Table 6 indicates the proportion of known-age females which bred in each age class. No females in their second or third year were recorded as having bred, after which the proportion breeding in each age group increased with age, up to a maximum of 80% of sows in their sixth year or older.

Modelling the population

Using our model a stable population was generated by Year 5, with a ratio of 15 adult males : 20 adult females. Table 7 shows the initial year's figures and the age structure of the stable population in Year 5. The model was found to be sensitive to changes in the initial sex ratio of cubs and the sex-specific mortality rates but insensitive to changes in initial numbers or numbers of cubs produced. The stable population generated closely

Table 6. Proportion of known-age females which produced cubs.

Year class	Animals which bred %	Sample size
Second	0	50
Third	0	33
Fourth	40.0	20
Fifth	66.7	9
Sixth or older	80.0	10

Table 7. Age structure of the model population in Years 1 and 5, when a stable population was achieved.

Year class	Year 1		Year 5	
	Males	Females	Males	Females
First	19	19	19	19
Second	6	4	7	7
Third	3	5	4	5
Fourth	1	1	2	3
Fifth	1	2	1	2
Sixth	1	3	1	2
Seventh	0	0	0	1
Total adult population	12	15	15	20

resembles the age and sex structure of the observed population in Bristol, and the population size was comparable to that observed, allowing for a number of animals in their seventh year or older, which in the trapping study were excluded from the total as animals of unknown age.

Discussion

For the most part, previous studies have concentrated on various aspects of the behavioural ecology of badgers living in high-density populations. The preceding paper in this volume, by Cheeseman *et al.* (p. 279), reported on a long-term capture–mark–recapture study, describing the dynamics of a high-density badger population over a period of time. Since our paper addressed the same problem, but concentrated on a population at lower density and in a contrasting habitat, the results of the present study are hereafter compared and contrasted with the previous paper. It should be noted, however, that although the recorded badger density in Bristol was lower than that observed by Cheeseman *et al.*, it should not be regarded as a low-density population, and the observed adult density (ranging from 4.4 to 7.5

adults per km^2) was higher than that reported in most of the studies summarized by Anderson & Trewhella (1985).

The direct enumeration technique used in this study is designed to give an accurate estimate of population size from capture records, simply by counting the number of marked animals present and adjusting this estimate on the basis of subsequent captures. Such a figure may also be referred to as a 'minimum number alive' estimate. In his review in this volume (p. 25) of the application of capture–mark–recapture methods to small mammal populations, Montgomery concluded that 'the fundamental failure of minimum number alive estimates is the underestimation of population size to an unknown and varying degree'. Such problems are likely to be most pronounced in short-lived species. In the present study, where trapping was over a long period and the study animal a relatively long-lived species, a very large proportion of the study population was captured and marked. Where a population such as ours is studied in detail over a long period, direct enumeration is an accurate and reliable technique. Such a view is reinforced by the close agreement found between the direct enumeration estimates of group size, and those obtained from the faeces labelling experiment. This interpretation is also supported by the rarity of unmarked badgers in the population; after 1981 only two animals were captured as unmarked adults, and no unmarked adults were recovered as road casualties on the study area.

The direct enumeration technique assumes that there is no emigration or immigration to the population. In the present study trapping took place around the outskirts of the study area, as did the recovery of corpses from road traffic deaths. Although group changes were common in the adult population, they were rare in younger animals and most movements were confined to the study area (S. Harris & W. J. Cresswell unpublished data). Therefore the calculations presented here were not significantly altered by emigration, and the rarity of unmarked animals on the study area suggests that immigration was uncommon. Since it was rare for younger animals to move, it was also justified to assume that animals first caught in their second year (yearlings) had been born in the group in which they were first caught.

No single obvious cause could be found to explain either the overall decrease in population size, or the group size fluctuations indicated in Table 1. Anderson & Trewhella (1985) used a series of mathematical models to predict overall trends in badger population dynamics. Their results suggested that in good habitats badger populations were likely to appear stable, but with detectable oscillations in cub density. It was difficult to relate the findings of the current study (concentrating as it did on a relatively short period of time) to such an overview. However, there was a considerable range in the numbers of cubs produced and the total population size was not stable. Cheeseman *et al.* (this volume) found some indication of density-dependent trends in cub and adult numbers, but in Bristol oscillations in cub production seemed entirely independent of the decrease in the adult population.

Past studies on badger reproduction have suggested that there is high perinatal mortality in this species (Wandeler & Graf 1982). The estimates of pre-capture mortality from this study (38.4%), and that of 36.6% by Cheeseman *et al.* (this volume), tend to support these observations. These latter authors found a significantly greater post-capture mortality rate in male cubs than in females, whereas in Bristol they were approximately equal. Both populations exhibited a preponderance of adult females in the total population, but while in Gloucestershire this was achieved by a greater mortality rate in male cubs, in Bristol the preponderance of adult sows was brought about by a sex-specific difference in mortality rates in older animals, particularly those in their fourth year and above. With our small sample size, the average mortality rate of adult males was not statistically significantly greater than that for females. However, it did represent a real phenomenon, which produced a statistically significantly greater proportion of females in the adult population. In addition, a comparison between the life tables and survivorship curves derived from the Bristol and Gloucestershire data revealed an overall higher mortality rate in Bristol badgers.

The validity of these observations was supported by the results of the modelling study. Obviously such a simple model had certain drawbacks: the greatest age an animal could reach was determined by the final step in the life table, and it is likely that older animals occur in the population than have so far been recorded in the known-age sample. Also the changes in mortality for older animals had little effect on the total population because of the small number of known-age animals available at that stage. However, since the parameters used in the model did produce a stable population of the right proportions, they suggest that the data presented here are likely to be correct and that the reported differences in sex-specific mortality rates are important. Although in this study consistently more males than females were produced as cubs (this difference was not statistically significant), the model only produced the correct stable population using an initial production of cubs on a 50:50 ratio (Anderson & Trewhella (1985), Cheeseman *et al.* (this volume) and Kruuk & Parish (1987) all record an equal sex ratio for cubs). It is likely, therefore, that our observed sex ratio was due to small sample size bias and had no biological significance. The most common observed mortality factor in both the Bristol and Gloucestershire populations was road traffic fatalities (although in the populations as a whole this may not have been the principal cause of death). From their third year onwards the badgers in Bristol, and particularly the males, were increasingly likely to change social group or undertake a longer-distance dispersal movement (S. Harris & W. J. Cresswell unpublished data). It was at this age that the increased rate of male mortality became apparent, and this could be associated with an increased risk from motor vehicles or intraspecific aggression either within or between social groups. Why there should be a higher mortality rate in sows in Bristol than in Gloucestershire is unclear; it may be that there is a greater (but unknown) risk to sows living in a less rigidly defined territorial system.

Not only did Bristol badgers suffer from a higher mortality rate than the Gloucestershire population, but Bristol sows also bred later in life. Although most authorities state that female badgers reach sexual maturity in their second year (Neal & Harrison 1958; Ahnlund 1980), and there have been occasional reports of yearling females producing cubs (S. Fargher & P. Morris unpublished; Ferris 1986; Cheeseman *et al.* this volume), in this study none of the sows in their second or third year was recorded as having bred. There was some evidence that in Bristol two or three litters of cubs per year could be produced by a single badger social group. In Gloucestershire also some groups were recorded as producing two or three litters (Cheeseman *et al.* this volume), whereas in Scotland, where badgers were living at a much lower density, there was evidence to suggest that only one sow in each group produced cubs (Kruuk & Parish 1987). To understand the underlying processes controlling badger population dynamics, one must investigate whether a mechanism for regulating badger numbers exists and if so, at what stage or stages in the life cycle such processes act. Cheeseman *et al.* (this volume) produced evidence to suggest that such a mechanism could exist, and that there appeared to be a density-dependent relationship between cub and adult numbers. They suggested that the time of implantation was likely to be the crucial period for such a regulatory mechanism, with the majority of females mating but only a proportion of these actually implanting their blastocysts. Kruuk & Parish (1987), however, could find no evidence for a density-dependent relationship in cub recruitment, since cub numbers were independent of both group and range sizes. They suggested that a combination of mortality and emigration was the process by which population size was stabilized. There is evidence to suggest that all the above population control mechanisms operate in Bristol, but in addition some factor(s) was/were responsible for delaying breeding in females in Bristol; further work is needed to determine the nature of these mechanisms and their role in population regulation.

At this stage it is difficult to make firm conclusions concerning the nature of badger population dynamics and the underlying factors that control them. Although the results of Kruuk & Parish (1987) were similar in some respects to the findings of the present study, theirs was a more behavioural approach and lacked the detailed figures on mortality and recruitment necessary to make a direct comparison possible. One is left, therefore, with only two long-term capture–mark–recapture studies on badgers, and both this paper and that of Cheeseman *et al.* (this volume) are interim reports of continuing studies. However, since the methodology in the two studies was standardized, we have an opportunity to compare the dynamics of a lower-density, more loosely organized badger population with a high-density, more rigidly organized badger population. The salient differences are as follows: there is a higher overall mortality rate in Bristol, particularly in adult females; in both populations there is the same preponderance of adult females in the population, but in Bristol it is brought about by a sex-specific difference in mortality of adults, rather than of cubs, as in Gloucestershire, and this preponderance of females in Bristol is most obvious in animals in their third year onwards. In both studies there were wide fluctuations in group sizes with sporadic cub production and breeding confined to the older females, particularly in Bristol, where no sows produced cubs before their fourth year. Comparative data from areas of medium to low badger density would be of great value in understanding the factors which influence badger population dynamics.

Acknowledgements

We would like to thank the many people who have helped with the collection of field data, and the Bristol City Council, the Avon and Somerset Constabulary and the Bristol Branch of the Royal Society for the Prevention of Cruelty to Animals, for their help in recovering badger corpses. We would also like to thank Dr C. L. Cheeseman for his advice and valuable discussions during the course of this study. The study was funded by the Ministry of Agriculture, Fisheries and Food, the Natural Environment Research Council, the Nature Conservancy Council, the Royal Society and the Royal Society for the Prevention of Cruelty to Animals; Harris would like to express his thanks to these organizations and Cresswell would like to thank the Natural Environment Research Council for postgraduate support.

References

Ahnlund, H. (1976). Age determination in the european badger, *Meles meles* L. *Z. Säugetierk.* **41**: 119–25.

Ahnlund, H. (1980). Sexual maturity and breeding season of the badger, *Meles meles* in Sweden. *J. Zool., Lond.* **190**: 77–95.

Anderson, R.M. & Trewhella, W. (1985). Population dynamics of the badger (*Meles meles*) and the epidemiology of bovine tuberculosis (*Mycobacterium bovis*). *Phil. Trans. R. Soc. (B)* **310**: 327–81.

Cheeseman, C.L. & Harris, S. (1982). Methods of marking Badgers (*Meles meles*). *J. Zool., Lond.* **197**: 289–92.

Cheeseman, C.L. & Mallinson, P.J. (1980). Radio tracking in the study of bovine tuberculosis in badgers. In *A Handbook on biotelemetry and radio tracking*: 649–56. (Eds Amlaner, C.J. & Macdonald, D.W.). Pergamon Press, Oxford.

Cresswell, W.J. & Harris, S. (In press). Foraging behaviour and home range utilisation in a suburban badger (*Meles meles*) population. *Mammal Rev.*

Ferris, C. (1986). Mating and early maturity of badgers in Kent. *J. Zool., Lond. (A)* **209**: 282.

Harris, S. (1982). Activity patterns and habitat utilization of badgers (*Meles meles*) in suburban Bristol: a radio tracking study. *Symp. zool. Soc. Lond.* No. 49: 301–23.

Harris, S. (1984). Ecology of urban badgers *Meles meles*: distribution in Britain and habitat selection, persecution, food and damage in the city of Bristol. *Biol. Conserv.* **28**: 349–75.

Harris, S. & Smith, G.C. (1987). Demography of two urban fox (*Vulpes vulpes*) populations. *J. appl. Ecol.* **24**: 75–86.

Kruuk, H., Gorman, M. & Parish, T. (1980). The use of ^{65}Zn for estimating populations of carnivores. *Oikos* **34**: 206–8.

Kruuk, H. & Parish, T. (1987). Changes in the size of groups and ranges of the European badger (*Meles meles* L.) in an area of Scotland. *J. Anim. Ecol.* **56**: 351–64.

Neal, E.G. & Harrison, R.J. (1958). Reproduction in the European badger (*Meles meles* L.). *Trans. zool. Soc. Lond.* **29**: 67–130.

Petrides, G.A. (1950). The determination of sex and age ratios in fur animals. *Am. Midl. Nat.* **43**: 355–82.

Wandeler, A.I. & Graf, M. (1982). Der Geschlechtszyklus weiblicher Dachse (*Meles meles* L.) in der Schweiz. *Revue suisse Zool.* **89**: 1009–16.

Symp. zool. Soc. Lond. (1987) No. 58: 313–328

The use of sociological data to explain the distribution and numbers of urban foxes (*Vulpes vulpes*) in England and Wales

Stephen HARRIS and
G. C. SMITH

Department of Zoology
University of Bristol
Woodland Road
Bristol BS8 1UG

Synopsis

Fox surveys in 14 English cities were undertaken to determine the total number of foxes present, measured as the number of fox family groups per square kilometre. These results were then combined with sociological data derived from both Moser & Scott's (1961) analysis of the social and economic differences in British towns and the 1981 Census in a series of regression analyses to form a number of predictive models. Predictions were attempted for 143 cities in England and Wales which in 1951 had a population of over 50,000. These predictions were compared with replies to questionnaires sent to Environmental Health Departments of all the Local Authorities to give a measure of success for the prediction for each town. A total of 77 predictions were considered to be correct and for a further 43 cities foxes were known to be rare or absent. Fox density estimates were thus considered to be accurate for 120 of the 143 cities. Predictions could not be made for five of the remaining cities, and the fox population estimates for the other 18 were considered to be too high. The important variables in the predictive equations and the possible reasons underlying the relative success and failure of the models used are discussed, together with their relevance to developing rabies control programmes.

Introduction

Foxes have been established in many British cities for over a quarter of a century; colonization of most of these urban areas probably began during or soon after the Second World War (Harris 1986). The distribution of urban foxes in Britain has been described by Macdonald & Newdick (1982) and Harris & Rayner (1986c). The latter authors showed that the change in housing policy in the 1930s, which produced extensive areas of low-density,

ZOOLOGICAL SYMPOSIUM No. 58
ISBN 0-19-854006-X

owner-occupied residential suburbs, provided an ideal habitat for foxes. Using a discriminant analysis, Harris & Rayner (1986c) found that simple sociological data could be used to explain the presence or absence of foxes in individual British cities.

To determine the actual numbers of foxes in a city, Harris (1981) developed a two-stage survey technique, in which the results of a spring count of the number of litters of fox cubs in an area of a few square kilometres were combined with the results of a sightings survey by school children throughout the entire urban area to produce an estimate of the total fox population. The results were presented as the number of fox family groups per square kilometre throughout the city. Despite the simplicity of the approach, surveys to check the accuracy of the technique showed that the results were remarkably reliable (Harris 1985).

Harris & Rayner (1986a) used the results of these surveys in nine English cities to analyse the habitat requirements of urban foxes. Although the cities surveyed differed greatly in character, the habitat variables they examined had remarkably consistent effects on the numbers and distribution of foxes in the various cities. Since these habitat requirements were so consistent between disparate urban areas, Harris & Rayner (1986b) used the results of the nine fox surveys to develop multivariate regression models which they suggested could be used to predict the distribution and numbers of foxes in other British cities. These models were based on a combination of sociological variables derived from the 1981 Census and habitat variables extracted from large-scale maps. Harris & Rayner (1986b) also described the value and application of their models in the event of an outbreak of rabies.

Throughout these earlier studies, the fox family group was used as the basic recording unit. Harris & Rayner (1986a) explained that the limited amount of data available suggested that the mean fox family group size was similar between cities, but that it was possible that the size of the fox family groups and/or their breeding success might vary in relation to different urban habitats within a single city. Subsequently Harris & Smith (1987) showed that in Bristol the productivity, age and sex structure of the fox population did not vary in areas of different fox density (and hence habitat quality). They argued therefore that the fox family group was a uniform recording unit and so could be used to compare fox densities in different types of urban habitat within a single city and probably also between cities.

Although Harris & Rayner (1986a,b) showed that a variety of habitat and sociological data had remarkably consistent correlations with urban fox numbers, and that these data could be used to predict the distribution and numbers of foxes within individual cities with a high degree of accuracy, their use of habitat data involved the collection of very time-consuming measurements from large-scale maps, and a degree of subjectivity, evident when other workers tried to apply their criteria to interpret the maps. To avoid any such biases, and to see if a model could be developed which was quicker to apply in the event of rabies, we decided to extend their line of investigation. Our aim was to use sociological data alone to develop models which could be used to classify and rank the suitability of British cities as a habitat for foxes, to predict the mean fox density in all British cities, and to explain the absence of foxes from many urban areas. Such models could then be used to identify those urban areas most at risk in an outbreak of rabies and to quantify the total number of foxes present. It was not our intention to produce reasons for the presence or numbers of urban foxes based upon the sociological data which were used, but only to use the data to produce a predictive model which gives accurate results.

Materials and methods

The fox surveys

The distribution and numbers of foxes in 14 cities were determined using the survey technique described by Harris (1981). These cities, with the year of their survey in brackets, were: Bath (1981), Birmingham (1982), Bournemouth (1981), Bristol (1979/1980), Cheltenham (1984), Coventry (1985/1986), Dudley (1982), Gloucester (1984), Leicester (1984), Nottingham (1985), Poole (1981), Solihull (1982), Walsall (1982) and Wolverhampton (1982).

Statistical techniques

A large amount of Census data is available for all the cities in England and Wales. Scottish cities have been excluded from this study since the Census asks for slightly different information (Rhind 1983) and only a few Scottish cities have urban foxes (Harris & Rayner 1986c). For the present analysis it was decided to concentrate on two main data sets. Census data covering the period when it is believed that foxes started colonizing the cities in large numbers have been collated by Moser & Scott (1961) in their analysis of the 157 British towns and cities which at that time had a population of over 50,000. To take into account any gross changes in the urban habitat since that time their data were combined with some information from the most recent (1981) Census (Office of Population Censuses and Surveys 1984). Hence for this analysis we were restricted to the 157 cities included in Moser & Scott (1961); of these we surveyed 14 to determine the number of foxes present, and were trying to develop predictive models for the remaining 143.

All the analyses were performed on a Honeywell mainframe computer with the latest version of Minitab, a statistical package supplied by Pennsylvania State University. A preliminary bivariate and multiple regression analysis was performed on 33 sociological variables for all 14 of the surveyed cities. First the data were analysed by multiple stepwise regression with both linear and logarithmically transformed data. This technique produced a series of multiple regression equations by eliminating variables within the equation that had low significance (a low *F*-statistic) and adding variables from the database to the equation that had high significance (a high *F*-statistic) until no variables had the appropriate level of significance for them to be either removed or added. The equations were valid at all stages of the analysis, but as the analysis continued they became more complex, with more variables, and explained a greater proportion of the variation in fox density. It is of paramount importance to note that the independent variables were probably intercorrelated, and that a variable which on its own in a bivariate analysis showed positive correlation with the dependent variable might show negative partial correlation in a multivariate analysis. This was especially likely where the independent variable (for example, human population size) had an indirect effect on the dependent variable (fox family density). Bivariate analyses can only show the sum of all of the correlations with the dependent variable, while multivariate analysis may show some of the lesser, partial correlations.

The results from this preliminary analysis were sufficient to persuade us to analyse the entire set of 60 variables used by Moser & Scott (1961), together with a measure of longitude and latitude derived from the map reference for each city, two measures of current rural fox density in the surrounding countryside and 19 variables chosen from the 1981 Census (some of which were the same variables as from the 1951 Census) as a measure of any changes in that city in the intervening 30 years. The measures of rural fox density were produced by using the original data of Macdonald, Bunce & Bacon (1981), where each of 32 land classes was given an estimated fox density based on averaging subjective estimates for ten 1 km squares chosen randomly for each land class. We rederived estimates for all 32 land classes by excluding those areas which fell within the limits of urbanization. Each city was then examined in turn to derive a mean rural fox density based on the type and number of each land class present around its boundary. Both the original estimates of Macdonald *et al.* (1981) and our rederived measures were then included in the analysis.

The variables which were found to be important in the stepwise regression were then used in a multiple regression analysis which permitted predictions for the remaining 143 cities listed in Moser & Scott (1961) to be made. The analyses were performed on both linear and logarithmically transformed data.

Two major problems were encountered. First, any model which was very accurate necessarily used a large number of variables. As the number of variables increased, the number of cities which could be predicted for decreased, since each variable for each city had to fall close to the mean of the same variable for the predictor cities. For example, if the maximum population for the 14 predictor cities was 600,000, and population size was one of the variables used in the model, then no prediction should be made for any city with a population greater than 600,000. To circumvent this problem the least important variables were systematically removed, resulting in increasingly inaccurate equations which predicted for a greater number of cities. When the 95% confidence limits of the predictions reached unacceptably high values or when most of the cities had received predictions, no more variables were removed. As some cities will never be available for prediction (namely those with extreme values for the most important variables) certain variables were systematically removed from the database and a new stepwise regression performed in an attempt to predict for these cities. An alternative method would be to include data for a city which was known to have no foxes present and this probably would help to increase the total number of cities for which predictions may be made. After investigation it was decided not to use the latter method since the problems associated with it outweighed the potential advantages. The major uncertainty inherent in such an approach is that it would be impossible to be sure whether the selected city was devoid of foxes for reasons of unsuitability of habitat or for any reasons independent of the available data.

The second major problem during the analysis was that of lack of compatibility between the two Censuses as a result of boundary changes. Although all the cities were included in both Censuses, a number of boundary changes, especially in Greater London, meant that 1981 Census data for the areas used in the analysis by Moser & Scott (1961) were not immediately available. Rather than attempt to recalculate the 1981 Census data for these areas to take account of the boundary changes, a number of stepwise regressions were performed solely on the data from Moser & Scott (1961) to obtain predictive equations for these cities.

The analyses resulted in a number of different equations which could be used to predict urban fox density. These equations then had to be assessed to find which was the most accurate. Approximate fox densities were calculated for each city by evaluating replies to questionnaires to the Environmental Health Departments of Local Authorities (Harris & Rayner 1986c), information collected from local natural history societies and published reports, and from a working knowledge of some of these areas (S. Harris unpublished data). Each equation was then used to produce as many predictions as possible and these values were compared with the expected density levels. The predictions were examined both geographically and by the groups of cities as determined by Moser & Scott (1961). Moser & Scott produced 14 groups of cities based on similarity of the principal components of their sociological data. More recently, Andrews (1971) and Grove

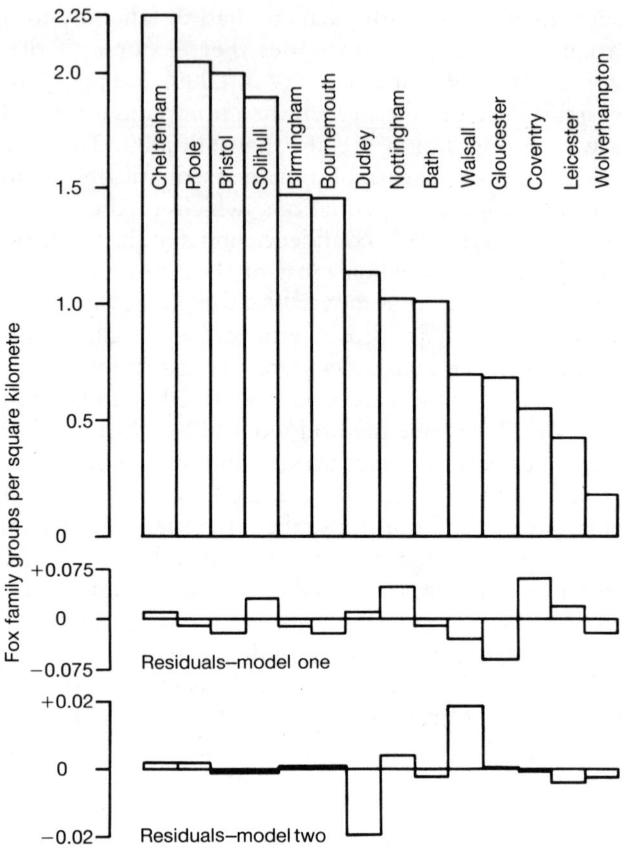

Fig. 1. Mean fox density (fox families per square kilometre) in each of the 14 predictor cities. The residuals of the two models used in the analyses are shown below, but in a different scale.

& Roberts (1980) have also calculated clusters of cities but these groupings were found to be less useful for our purposes and so are not discussed here.

Results

The detailed results from some of the surveys have been given by Harris (1981) and Harris & Rayner (1986a). For the following analyses, only the mean fox density for each of the 14 cities has been used (Fig. 1).

Seven predictive models were finally produced which fulfilled the necessary requisites of predicting mean fox densities for most cities and producing predictions in the correct range of values. Evaluation of the seven models showed that two were the most useful, and so only these two will be reported and discussed in any detail. The first model used a maximum of

Table 1. Summary of the first multiple regression model, which is based on data from Moser & Scott (1961) and the 1981 Census. The 95% confidence limits are ± fox family groups.

Stage	Variable	Per cent explained variation	95% confidence limits	Number of cities predicted
1	X1: Log(% private households with six or more persons; 1981)			
	X2: Population (thousands); 1951			
	X3: New houses built by Local Authorities per 1,000 population; 1945–58	73.0%	±0.8	132
2	X4: Social Survey J-index; 1954[a]			
	X5: Log(% households owner-occupied; 1981)	91.4%	±0.5	118
3	X6: 1955 General Election: % voting left	95.8%	±0.4	61
4	X7: Log(% population aged 0–14; 1981)	98.4%	±0.3	32
5	X8: Log(% Social Class I+II; 1981)[b]	99.3%	±0.16	17

Equations used:

Stage 1: Fox Density = 2.44 − 2.09X1 + 0.00198X2 + 0.0269X3

Stage 2: Fox Density = 4.38 − 1.85X1 + 0.00218X2 + 0.0367X3 + 0.0880X4 − 0.800X5

Stage 3: Fox Density = 5.55 − 1.60X1 + 0.00232X2 + 0.0460X3 + 0.0644X4 − 0.908X5 − 0.0302X6

Stage 4: Fox Density = 4.93 − 2.09X1 + 0.00223X2 + 0.0393X3 + 0.0608X4 − 1.230X5 − 0.0422X6 + 1.99X7

Stage 5: Fox Density = 5.62 − 2.53X1 + 0.00219X2 + 0.0388X3 + 0.0460X4 − 1.080X5 − 0.0582X6 + 3.05X7 − 0.566X8

[a] Social Survey J-index: the percentage of voters on the Electoral Register who are jurors.
[b] Social Class I+II: the percentage of occupied and retired males in social classes I (professional occupations) and II (intermediate occupations).

eight variables and predicted for 132/143 cities, while the second model used a maximum of nine variables and also predicted for 132 cities. The two models are summarized in Tables 1 and 2 respectively. Both models were split up into five different equations with each stage predicting with greater accuracy for fewer cities, since a number of cities had habitat variables outside the range of the predictor cities. The 95% confidence limits improved from approximately ± 0.8 fox families per square kilometre to ± 0.16 families in the first model and from ± 0.6 to ± 0.04 families per square kilometre in the second model. As the confidence limits were reduced in each succeeding equation the number of cities for which valid predictions could be made decreased to only 17 and 21 respectively. The residuals of the most accurate equation for each model are shown for the 14 predictor cities in Fig. 1. No pattern was found to be present in the residuals of any of the ten equations used in the two models and so all were performing equally well in low-, medium- or high-density towns.

Analysing both models in detail showed the demographical areas in which each model was working best. The 14 groups of cities derived by Moser & Scott are shown in Table 3. The first model predicted accurately

Table 2. Summary of the second multiple regression model, which is based entirely on data from Moser & Scott (1961). The 95% confidence limits are ± fox family groups.

Stage	Variable	Per cent explained variation	95% confidence limits	Number of cities predicted
1	Y1: Per capita retail sales; 1950			
	Y2: % of population working in insurance, banking and finance; 1951			
	Y3: Tuberculosis notification rate; 1957	86.9%	±0.6	132
2	Y4: Infant mortality rate 1950–52	93.4%	±0.4	119
3	Y5: Mortality rate, other cancers; 1957[a]	95.2%	±0.4	102
4	Y6: % of private households with six or more persons; 1951			
	Y7: New houses built by Local Authorities as % of total new houses built 1945–58			
	Y8: % Social class IV+V; 1951[b]	99.7%	±0.1	34
5	Y9: Total change of population (1951–58) as % of population 1951	99.9%	±0.04	21

Equations used:

Stage 1: $\text{Fox Density} = 3.17 - 0.0181Y1 + 1.06Y2 - 0.00988Y3$

Stage 2: $\text{Fox Density} = 2.24 - 0.0206Y1 + 1.22Y2 + 0.01120Y3 + 0.0393Y4$

Stage 3: $\text{Fox Density} = 3.01 - 0.0225Y1 + 1.26Y2 - 0.01100Y3 + 0.0562Y4 - 0.01040Y5$

Stage 4: $\text{Fox Density} = 2.52 - 0.0201Y1 + 1.30Y2 - 0.01360Y3 + 0.0346Y4 - 0.00940Y5 + 0.127Y6 - 0.01050Y7 + 0.0222Y8$

Stage 5: $\text{Fox Density} = 2.37 - 0.0194Y1 + 1.290Y2 - 0.0133Y3 + 0.0315Y4 - 0.00919Y5 + 0.127Y6 - 0.00963Y7 + 0.0236Y8 + 0.00485Y9$

[a] Mortality rate, other cancers: deaths from cancers (malignant neoplasm) other than cancer of the lung or bronchus; the ratio of local to national rate is adjusted for sex and age differences.
[b] Social Class IV+V: percentage of occupied and retired males in social classes IV (partly skilled occupations) and V (unskilled occupations).

for Groups 1, 2, 6, 12, and for three towns in Group 9, poorly for Group 4 and did not produce many predictions for Groups 9, 10, 11, and 13 (see Table 4). The remaining groups had both good and poor predictions. A prediction was considered to be accurate if it fell within the estimated range for that particular city. The second model produced its best predictions for Groups 9, 10 and 11 and these were used to supplement the predictions from the first model (see Table 4). Geographical analysis showed that the first model produced good results for cities in the south and in the west Midlands, but not for cities in East Anglia, the north, and east Midland towns (Fig. 2). This result did not simply reflect the geographical distribution of the groups of cities for which the first model predicted well. The second model predicted well for cities in the south-east, and in Greater London this model produced good results for nearly all urban areas in the south and west (Fig. 2). Using both these models to supplement each other produced predicted fox densities which we considered to be accurate for 77 cities. A further 43 cities were believed to have few or no urban foxes (Harris &

Table 3. Classification of the towns used by Moser & Scott (1961); see text for a fuller explanation.

Group	Classification	Total number of cities	Number of predictor cities
1	Mainly seaside resorts	10	1
2	Mainly spas, professional and administrative centres	10	3
3	Mainly commercial centres with some industry	16	2
4	Includes most of the traditional railway centres	14	1
5	Many of the large ports and two Black Country towns	10	2
6	Mainly textile centres of Yorkshire and Lancashire	16	2
7	Industrial towns of north-east seaboard and mining towns of Wales	11	0
8	The more recent metal manufacturing towns	14	2
9	Mainly 'exclusive' residential suburbs	9	0
10	Mainly older mixed residential suburbs	12	0
11	Mainly newer mixed residential suburbs	11	1
12	Light industry suburbs, national defence centres and towns within the sphere of influence of large conurbations	9	0
13	Mainly older working class and industrial suburbs	10	0
14	Mainly newer industrial suburbs	3	0
—	Towns not allocated to any group	2	0
Total		157	14

Rayner 1986c), and so any predictions for these areas were not required. This resulted in a total of 120 cities out of 143 which had fox density estimates which were considered to be accurate (Tables 5 and 6). Of the remaining 23 cities, predictions could not be made for five, and for the other 18 the predictions were considered to be overestimates (Table 7). There were no predictions considered to be underestimates. The overestimated cities were mainly in the north-east of London and the Midlands. These were all places where foxes were believed to be present, but not in the numbers the models suggest.

Discussion

To evaluate the accuracy of each of the models was of paramount importance. Obviously the only method of being entirely sure of their accuracy was to measure the actual fox density using the same technique as applied to each of the 14 predictor cities. Since doing this for 143 urban areas was impractical, another method of testing the predictions was required. Harris & Rayner (1986c) sent questionnaires to all Environmental Health Depart-

Table 4. A breakdown of the success of the predictions for each of the 14 groups of cities using both models combined.

Group	Number of correct predictions	Number known to have very few or no foxes[a]	Number of incorrect predictions	Number not predicted	Total number of cities[b]
1	6	3	0	0	9
2	6	1	0	0	7
3	6	6	2	0	14
4	3	6	3	1	13
5	3	5	0	0	8
6	11	3	0	0	14
7	3	8	0	0	11
8	4	6	0	2	12
9	7	0	2	0	9
10	8	1	3	0	12
11	10	0	0	0	10
12	7	0	2	0	9
13	3	1	6	0	10
14	0	2	0	1	3
—	0	1	0	1	2
Total	77	43	18	5	143

[a] This figure denotes cities known to have very few foxes, but in which the models suggest foxes should be present, and those cities known to have no foxes and for which predictions are not available. It does not include those cities where the models correctly predict zero foxes; these are included in the column for correct predictions.
[b] This excludes the 14 predictor cities.

ments in England and Wales and the information accrued from these results and data from local naturalists was sufficient to estimate an approximate fox density for each of the towns. It was this estimate which was used to measure the success of each prediction. Although this method had its shortcomings, a further, *post hoc*, justification of its validity is that other absolute measures of fox density support our predictions. Page (1981) calculated that there was about one fox family group per square kilometre in part of the London Borough of Hillingdon. His survey area corresponds to Ruislip–Northwood and Uxbridge in this paper, and his results agree well with the mean of the predictions for these two urban areas given here. Similarly, the predicted fox densities in parts of south and east London correspond well with the densities determined by surveys in the early 1970s (S. Harris unpublished data).

The first equation, using both sets of data, employed a maximum of eight variables. Four of these factors (X1, X4, X6 and X8—Table 1) were measures of social class, two variables (X3 and X5) related to the housing quality and the remaining two (X2 and X7) referred to the population size and structure. In the second equation, social class was represented by four factors (Y1, Y2, Y6 and Y8—Table 2), housing quality by one variable (Y7) and population size by one (Y9). In addition three factors (Y3, Y4 and Y5)

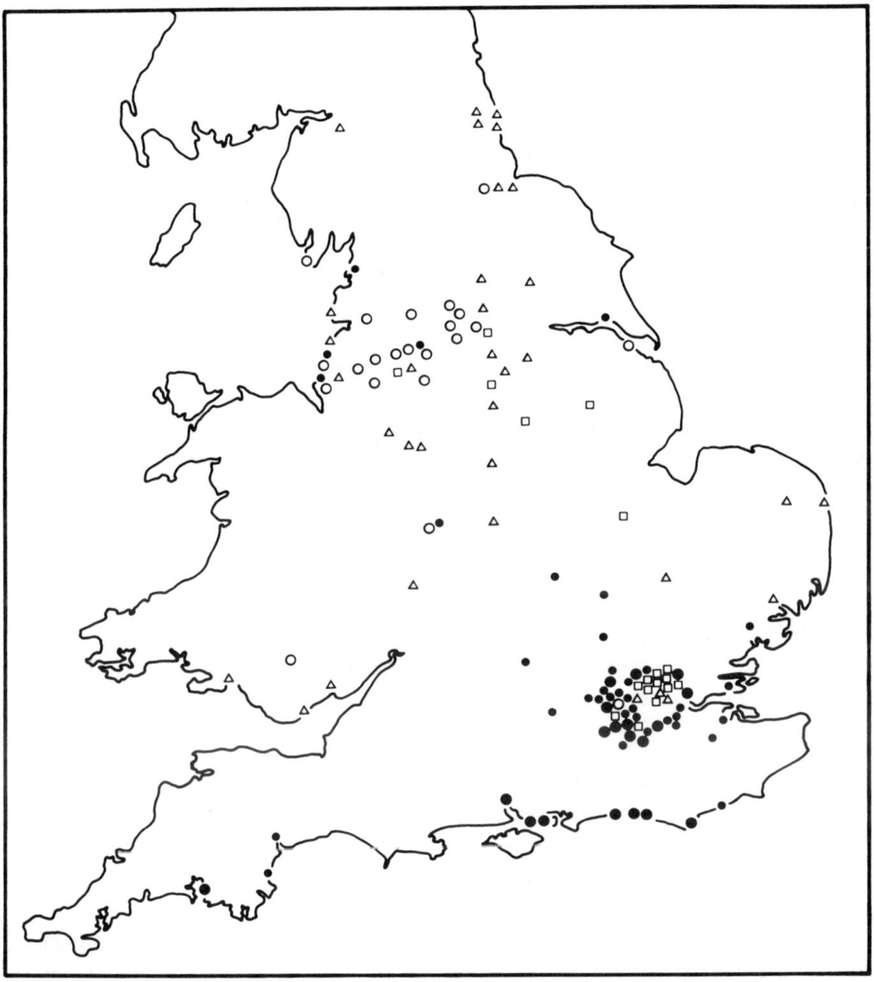

Fig. 2. Performance of the two models for English and Welsh cities. Symbols as follows: ● fox density up to 1.5 family groups per square kilometre correctly predicted; ⬤ fox density above 1.5 family groups per square kilometre correctly predicted; ○ foxes correctly predicted to be rare or absent; △ foxes believed to be rare or absent but the models suggest the habitat is suitable for foxes; □ cities in which foxes are believed to be present but for which the predictions are believed to be overestimates or for which it was not possible to predict the number of foxes present.

referred to the health of the population and so can be included amongst the social class variables. Most of these variables obviously fall into the above categories but some may require an explanation. The social survey J-index is the percentage of voters on the electoral register who are jurors and is known to sociologists as a social stratification factor, such that high values correspond with a high social class (Moser & Scott 1961). *Per capita* retail

Table 5. Predicted mean fox density[a] (number of fox families per square kilometre) ± 95% confidence limits for those cities with foxes present, arranged in order of increasing density.

City	Group	Density	City	Group	Density
Rochdale	6	0.1±0.2	Chislehurst & Sidcup	11	1.4±0.1
Enfield	12	0.1±0.6	Ruislip-Northwood	11	1.4±0.3
Crosby	10	0.2±0.1	Exeter	2	1.5±0.2
Slough	12	0.2±0.2	Colchester	2	1.5±0.3
Bedford	2	0.3±0.4	Southend-on-Sea	2	1.5±0.3
Brentford & Chiswick	13	0.3±0.4	Maidstone	2	1.5±0.4
Luton	12	0.3±0.4	Ealing	10	1.5±0.5
Reading	3	0.3±0.4	Epsom & Ewell	9	1.5±0.5
Lancaster	3	0.3±1.0	Southampton	3	1.6±0.2
Hull	5	0.4±0.4	Heston & Isleworth	10	1.7±0.5
Gillingham	12	0.6±0.4	Worthing	1	1.8±0.4
Willesden	13	0.6±0.4	Hove	1	1.9±0.4
Croydon	10	0.7±0.4	Esher	9	1.9±0.5
West Bromwich	7	0.7±0.5	Surbiton	10	2.0±0.1
Oxford	2	0.8±0.2	Brighton	1	2.0±0.2
Wallasey	10	0.8±0.2	Harrow	11	2.1±0.5
Uxbridge	12	0.8±0.4	Gosport	12	2.1±0.6
Merton & Morden	11	0.8±1.0	Eastbourne	1	2.2±0.2
Northampton	3	0.9±0.4	Finchley	9	2.2±0.4
Bexley	11	0.9±0.5	Hornchurch	11	2.2±0.4
Wimbledon	10	1.0±0.6	Sutton & Cheam	9	2.2±0.5
Hendon	10	1.1±0.5	Plymouth	3	2.3±0.3
Watford	12	1.2±0.2	Chigwell	11	2.4±0.5
Hastings	1	1.2±0.4	Portsmouth	3	2.5±0.3
Bromley	9	1.3±0.3	Carshalton	11	2.6±0.4
Orpington	11	1.3±0.4	Beckenham	9	2.7±0.5
Torquay	1	1.3±0.4	Coulsdon & Purley	9	2.8±0.6
Wembley	11	1.3±0.5			

[a] This mean figure relates to the whole of the urban area as defined in the 1951 Census, and within that urban area there will be regions with both higher and lower fox densities.

sales of an area are a direct measure of the amount of shopping in that area, so that high retail sales indicate a high number, or at least an extensive area, of shops, and/or a prosperous community. Most of the variables in the multivariate equations correlated in the same direction as the bivariate correlation, but seven had reversed signs. In Model 1 these were X3, X5, X8 and X9 and in Model 2, Y4, Y6 and Y8. This showed that these variables did not have a simple effect on fox density but rather that they were themselves composed of a number of separate and distinct factors. We will make no attempt to explain these, or other, variables used in the model, since the actual reasons for their inclusion may be complex and hard to ascertain, because each variable may be composed of more than one relationship with fox density, may be acting as a surrogate for other more meaningful variables or may simply be present as a result of a random correlation with the 14 points of the dependent variable.

Table 6. Performance of the models for cities believed to have few or no foxes present.

City	Group	City	Group
Zero fox density correctly predicted by models:			
City	Group	City	Group
Barrow-in-Furness	4	Huddersfield	6
Birkenhead	5	Keighley	6
Bolton	6	Merthyr Tydfil	7
Bootle	8	Oldbury	8
Bradford	6	Oldham	6
Burnley	6	Preston	6
Bury	6	St. Helens	8
Darlington	4	Stockport	4
Dewsbury	6	Southall	13
Grimsby	5	Warrington	7
Halifax	6	Wigan	8
Zero fox density not correctly predicted by models:			
City	Group	City	Group
Acton	13	Liverpool	5
Barking	14	Manchester	6
Barnsley	7	Middlesbrough	8
Blackpool	1	Newcastle-under-Lyme	4
Cambridge	2	Newcastle-upon-Tyne	5
Cardiff	3	Newport	5
Carlisle	4	Norwich	3
Chesterfield	4	Nuneaton	8
Crewe	4	Rotherham	8
Dagenham	14	Southport	1
Derby	4	South Shields	7
Doncaster	4	Stockton-on-Tees	8
Gateshead	7	Stoke-on-Trent	8
Great Yarmouth	3	Sunderland	7
Harrogate	1	Swansea	5
Ipswich	3	Worcester	3
Leeds	6	York	3
Cities believed to have few or no foxes present for which it was not possible to predict a fox density with the two models:			
City	Group	City	Group
Blackburn	6	Tynemouth	5
City of London	—	West Ham	7
Rhondda	7	West Hartlepool	7
Salford	7	Wood Green	10
Scunthorpe	8		

The two models combined produced estimates indicating that foxes were present in 43 cities which in fact were known to contain few or no foxes. These included some coastal towns of Wales, cities in East Anglia and some cities in north-east England and parts of the Midlands. These cities can in effect be taken to have zero fox density (Harris & Rayner 1986c). In some of these areas, notably East Anglia, the north of England and parts of south Wales, rural fox densities are lower than in the rest of the country and some

Table 7. Cities in which foxes are present but for which either the predictions are believed to be incorrect or no predictions were made.

Predictions believed to be overestimates; the models suggest densities up to one fox family group per square kilometre may be sustained:

City	Group	City	Group
East Ham	13	Peterborough	3
Edmonton	13	Tottenham	13

Predictions believed to be overestimates; the models suggest densities above one fox family group per square kilometre may be sustained:

City	Group	City	Group
Hornsey	10	Sheffield	4
Ilford	10	Southgate	9
Leyton	13	Stretford	13
Lincoln	3	Twickenham	10
Mansfield	4	Wakefield	4
Mitcham	12	Walthamstow	13
Romford	12	Wanstead & Woodford	9

Cities believed to have foxes for which it was not possible to predict a fox density with the two models:

City	Group	City	Group
Hayes & Harlington	14	Swindon	4
Huyton with Roby	—	Thurrock	8
Smethwick	8		

of these areas also had considerably fewer foxes in the recent past. In their analysis, Harris & Rayner (1986c) argued that the rarity of rural foxes in some areas might explain the absence of urban foxes in adjacent cities. However, of the 59 variables used in the seven models, rural fox density, based on the estimates of Macdonald *et al.* (1981), did not appear. Since their data were based on estimated rather than surveyed fox densities, it seems likely that improved rural fox density estimates would increase the accuracy of our models. Neither did longitude and latitude appear in the two models used. This is because the location of cities with foxes present appears to relate to both longitude and latitude in a complex manner, and such complex relationships were not examined in the analyses.

Using just the two models we were able to predict mean fox densities for most cities in England and a few in Wales. This is of obvious benefit to any rabies control programme, since the models identify the cities where foxes are most numerous and, therefore, those in which the problems in an outbreak of rabies would be greater. There are, however, some limitations to our approach. We have confined the analyses to urban areas with a minimum population of 50,000, since Harris & Rayner (1986a) found that in small urban areas, and the periphery of large urban areas, the correlations between fox numbers and habitat were less pronounced. This they attributed to foxes living in the rural surroundings moving into the urban area at night to forage; such animals would affect the fox population estimates. It is possible that a more detailed predictive model may be better able to cope with such variation. Also, since our approach has been based on the earlier analyses of Moser & Scott (1961), we can only try to predict fox densities for the cities included in their work. This has been done for 120 of 143 cities. However, today there are 20 cities or towns in Scotland and 207 urban areas, cities or towns in England and Wales with a population in excess of 50,000 (1981 Census). We have surveyed or predicted fox densities for 59% of these (134 out of 227). To predict fox densities for the remaining areas will require a different approach.

It is also important to remember that our predictions are not directly measuring fox density, but rather indicate the carrying capacity of the city. Not every city will have as large a fox population as it is capable of carrying, even though the habitat is suitable. This can be for a variety of reasons. We have already argued that rural fox density during the colonization period must be an important factor. In the present analyses, the predictions for a total of 61 cities were considered to be overestimates. These included 43 positive predictions where the questionnaire survey had indicated that the fox population was in effect zero, and 18 overestimates for cities where foxes were known to be present but not in the numbers indicated by the models. For some of these urban areas fox density may be increasing and may attain predicted carrying capacity in the near future. For others, the fox density may already be at the predicted carrying capacity, and the numbers indicated by the questionnaire survey may in fact be an underestimate. Further information from such cities may indicate that the fox density predicted by the models is actually correct.

Finally, the predictions are only mean values for the whole city. They enable cities with large numbers of foxes to be identified and the total fox population estimated, but they supply no information on the distribution and relative densities of foxes within the urban area. However, since these analyses have shown that simple sociological data can be successful in predicting the mean fox density in individual cities, further investigations may show that sociological data can be used to predict the distribution and numbers of foxes in small areas within an urban area. If this is so, then predicting the distribution and numbers of foxes in any city during an outbreak of rabies would be a very quick and simple process, only requiring access to the Census data.

Acknowledgements

S. Harris is grateful to the Ministry of Agriculture, Fisheries and Food, the Natural Environment Research Council and the Nature Conservancy

Council for funding this work. Additional grants to help with the fox surveys were provided by the Manpower Services Commission and the Bristol Branch of the Royal Society for the Prevention of Cruelty to Animals. G. C. Smith is grateful to the Science and Engineering Research Council for a studentship. Both authors are grateful to the many people who have assisted with the fox surveys and with collating the survey data, and to the Education Departments of the City of Coventry, the County Councils of Avon, Dorset, Gloucestershire, Leicestershire and Nottinghamshire and the Metropolitan Boroughs of Birmingham, Dudley, Sandwell, Solihull, Walsall and Wolverhampton for their participation in the fox surveys.

References

Andrews, H.F. (1971). A cluster analysis of British towns. *Urban Stud.* **8**: 271–84.

Grove, D.M. & Roberts, C.A. (1980). Principal component and cluster analysis of 185 large towns in England and Wales. *Urban Stud.* **17**: 77–82.

Harris, S. (1981). An estimation of the number of foxes (*Vulpes vulpes*) in the city of Bristol, and some possible factors affecting their distribution. *J. appl. Ecol.* **18**: 455–65.

Harris, S. (1985). Surveying the urban fox. *Biologist, Inst. Biol.* **32**: 259–64.

Harris, S. (1986). *Urban foxes*. Whittet Books, London.

Harris, S. & Rayner, J.M.V. (1986a). Urban fox (*Vulpes vulpes*) population estimates and habitat requirements in several British cities. *J. Anim. Ecol.* **55**: 575–91.

Harris, S. & Rayner, J.M.V. (1986b). Models for predicting urban fox (*Vulpes vulpes*) numbers in British cities and their application for rabies control. *J. Anim. Ecol.* **55**: 593–603.

Harris, S. & Rayner, J.M.V. (1986c). A discriminant analysis of the current distribution of urban foxes (*Vulpes vulpes*) in Britain. *J. Anim. Ecol.* **55**: 605–11.

Harris, S. & Smith, G.C. (1987). Demography of two urban fox (*Vulpes vulpes*) populations. *J. appl. Ecol.* **24**: 75–86.

Macdonald, D.W., Bunce, R.G.H. & Bacon, P.J. (1981). Fox populations, habitat characterization and rabies control. *J. Biogeogr.* **8**: 145–51.

Macdonald, D.W. & Newdick, M.T. (1982). The distribution and ecology of foxes, *Vulpes vulpes* (L.), in urban areas. In *Urban ecology*: 123–35. (Eds Bornkamm, R., Lee, J.A. & Seaward, M.R.D.). Blackwell Scientific Publications, Oxford.

Moser, C.A. & Scott, W. (1961). *British towns: A statistical study of their social and economic differences*. Oliver & Boyd, Edinburgh.

Office of Population Censuses and Surveys (1984). *Census 1981: Key statistics for urban areas Great Britain*. Her Majesty's Stationery Office, London.

Page, R.J.C. (1981). Dispersal and population density of the fox (*Vulpes vulpes*) in an area of London. *J. Zool., Lond.* **194**: 485–91.

Rhind, D. (1983). (Ed.) *A census user's handbook*. Methuen & Co., London.

Symp. zool. Soc. Lond. (1987) No. 58: 329–342

Population dynamics of mammalian diseases

Denis MOLLISON

Dept of Actuarial Mathematics & Statistics
Heriot-Watt University
Riccarton
Edinburgh EH14 4AS

Synopsis

Mathematical models for the dynamics of animal populations, even without the added complications of disease, inevitably require many simplifying assumptions, often on the basis of very sketchy data. A dissection of epidemic models into their basic ecological components, such as the generation gap of the disease, the dependence of infectivity on population density, and the dispersal distribution of contacts, helps us to assess which conclusions are likely to be robust, and which are highly sensitive to modelling details.

In qualitative terms at least, mathematical models can help us understand the broad features of epidemics. In the initial establishment of disease in a new population, numbers are low and the effects of chance play an important role, so that stochastic models are required. In considering the velocity of subsequent spread, simpler deterministic models may suffice. Where endemic conditions are established, the pattern of infection is often irregular, suggesting the need again for stochastic models which can, for instance, explain patchy behaviour in a homogeneous environment.

Long-distance infectious contacts, though individually of low probability, can contribute significantly to the advance of an epidemic, as in the case of airborne diseases such as foot-and-mouth, or of the role of the annual dispersal of young male foxes in the spread of rabies. The possibility of such contacts is of particular importance when we consider attempts to control the advance of a disease by barriers.

One of the most crucial, and controversial, problems is in evaluating strategies for eliminating endemic diseases, such as fox rabies or tuberculosis in badgers. Mathematical results on this need to be treated with caution, as they depend heavily on the assumptions made as to how infectivity will change under the strategies considered. For instance, reducing population density by killing a certain proportion of the population could actually increase infectivity if the social structure of the population is disrupted.

ZOOLOGICAL SYMPOSIUM No. 58
ISBN 0-19-854006-X

Introduction

There are a number of basic reasons for studying mammalian diseases; it is a particular pleasure to speak on the modelling of diseases at this symposium, where our concern is primarily for the wild species that suffer from them, rather than anthropocentric interest in protecting ourselves and our livestock, or the ethically dubious use of diseases to 'control' wild species.

A recent Zoological Society symposium (Edwards & McDonnell 1982) gave detailed consideration to a wide range of animal diseases, in the context of conservation, including several discussions of the practical problems of management and management attitudes. What I shall attempt here is a brief discussion of the modelling of disease, its uses and limitations. I shall try to avoid repetition of a recent survey of the closely similar modelling problems of invasions (Mollison 1986), or of more detailed treatments of sensitivity to modelling details (Mollison 1984, 1985) and of spatial models (Mollison & Kuulasmaa 1985).

I shall here take disease as given, but it is important to note that severe disease is ecologically unnatural (Jones 1982) (on an evolutionary timescale it makes little sense for a disease to harm its host too severely). Severe epidemics and endemics are thus usually associated with stressed conditions, or the introduction of a disease to a new population (immunologically new, not necessarily a different species), and it is not surprising that, in historic times, man has usually been responsible for their occurrence, whether accidentally or intentionally (including many in human populations: Crosby 1986). Man has also often created suitable conditions for the spread of epidemic disease through his intensive cultivation of plants and animals.

Examples of epidemics accidentally caused by man include the spread of rinderpest through Africa, which is thought to have been introduced by zebu cattle brought by Italians to Ethiopia in 1889, and the current European outbreak of rabies, whose origins seem to be associated with the stressed conditions of eastern Europe during the 1939–45 war.

Myxomatosis furnishes a good example of a deliberate introduction. This is a mild disease in the rabbits (*Sylvilagus brasiliensis*) of South America among whom it has evolved, but severe in the British and Australian populations of *Oryctolagus cuniculus*, who had no experience of the disease. It is interesting to note how the latter populations and the disease have co-evolved subsequently. Within 25 years there were very significant adaptive changes, both in increased resistance among the rabbits and in decreased virulence in the virus (Fenner & Myers 1978; Ross 1982).

I shall survey briefly models for the various stages of epidemic outbreaks (arrival, establishment, spread), and for endemic disease, mainly in the context of the simpler 'single-species' diseases such as rinderpest, rabies and myxomatosis, and will then discuss the crucial problems of control. First,

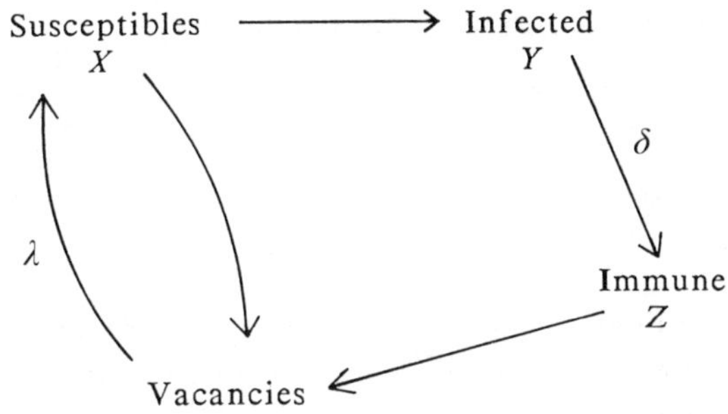

Fig. 1. Simple flow diagram for a non-fatal disease.

however, I shall discuss epidemic models more generally, the main theme being the need to relate modelling conclusions to the assumptions on which they are based.

Models

The first stage in modelling is to describe the logical structure of the situation. For instance, the flow diagram of Fig. 1 shows the possible transitions in a simple model for a non-fatal disease. It is possible to draw some useful conclusions already at this stage, even before we make specific assumptions about such important aspects as the contact rates between infectious and susceptible individuals.

As an example, consider when the population and disease are both at equilibrium levels. Then the rate of recruitment to the population as a whole will be $\lambda = N/T$, where N is the population total, $= X+Y+Z$, and T the mean lifetime. Hence the equilibrium flow rate into the infectious class must be pN/T, where p is the lifetime probability of catching the disease. But the equilibrium flow rate out of the infectious class is $\delta = Y/\tau$, where τ is the mean duration of the infection. Equating these, we have $pN/T = Y/\tau$, and have thus shown that in equilibrium the mean proportion with the disease at any one time, Y/N, is equal to $p\tau/T$. A similar result can be obtained for a model for fatal diseases such as rabies (Mollison 1984, 1985): namely that in equilibrium Y/X is equal to $r_* \tau$, where r_* is the equilibrium *per capita* net growth rate.

In order to proceed to a specific mathematical model, we have to give each component a specific mathematical description. This description can only be a rough approximation to ecological reality, and the behaviour of the model may be quite sensitive to the form we choose. This is a quite general problem in mathematical biology; to illustrate it, consider the following two examples.

First, consider the behaviour of a stone thrown vertically upwards; its height $y(t)$ at time t, according to Newtonian physics, is described by the equation

$$\ddot{y} = -g$$

This exactly expresses Newton's law, and its solution

$$y(t) = y(0) + \dot{y}(0)\,t - \tfrac{1}{2}g\,t^2$$

gives a correspondingly almost exact description of the position of the stone. (And further terms, e.g. allowing for air resistance, can be added if great accuracy is required.)

In contrast, consider the well-known logistic model for population growth

$$\dot{y} = r\,y\,(1 - y/K)$$

where r is the *per capita* net growth rate at low density, and K the carrying capacity; from which we obtain

$$y(t) = K \;/\; [1 + e^{-rt}\,(K/y(0) - 1)]$$

The derivation of this solution is mathematically almost as straightforward as in the preceding example. Like that solution, it specifies the value of $y(t)$ in terms of certain parameters (here r and K, there g), and initial values (here $y(0)$, there both position $y(0)$ and velocity $\dot{y}(0)$).

The great difference between these two is that the logistic equation cannot pretend to be an exact description of population growth in the way that the first equation describes the thrown stone. For instance, it is a reasonable assumption that the *per capita* net growth rate declines from a value r at low densities to zero at the carrying capacity, but there is no particular reason why the shape of this decline should be linear, as the term $r(1-y/K)$ assumes: for instance, the form $r(1-(y/K)^z)$ has been used in modelling fish populations, with z typically between 2 and 3 (May 1981). For many purposes the exact form may not matter, but it can make the difference between stable and cyclic behaviour when incorporated in an epidemic model (Mollison 1984, 1985). More basically, the concept of a constant carrying capacity is an oversimplification, as is the idea that a population grows in a smooth and continuous manner, rather than in discrete, stochastic steps, often with a pronounced seasonal character.

The answer to this problem is not to build complex models with a dozen or more parameters: this only serves to disguise crucial assumptions more effectively. It is better to recognize that simplifying assumptions are inevitable in ecological models, and to aim for a minimal set of basic components, where possible analysing how each one contributes to the model's behaviour.

For epidemic models, basic components include the following. First, the generation gap of the disease, i.e. the time between an individual's acquiring and passing on the disease: for some aspects, such as the equilibrium proportion of infected considered above, only its mean value matters. However, the stability of equilibrium conditions will depend on its distribution; the most common mathematical formulations (differential equations and Markov processes) assume the generation gap to be distributed exponentially, or as the sum of exponentials; while discrete time models assume a constant generation gap, which tends to give less stable behaviour.

Secondly, the infection term: this is most informatively expressed in relation to the mean number of potentially infectious contacts made by an individual, the reproductive ratio R_0; this value is crucial in determining whether the disease will spread, and with the common 'βXY' formulation (for the overall rate of contacts between the infectious number Y and the susceptible number X) it is too easy to slip in ecologically implausible assumptions as to how the number of contacts varies, for instance with population density (Mollison 1984, 1985). Thirdly, the dispersal distribution of contacts made by an individual: for animal populations a local spatial distribution is often appropriate, distinguishing between contacts outwith and within family groups where these exist.

Epidemics

Several examples of the introduction and evolution of diseases have already been given in the Introduction. Some simple calculations on the dependence of the probability of chance arrival on the dispersal distribution, mainly in the context of organisms crossing ocean gaps, are presented by MacArthur & Wilson (1967) and Mollison (1986). The shape of the dispersal distribution matters, as well as its mean: a successful invasion may result from a very small proportion of organisms travelling much greater distances than average. Not surprisingly, the best examples of natural long-distance movement are for windborne infections such as foot-and-mouth disease and various diseases carried by small insects (Smith 1982; Pedgley 1983).

Whether a disease becomes established once it has arrived depends almost exclusively on R_0, the mean number of contacts made by an infectious individual. The classic threshold theorem of Whittle (1955), extending the original deterministic theorem of Kermack & McKendrick (1927), states that for a homogeneously mixing population the disease has a non-zero chance of becoming established if and only if R_0 is greater than its 'threshold' value of 1. Where, as in most animal populations, mixing is local rather than homogeneous the threshold value of R_0 is somewhat greater than 1 (Mollison & Kuulasmaa 1985).

These theoretical results are clear and extremely robust: that is, they depend hardly at all upon other details of the model. The problem lies in estimating the value of R_o, even in an undisturbed population; or worse, in predicting its new value under some control strategy (see below pp. 339–40).

Once a disease becomes established, it commonly spreads as an epidemic 'wave' at a velocity depending chiefly on R_o, on the mean generation gap τ of the disease, and on the dispersal (or 'contact') distribution of the disease in the given population. Stochastic and deterministic models for spread are surveyed by Mollison (1977). Deterministic models are usually adequate where numbers are large, and it might be thought that this would apply to the spread of epidemics once they have become established. However, this does not appear to be the case, for two reasons. First, the number with whom an individual interacts remains relatively small; and secondly, long-distance dispersal by a very small proportion of the population can have a significant effect on the velocity and manner of spread: if the dispersal distribution is long-tailed, 'great leaps forward' caused by single contacts can be a major factor in the advance of the epidemic.

For short-tailed dispersal distributions (the exact condition is that the dispersal probability should fall off at least exponentially with distance), relatively steady progress may be expected, with a velocity of the order of $R_o\sigma/(2\tau)$ (Mollison & Kuulasmaa 1985), where σ is the root mean square dispersal distance. For instance for fox rabies, taking $R_o \sim 5$, $\sigma \approx 1$–2 km, and $\tau \approx 0.1$ year, this formula gives a velocity of 25–50 km/year; this figure, which should be increased somewhat to allow for the effect of individual 'great leaps forward', such as may particularly occur during the annual dispersal of juvenile males, is in reasonable agreement with data for the spread of rabies in Europe (Macdonald & Voigt 1985; Ball 1985b, see Fig. 2).

Endemics

A disease is said to be endemic when it persists in a population. Whether endemicity is possible again depends principally on the transmissibility of the disease as measured by the reproductive ratio R_o (though possibly with a higher threshold value), but will also depend on other factors, in particular the requirement for a sufficiently large and widely spread population to sustain the disease.

Where a disease affects several species, we need to distinguish between those species which can maintain the disease on their own (maintenance or reservoir species), and others in which the disease would die out if eliminated in the maintenance species. For instance, while rabies affects many species, in each area there are usually a small number of maintenance species: e.g. the red fox (*Vulpes vulpes*) in most of Europe, the red fox and

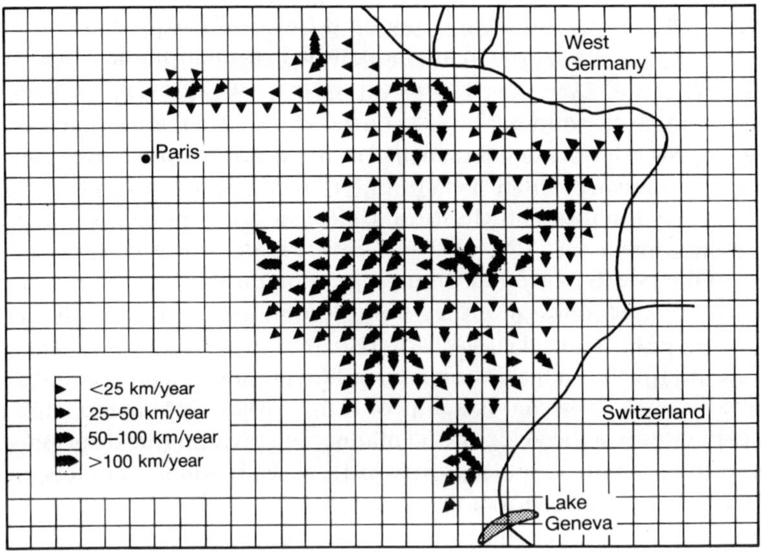

Fig. 2. Estimated front-wave velocities for French fox rabies (from Ball 1985b; grid squares are 20 × 20 km).

striped skunk (*Mephitis mephitis*) in eastern Canada, the vampire bat (*Desmodus rotundus*) in South America; though a wider complex of jackals, foxes and other small carnivores is implicated in South Africa (Macdonald & Voigt 1985). It is not always clear which is the maintenance species. For instance, it was thought that wildebeest played this role for rinderpest in Kenya, but it died out among them once cattle were vaccinated, showing that maintenance had relied on the latter; whether the cattle on their own could have maintained the disease, or whether it required the two interacting species, remains an open question (Plowright 1982).

The endemic phase is the most difficult to model, particularly for diseases which appreciably depress population numbers, where we must allow for density-dependent population growth, as well as the basic components of infection term, generation gap and dispersal distribution.

Consider first non-fatal diseases, for which we have already seen that the equilibrium proportion of infected individuals is $p_y = p\tau/T$ (p. 331 above). Homogeneous mixing models yield other apparently robust results on endemic equilibrium. For instance, the following simple argument suggests that the equilibrium proportion of susceptibles p_x is equal to $1/R_o$ (Dietz 1975). In equilibrium the mean number of effective contacts per infected must = 1; thus the proportion of contacts that are effective must average $1/R_o$. But if we assume homogeneous mixing, the probability that a contact is effective is simply equal to the proportion of susceptibles in the population, p_x.

Since p_x is much easier to estimate than R_o, this relation has mostly been used inverted, i.e. $1/p_x$ is used as an estimate for R_o (e.g. Anderson & May (1982) estimate R_o for a variety of human 'childhood diseases' in this way). However, this relation depends strongly on the assumption of homogeneous mixing. For instance, either for a population consisting of groups relatively isolated from each other (Mollison 1985), or for a population with age-dependent contact rate (Dietz & Schenzle 1985), it is possible to have R_o substantially less than $1/p_x$.

Returning to the case of diseases which are fatal, or at least significantly depress the population density, a similar argument to that of Dietz (1975) suggests a similar conclusion for susceptibles as a proportion of the carrying capacity, namely that $p_{\acute{x}}$, defined as X/K, $= 1/R_o$. However, this result depends not only on homogeneous mixing, but also on the assumption that the contact rate falls proportionately as the population density is reduced—an ecological assumption that seems particularly doubtful when the socially disruptive effect of severe disease is considered.

In simulations of a spatial model for a fatal disease, with parameters appropriate to fox rabies, Mollison & Kuulasmaa (1985) found that the population numbers were not depressed nearly as much as the simpler homogeneous mixing model suggests: in the spatial model, the disease appears to maintain itself through wandering foci, mostly well separated from each other, so that numbers are only held down locally at any one time. For instance, Fig. 3 shows a simulation where homogeneous mixing would predict $p_{\acute{x}} = 25\%$, but the actual value is approximately 80%.

Such simulations show reasonable agreement with data on fox rabies. Detailed data analysis by Sayers, Ross, Saengcharoenrat & Mansourian (1985) has identified patterns of wandering foci; and the general pattern is evident from data on case occurrences such as are shown in Fig. 4 (for comparison with Fig. 3 it is probably fairest to think of the simulation's inter-site distances as representing a typical contact distance, say ~ 2 km, so that the simulation represents an area of the order of 100×100 km).

As regards the proportion of infected, the result $Y/X = r_* \tau$ was quoted earlier (p. 331). We might have expected that for $R_o >> 1$, the population would be so reduced that the net *per capita* growth rate r_* would be near its maximum value, r, and hence $Y/X \approx r\tau$. But if the population is only reduced locally at any one time, r_* can be much smaller, even in heavily endemic conditions (e.g. only $= .3r$ in the simulation illustrated in Fig. 3), and the proportion of infected in consequence also much smaller than predicted by a homogeneous mixing model. This has some bearing on the interesting question as to how large an area is needed to sustain endemicity: if the equilibrium number of infected is sufficiently small, chance fluctuations

Fig. 3. Simulation of stochastic spatial endemic model (from Mollison & Kuulasmaa 1985). Filled circles, infectious sites; open circles, incubating sites; dots, susceptible sites; vacant sites are left blank.

will extinguish the disease. (Note that the answer to this question cannot be simply an area; the shape of the habitat can be important too, as the yellow fever example of the next section illustrates.)

One other feature of endemic diseases is that if we look at cases in a specific small area we often find a pattern of recurrent outbreaks at fairly regular intervals, typically four to five years in the case of fox rabies (Macdonald & Voigt 1985). This behaviour can be explained by the spatial model, where we would expect the intervals between successive visits by wandering foci to have an average value equal to the 'turnover' period, $\tau/p_{\hat{y}}$, where $p_{\hat{y}}$ is the equilibrium proportion of sites infected at any one time, i.e. Y/K (Mollison

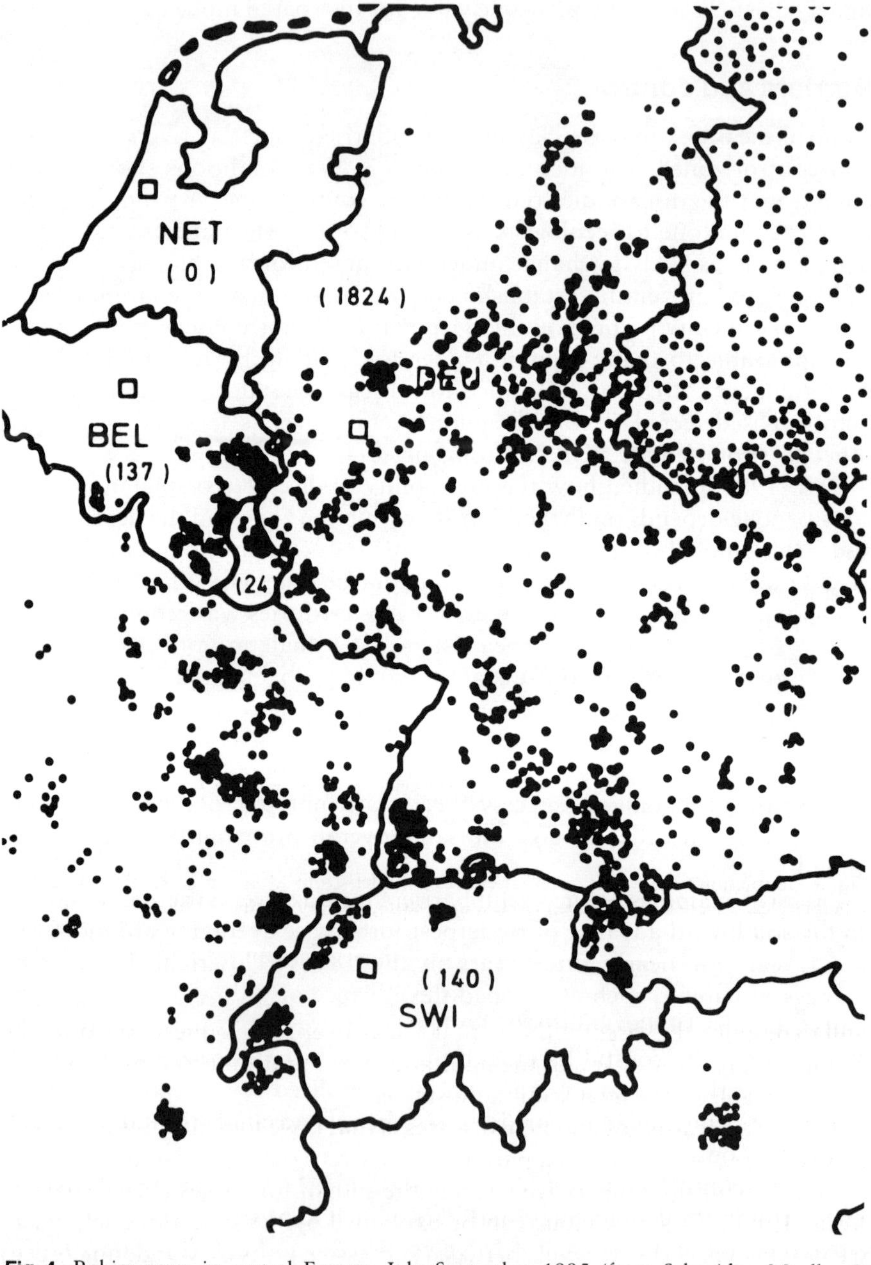

Fig. 4. Rabies cases in central Europe, July–September 1985 (from Schneider, Mueller & Hohnsbeen 1985). The figures in brackets indicate the total numbers of cases in each country during the period.

1986); for fox rabies, this period will be four to five years if $p_{\hat{y}}$ is approximately 2%, a reasonably plausible value for the spatial model.

Barriers and control

Control measures may be classified into barriers, such as rivers or fences, and control zones. The idea of the latter is that conditions should be so altered that the disease dies out within the control zone. A major problem therefore, and one for ecologists as much as for mathematicians, is to estimate how R_0 is likely to change under a proposed control strategy.

As regards preventing establishment, we should first note an important practical difficulty pointed out by Bacon (1981) in the context of fox rabies, but applicable to many wildlife epidemics: judging by detection rates in rabies endemic areas, the distance the disease would be likely to spread before it is detected is considerable, several times as large as the radius officially proposed for a UK control zone.

Ball (1985a) studied how the probability of disease crossing an annular control zone depends on the level of R_0 within the zone and the depth of the zone, for a model of family groups making only within-group and nearest-neighbour contacts. For instance, for a control zone with R_0 half its threshold value, a depth of approximately seven territories was required to give 80% probability of success. (The existence of a small proportion of individuals dispersing relatively long distances would probably greatly increase the required depth of the control zone.)

The problem of stopping the spread of an established epidemic is similar, except that in as much as the barrier may have to be much longer, we need to make the probability of success of each 'attempt' to cross the barrier that much smaller in order to give the same overall probability of preventing spread.

Practical examples of successful barriers include rivers: the Zambesi held up the southward advance of rinderpest (otherwise averaging around 1000 km a year) for nearly three years in the 1890s (Plowright 1982); and changes of habitat, such as stopped the advance of yellow fever in Central America in the 1950s (Smith 1971). The latter example is particularly interesting in that the coastal habitat through which the disease had advanced was sufficiently close to a mathematician's idealized one-dimensional space that the epidemic failed to spread back when its advance was stopped, and simply died out.

Human control methods have included physical barriers, such as the fence around the Kruger National Park which kept trypanosomiasis from spreading into cattle outside (Molyneux 1982); and control zones, as for instance in the defence from rabies of Denmark by killing foxes, and of Swiss valleys by vaccinating them (Steck 1982).

Control zones where the population density is reduced by killing are not necessarily as effective as simple mathematical models suggest; because of social disruption, R_0 may not fall in proportion to population density, as suggested (e.g.) for fox rabies by Anderson, Jackson, May & Smith (1981) (see Mollison 1984, 1985). The effects of vaccination are more predictable since social behaviour is altered little if at all, though we must remember the difficulty of estimating R_0 in the first place (see preceding section) (Mollison 1984, 1985; Bacon & Macdonald 1980). We should note also that the effect of culling or vaccinating a specific proportion of the population will be enhanced if we can bias our intervention towards those sections of the population that make the most contacts or disperse furthest (and reduced if our bias goes the other way).

These final examples again illustrate how difficult it is to obtain useful quantitative predictions from mathematical models of disease. This is why it is important to study the simplest possible models consistent with the basic features of a disease. If models are to be more of a help than a hindrance, we must understand how their conclusions follow from their assumptions, and whether or not they are sensitive to small variations in those assumptions.

As I hope I have shown, such simple models can be useful in clarifying the logic of the situation, in qualitatively describing the population dynamics of disease, and in identifying gaps in our ecological knowledge. For the control of disease the most crucial of these relate to its reproductive ratio, R_0. How do we estimate R_0? How does it depend on population density and habitat? And how will it change under a particular control strategy?

References

Anderson, R.M. & May, R.M. (1982). Directly transmitted infectious diseases: control by vaccination. *Science, N.Y.* **215**: 1053–60.

Anderson, R.M., Jackson, H.C., May, R.M. & Smith, A.D.M. (1981). Population dynamics of fox rabies in Europe. *Nature, Lond.* **289**: 765–71.

Bacon, P.J. (1981). The consequences of unreported fox rabies. *J. env. Mgmt* **13**: 195–200.

Bacon, P.J. & Macdonald, D.W. (1980). To control rabies: vaccinate foxes. *New Scient.* **87**: 640–5.

Ball, F. (1985a). Spatial models for the spread and control of rabies incorporating group size. In *The population dynamics of rabies in wildlife*: 197–222. (Ed. Bacon, P.J.). Academic Press, London.

Ball, F. (1985b). Front-wave velocity and fox habitat heterogeneity. In *The population dynamics of rabies in wildlife*: 255–89. (Ed. Bacon, P.J.). Academic Press, London.

Crosby, A.W. (1986). *Ecological imperialism: the biological expansion of Europe, 900–1900*. Cambridge University Press, Cambridge.

Dietz, K. (1975). Transmission and control of arbovirus diseases. In *Epidemiology*: 104–21. (Eds Ludwig, D. & Cooke, K.L.). SIAM, Philadelphia.

Dietz, K. & Schenzle, D. (1985). Mathematical models for infectious disease statistics. In *A celebration of statistics: the ISI centenary volume*: 167–204. (Eds Atkinson, A.C. & Fienberg, S.E.). Springer-Verlag, New York.

Edwards, M.A. & McDonnell, U. (Eds) (1982). *Animal disease in relation to animal conservation*. Academic Press, London and New York. (*Symp. zool. Soc. Lond.* No. 50.)

Fenner, F. & Myers, K. (1978). Myxoma virus and myxomatosis in retrospect: the first quarter century of a new disease. In *Viruses and environment*: 539–70. (Eds Kurstak, E. & Maramorosch, K.). Academic Press, New York and London.

Jones, D.M. (1982). Conservation in relation to animal disease in Africa and Asia. *Symp. zool. Soc. Lond.* No. 50: 271–85.

Kermack, W.O. & McKendrick, A.G. (1927). Contributions to the mathematical theory of epidemics. *Proc. R. Soc.* (A) **115**: 700–21.

MacArthur, R.M. & Wilson, E.O. (1967). *The theory of island biogeography*. Princeton Univ. Press, New Jersey.

Macdonald, D.W. & Voigt, D.R. (1985). The biological basis of rabies models. In *The population dynamics of rabies in wildlife*: 71–108. (Ed. Bacon, P.J.). Academic Press, London.

May, R.M. (1981). Mathematical models in whaling and fisheries management. In *Some mathematical questions in biology: Lectures on mathematics in the life sciences* **13**: 1–64. (Ed. Oster, G.F.). Am. Math. Soc., Providence, Rhode Island.

Mollison, D. (1977). Spatial contact models for ecological and epidemic spread (*with Discussion*). *J. R. statist. Soc.* **B39**: 283–326.

Mollison, D. (1984). Simplifying simple epidemic models. *Nature, Lond.* **310**: 224–5.

Mollison, D. (1985). Sensitivity analysis of simple endemic models. In *The population dynamics of rabies in wildlife*: 223–34. (Ed. Bacon, P.J.). Academic Press, London.

Mollison, D. (1986). Modelling ecological invasions: chance, explanation, prediction. *Phil. Trans. R. Soc.* (B) **314**: 675–93.

Mollison, D. & Kuulasmaa, K. (1985). Spatial epidemic models: theory and simulations. In *The population dynamics of rabies in wildlife*: 291–309. (Ed. Bacon, P.J.). Academic Press, London.

Molyneux, D.H. (1982). Trypanosomes, trypanosomiasis and tsetse control: impact on wildlife and its conservation. *Symp. zool. Soc. Lond.* No. 50: 29–55.

Pedgley, D.E. (1983). Windborne spread of insect-transmitted diseases of animals and man. *Phil. Trans. R. Soc.* (B) **302**: 463–70.

Plowright, W. (1982). The effects of rinderpest and rinderpest control on wildlife in Africa. *Symp. zool. Soc. Lond.* No. **50**: 1–28.

Ross, J. (1982). Myxomatosis: the natural evolution of the disease. *Symp. zool. Soc. Lond.* No. 50: 77–95.

Sayers, B.McA., Ross, A.J., Saengcharoenrat, P. & Mansourian, B.G. (1985). Pattern analysis of the case occurrences of fox rabies in Europe. In *The population dynamics of rabies in wildlife*: 235–54. (Ed. Bacon, P.J.). Academic Press, London.

Schneider, L.G., Mueller, W.W. & Hohnsbeen, K.-P. (1985). Map of rabies cases in Europe, 3rd quarter 1985. *Rabies Bulletin Europe* 9 (3), Annex 1. WHO Collaborating Centre for Rabies Surveillance and Research, Tubingen.

Smith, C.E.G. (1971). Human and animal ecological concepts behind the distribution, behaviour and control of yellow fever. *Bull. Soc. Path. Exot.* **64**: 683–94.

Smith, C.E.G. (1982). Major factors in the spread of infections. *Symp. zool. Soc. Lond.* No. 50: 207–35.

Steck, F. (1982). Rabies in wildlife. *Symp. zool. Soc. Lond.* No. 50: 57–75.

Whittle, P. (1955). The outcome of a stochastic epidemic—a note on Bailey's paper. *Biometrika* **42**: 116–22.

Symp. zool. Soc. Lond. (1987) No. 58: 343–346

Summary

M. L. GORMAN

Culterty Field Station
University of Aberdeen
Newburgh, Ellon
Aberdeenshire AB4 0AA

A striking feature of this symposium has been that the majority of the papers have dealt with rare and endangered species, or with species that are cropped as a resource, or with species that through their impact on the production of food or timber, or as vectors of human disease, are regarded as pests. A central theme has been the problems associated with estimating and monitoring population sizes and the elucidation of population dynamics.

As human populations continue to grow, and as pressure on land increases throughout the world, other species of mammals will inevitably decline in numbers, and many are likely to become extinct in the not too distant future. Equally, as our own needs and numbers multiply, many species with which in the past we have found it relatively easy to share space and resources will increasingly come to be regarded as pests and competitors.

Here in Britain we have a woeful record as far as our relationships with mammals are concerned and a significant part of our natural mammalian fauna has been lost as a result of habitat destruction and straightforward persecution. In the past our carnivores were particularly hard hit; a number are extinct and others including the polecat and wildcat were taken to the very brink by sporting interests but happily are now increasing thanks to changes in land-use and attitudes to wildlife.

Other species have suffered in more recent times, most notably several species of bats that are near extinction as a result of loss of roost sites and of poisoning by insecticides. The magnitude of this problem is clearly exemplified by the paper on greater horseshoe bats by Stebbings & Arnold.

Agricultural policy over the last few decades has led to the destruction of natural and ancient man-made habitats on a huge scale. Such changes in and intensification of land-use have undoubtedly had serious impacts on some of our remaining mammal species; in many parts of Britain, for example, it is becoming difficult to see brown hares.

The great irony is that having destroyed so much we are now faced with the very real problems of over-production of food in Europe and, as I write

ZOOLOGICAL SYMPOSIUM No. 58
ISBN 0-19-854006-X

these comments, leading agricultural economists are telling of the need to take as much as 10% of agricultural land out of food production.

Such profound changes in land-use may be the first good news for British wildlife for generations. It is likely that much of the land released from food production will be used for afforestation. This will undoubtedly be of benefit to woodland carnivores and we can confidently expect that the wildcat, pine-marten and polecat will continue to increase in numbers and to expand their ranges. Equally both red and roe deer thrive in coniferous plantations and as the papers by Staines and Ratcliffe show are increasing rapidly in those habitats. Increasingly we can expect that deer will come into conflict with forestry production and that rational management plans will be required.

Other changes in the pattern of food production may not be so benign. The farming of salmon (and in future other species of marine fish) in the sea-lochs of western Scotland and the Northern Isles is rapidly growing. These same areas support some of the most important populations of otters and seals to be found in Europe. What the ultimate impact of the industry will be is unclear, but it is likely to be detrimental. Disturbance; persecution; poisoning by the chemicals used to prevent fouling of nets and to control fish ectoparasites; changes in benthic communities subject to a constant rain of fish excrement; all are likely to be increasingly important.

In such a world the monitoring of population sizes and structure is likely to become increasingly important if we are to effectively manage susceptible, pest or crop species.

To do so is a very diverse problem when we are dealing with species as different as woodland rodents and pelagic whales. As the papers in this symposium so eloquently show, the principles are likely to be similar from species to species but the practicalities and techniques required differ enormously.

Whatever the method used, unless monitoring is precise, and capable of accurately detecting relatively small changes in numbers and structure, populations may fluctuate quite considerably before we can be statistically confident that any change has taken place. This could be extremely serious when we are dealing with relatively rare species.

Otters are a case in point. The paper by Conroy & French shows clearly that although spraint surveys are a good means of detecting the presence of otters in an area they give only the very crudest measure of otter abundance. Indeed, spraint density itself has to change by at least 30% before the change can be detected. Even then we have no real idea of what that means in terms of otter numbers. One really must question the value of extensive spraint surveys such as those carried out around Sullom Voe in Shetland where the accuracy is such that only very major losses of otters are ever likely to be detected with any certainty.

However good population monitoring may be, it is little more than an academic exercise unless remedial action can be taken to redress undesirable population fluctuations, whatever their direction. There will be little joy to be gained from documenting in great detail the final demise of a species. Technically difficult and expensive though monitoring may be, it is a relatively simple problem compared to the problem of population management.

Where declines in populations are due to over-exploitation of a species the design of a sustainable cropping programme may well be biologically easy but often it will be politically difficult to implement. In other circumstances the causes of population changes may be obscure and extremely complex. In such cases the difficulties of solving the problem will be enormous, if not impossible. If there is a lesson to be learned it is surely that monitoring for its own sake is likely to be a sterile occupation, its only redeeming feature being a warm feeling that we are doing something worthwhile. For every pound spent on monitoring, several pounds must be spent on fundamental research into the population biology of the species in queston. Only then *may* we be in a position to implement effective management should that become necessary. The work on otters reported by Kruuk, Conroy & Moorhouse is a case in point. Their approach is much more likely to produce data of lasting value, and of relevance to otter conservation, than is any amount of intensive and detailed spraint monitoring.

It is perhaps ironic, but inevitable, that some of the best data we have on the population dynamics of British mammals have come from the study of such pest species as the rabbit, the coypu and the West Country badger. To collect good population data is expensive in time and money and only rarely shall we see financial input on the necessary scale, except for research into economically important mammals or those implicated in the spread of disease. We can only be grateful that those workers charged with dealing with these species have so fully exploited the opportunities presented to them.

And what of the future? The major problem of population ecology is still that of population regulation and it is central to the preservation of rare species, to the control of pests and to the sustained cropping of economically important species. The measurement of population size and the elucidation of population dynamics are prerequisites to any understanding of population processes.

Undoubtedly, as shown so clearly in the papers presented here, the study of wild mammals will continue to be the province of gifted field workers. Increasingly, however, they will be required to work with complex technological aids and, more importantly, to be highly numerate and competent in a range of statistical and modelling skills, or failing that to be aware of the need to work in the closest collaboration with colleagues with such skills.

For me one of the most pleasurable aspects of this excellent symposium was to sit through a number of papers of very high quality given by young graduate students. I was much impressed not only with the quality of their work but also by the sheer professionalism with which they presented their papers. With young people of this calibre in our profession we can be confident that, given the necessary financial support, the study of mammalian populations in this country will continue in the fine tradition and to the high standards set by earlier workers.

Index

Aonyx capensis 248
Apodemus sylvaticus 44, 48

badgers 140, 279–94, 295–311
  age determination 296
  age structure 282, 285–6, 298–9, 302–3
  body weight 283, 290
  breeding 287–90, 292–3, 305, 309
  group size 285, 292, 298, 301–2
  litter size 293
  marking 281
  mortality 282, 286–7, 292, 298–9, 303–4, 308–9, 310
  population dynamics 280
  population model 299, 305–6, 308
  population size 140, 282, 284–5, 291, 297–8, 300–1
  sex ratio 305, 308
  territories 283
  tuberculosis 280, 293
  trapping 281, 297
Balaena mysticetus 237–40
Balaenoptera acutorostrata 227–31
bats, greater horseshoe 7–24, 343
  breeding 18–20, 23
  colony, definition of 10
  density 22–3
  distribution 9–11
  hibernation 21–2
  historical population changes 11–12, 24
  mark–recapture 9, 12–14
  marking 8–9
  recruitment 18
  sex ratio 14–15
  survival 15–18, 23
bats, vampire 335
blackgame 80
Bonasa umbellus 95
bovine tuberculosis 280, 293
butterfish 264, 270–1, 273
caiman 127
Calluna spp. 168
Canis familiaris 127
Capreolus capreolus 132–48, 168, 169, 172–3, 177, 178; *see also* deer
capture–mark–recapture (CMR) 25–57, 137–8, 307; *see also* population estimation
  advantages of 27
  age dependent survival 41–2
  badgers 281, 297–8
  bats 9, 12–14
  guidelines 47–9
  loss of markings 37
  problems 44–7
  seals 182, 195, 202–8
  sequential 31–3
  test for heterogeneity 38–41
  trap response 37
  whales 231–5
capybara 115–30
  group size 116, 118–20, 125–6, 127–8
  group stability 118, 120–1, 128
  harvesting 128–9
  marking 117
  predation 127
  reproductive success 126
  territorial behaviour 121–5
Cayman crocodilus 127
Census data 314, 315, 316–17
Cervus elaphus 131–52, 153–65, 167, 176; *see also* deer
Cervus nippon 168, 169, 173–4, 177; *see also* deer
coccidia 72
conservation strategy 7–8
correlogram 82–4
  detrending 82
coypu 99–113
  distribution 103
  population estimates 101–2, 106
  mortality 112

coypu—*cont.*
  recruitment 105–6, 111–12
  reproduction 107–8, 111
  trapping 100–1, 103–5, 112–13

Dama dama 168, 173, 178; *see also* deer
deer
  classification 133–4
  culling 156–7, 160, 161–2, 175–7
  fallow 168, 173, 178
  habitat 132
  muntjac 168
  population estimation, direct 134–9, 148, 169–70, 172, 176
  population estimation, indirect 140–3, 148
  red 131–52, 153–65, 167, 176
  roe 131–52, 168, 169, 172–3, 177, 178
  sika 168, 169, 173–4, 177
  white-tailed 140
Desmodus rotundus 335
disease
  control 339–40
  endemics 334–9
  epidemics 333–4
  models 331–9
  myxomatosis 68–9, 73–4, 90, 330
  rabies 327, 330, 334–5, 336, 339, 340
  rinderpest 330, 335, 339
  trypanosomiasis 339
  tuberculosis 280, 293
dogs, feral 127

Eimeria steidia 72
eelpout 264, 270–1, 273
Erica tetralix 168
Eschrichtius robustus 235–7

faecal accumulation rate 141–3, 144
faecal standing crop 141–3, 144
Felis concolor 127
flatfish 264
Forestry Commission 169–71, 175–6
fox, red 81, 296, 313–28
  densities 324
  population estimation 314, 315, 322
  population prediction 318–27
  rabies 327, 334, 336, 339, 340
  urban 313–15

Gadoidea 264, 270–1, 273
game bag records 92
Game Conservancy 81
group size
  badgers 285, 298, 301–2
  capybara 116, 118–20, 125–6, 127–8
  whales 230
grouse
  red 80, 95
  ruffed 95

Halichoerus grypus 181–91; *see also* seals
hares
  brown 80, 84–7, 92
  mountain 80, 84, 89–90, 92–5
  snowshoe 87–9, 90, 95
Hydrochoerus hydrochaeris 115–30; *see also* capybara

jaguar 127
Jolly–Seber model 26, 27, 28, 35, 41, 42, 43–4, 45–7, 50, 51, 233

Lagopus lagopus 80, 95
Larix sp. 173
Lepus americanus 87–9, 90, 95
Lepus europaeus 80, 84–7, 92
Lepus timidus 80, 84, 89–90, 92–5
Leslie–Chitty model 34
Lincoln Index 28, 29–30, 49
Lindane, effects on bats 11, 24
Lutra canadensis 264, 276
Lutra lutra 247–62, 263–78; *see also* otters
Lutra provocax 248
lynx 79
Lynx canadensis 79

marking
  badgers 281

marking—*cont.*
bats, greater horseshoe 8–9
capybara 117
loss of 37
rabbits 60
whales 231–2
Megaptera novaeangliae 231–5
Meles meles 140, 279–94, 295–311; *see also* badgers
Mephitis mephitis 335
Mirounga leonina 211–23; *see also* seals
models 329–42
differential equations 333
disease 331–9
Leslie–Matrix model 155–6
Markov processes 333
population estimation 316–17, 318–27
population simulations 102, 107–8, 170–1, 175, 299, 305–6, 308
technique evaluation 195–6
Monachus monachus 193–211; *see also* seals
multivariate analysis 254–5
multiple regression 315–16
Muntiacus reevesi 168; *see also* deer
Mustela nivalis 80
Myocastor coypus 99–113; *see also* coypu
myxomatosis 68–9, 73–4, 90, 330

New Forest 168–9, 176

Odocoileus virginianus 140
oil spills 256–7
Oryctolagus cuniculus 59–77, 80, 330; *see also* rabbits
otters 247–62, 263–78
birth season 266–8, 275
condition 266, 269–70, 274–5
counts 253–5
diet and feeding 270–3
litter size 268–9, 275–6
mortality 270, 274
spraint counts 250–60

Panthera onca 127
partridge 80
Perdix perdix 80
Phoca vitulina 181–91; *see also* seals
pine marten 344
Pinus nigra 173
Pinus sylvestris 143, 173, 177
Pholis gunnellus 264, 270–1, 273
Plecotus auritus, survival 23
polecat 344
population estimation 12–14, 25–57, 101–2; *see also* capture–mark–recapture
acoustic recording 239–40
aerial survey 183–5
badgers 282, 297–8, 300–2
Bailey's correction 34, 42
bats 13–14
camera traps 196–201
cervids 134–43, 154–5, 157–8, 169–70
choice of technique 43–4
correction model 212–14
direct censusing 134–9, 154–5
foxes 314, 315, 317
frequency-of-capture models 33–4
ground counting 182–3
indirect censusing 140–3
Jolly–Seber 26, 27, 28, 35, 41, 42, 43–4, 45–7, 50, 51, 233
Leslie–Chitty model 34
Leslie–Matrix model 155–6
Lincoln Index 28, 29–30, 49
maximum likelihood (ML) 26–7, 28, 30–1, 186–90, 204–6
minimum number alive (MNA) 27, 28, 44–7, 50, 51, 61, 297, 307
models 316–17, 318–27
otters 247–60
population closure 36, 208
rabbits 61, 62
radio-isotopes 298, 301–2
seals 182–90, 193–209, 211–23 211–23
sequential captures 31–3
subsampling 183
transects 139, 227–31, 241
vantage point counts 138–9, 157–8, 160, 236–7, 238–9
whales 225–42
population cycles 79–81, 82–96
puma 127

rabbits 59–77, 80, 330
  cause of mortality 68–9, 72–4
  marking 60
  myxomatosis 68–9, 73–4, 90
  population cycles 80, 87
  sex ratio 65
  survival 60, 61, 64–8, 70–2
rabies 327, 330, 334–5, 336, 339, 340
radio-isotopes 140, 298, 301–2
Red Deer Commission 136, 137, 146
Rhinolophus ferrumequinum 8–23, 343; *see also* bats
Rhum 176
rinderpest 330, 335, 339
rockling 264, 270–1, 273

salmon 344
Scolopax rusticola 80
Scots pine 143, 173, 177
seals 181–91, 193–210, 211–23
  censusing 182–3, 195–201, 202–8, 217–18
  common 181
  elephant 211–23
  grey 181–90
  monk 193–209
  pup production 183–5, 219–22
seascorpion 264
simulations 156–7; *see also* modelling
skunk, striped 335
spectral analysis 82
stress 8
Sylvilagus brasiliensis 330

Taurulus bubalis 264
Taxidea taxus 299
Tetrao tetrix 80
time series analysis 81–4, 90
trapping
  badgers 281, 297
  coypu 100–1, 103–5, 112–13
  response to 37
trypanosomiasis 339
Tsuga heterophylla 173
tuberculosis 280, 293

Vulpes vulpes 81, 296, 313–28; *see also* fox

weasels 80
whales
  bowhead 237–40, 241
  California gray 235–7, 241
  group size 230
  humpback 232–5, 241
  minke 227–31, 241
  right 241
wildcat 344
woodcock 80

Yellowstone National Park 176

Zoarces viviparus 264, 270–1, 273